RNA Genetics

Volume II
Retroviruses, Viroids, and RNA Recombination

Editors

Esteban Domingo
Scientist
Instituto de Biologia Molecular
Facultad de Ciencias
Universidad Autonoma de Madrid
Canto Blanco, Madrid, Spain

John J. Holland
Professor
Department of Biology
University of California, San Diego
La Jolla, California

Paul Ahlquist
Associate Professor
Institute for Molecular Virology
and
Department of Plant Pathology
University of Wisconsin-Madison
Madison, Wisconsin

CRC Press is an imprint of the
Taylor & Francis Group, an **informa** business

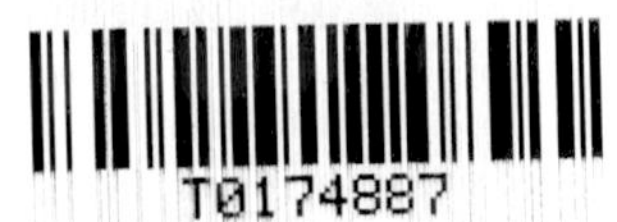

CRC Press
Taylor & Francis Group
6000 Broken Sound Parkway NW, Suite 300
Boca Raton, FL 33487-2742

First issued in paperback 2020

© 1988 by Taylor & Francis Group, LLC
CRC Press is an imprint of Taylor & Francis Group, an Informa business

No claim to original U.S. Government works

ISBN-13: 978-1-315-89733-2 (hbk)
ISBN-13: 978-0-367-65746-8 (pbk)

This book contains information obtained from authentic and highly regarded sources. Reasonable efforts have been made to publish reliable data and information, but the author and publisher cannot assume responsibility for the validity of all materials or the consequences of their use. The authors and publishers have attempted to trace the copyright holders of all material reproduced in this publication and apologize to copyright holders if permission to publish in this form has not been obtained. If any copyright material has not been acknowledged please write and let us know so we may rectify in any future reprint.

Except as permitted under U.S. Copyright Law, no part of this book may be reprinted, reproduced, transmitted, or utilized in any form by any electronic, mechanical, or other means, now known or hereafter invented, including photocopying, microfilming, and recording, or in any information storage or retrieval system, without written permission from the publishers.

For permission to photocopy or use material electronically from this work, please access www.copyright.com (http://www.copyright.com/) or contact the Copyright Clearance Center, Inc. (CCC), 222 Rosewood Drive, Danvers, MA 01923, 978-750-8400. CCC is a not-for-profit organization that provides licenses and registration for a variety of users. For organizations that have been granted a photocopy license by the CCC, a separate system of payment has been arranged.

Trademark Notice: Product or corporate names may be trademarks or registered trademarks, and are used only for identification and explanation without intent to infringe.

Library of Congress Cataloging-in-Publication Data

RNA genetics.

Includes biblioigraphies and index.
Contents: v. 1. RNA-directed virus replication --
v. 2. Retroviruses, viroids, and RNA recombination --
v. 3. Variability of RNA genomes.
1. Viruses, RNA. 2. Viral genetics. I. Domingo,
Esteban. II. Holland, John J. III. Ahlquist, Paul.
[DNLM: 1. RNA--genetics. QU 58 R6273]
QR395.R57 1988 574.87'3283 84-22432
ISBN 0-8493-6666-6 (vol. 1)
ISBN 0-8493-6667-4 (vol. 2)
ISBN 0-8493-6668-2 (vol. 3)

A Library of Congress record exists under LC control number: 87022432

Publisher's Note
The publisher has gone to great lengths to ensure the quality of this reprint but points out that some imperfections in the original copies may be apparent.

Disclaimer
The publisher has made every effort to trace copyright holders and welcomes correspondence from those they have been unable to contact.

Visit the Taylor & Francis Web site at
http://www.taylorandfrancis.com

and the CRC Press Web site at
http://www.crcpress.com

THE EDITORS

Esteban Domingo was born in Barcelona, Spain (1943), where he attended the University of the city, receiving a B.Sc. in Chemistry (1985) and a Ph.D. in Biochemistry (1969). Subsequently he worked as postdoctoral fellow with Robert C. Warner at the University of California, Irvine (1970 to 1973) and with Charles Weissmann at the University of Zürich (1974 to 1976). Work with Weissmann on phage Qβ led to the measurements of mutation rates and the demostration of extensive genetic heterogeneity in phage populations.

He is presently staff scientist of the Consejo Superior de Investigaciones Científicas at the Virology Unit of the Centro de Biología Molecular and Professor of Molecular Biology at the Universidad Autónoma in Madrid. For the last 10 years his group has contributed to research on the genetics of foot-and-mouth diseases virus (FMDV). They have characterized the extreme genetic heterogeneity of this virus and have established cell lines persistently infected with FMDV. Dr. Domingo has been adivser to international organizations on FMDV and problems of research in biology and medicine. His current interests include the relevance of FMDV variability to viral pathogenesis and the design of new vaccines, subjects on which his group collaborates with several laboratories in Europe and America.

John J. Holland, Ph.D., is Professor of Biology at the University of California at San Diego, La Jolla, California. He received his Ph.D. in Microbiology at U.C.L.A., did postdoctoral work, and was Assistant Professor in the Department of Microbiology, University of Minnesota. He was Assistant Professor and Associate Professor of Microbiology at the University of Washington in Seattle and Professor and Chairman of the Department of Molecular Biology and Biochemistry at the University of California at Irvine before moving to the University of California at San Diego, and spent a sabbatical year as a visiting scientist at the University of Geneva, Switzerland. He has published numerous papers in the field of virology, and presently, he and his colleagues do research on rapid evolution of RNA viruses, virus-immunocyte interactions, defective interfering particles, persistent infections, and related areas.

Paul Ahlquist, Ph.D., is Associate Professor of Molecular Virology and Plant Pathology at the University of Wisconsin, Madison. He is a member of the American Society for Virology and currently serves on the executive committee of the International Committee on Taxonomy of Viruses. His research interests include virus genome structure, organization and evolution, RNA replication and gene expression mechanisms, and viral gene functions.

CONTRIBUTORS

Jef D. Boeke, Ph.D.
Assistant Professor
Department of Molecular Biology and Genetics
Johns Hopkins University School of Medicine
Baltimore, Maryland

Jean-Marc Bonneville, Ph.D.
Friedrich Miescher Institut
Basel, Switzerland

Valerie Bosch, Ph.D.
Research Scientist
Zentrum für Molekulare Biologie
Universität Heidelberg
Heidelberg, West Germany

George Bruening, Ph.D.
Professor
Department of Biochemistry and Biophysics
University of California at Davis
Davis, California

Jamal M. Buzayan, Ph.D.
Postdoctoral Researcher
Department of Plant Pathology
College of Agriculture and Environmental Sciences
University of California at Davis
Davis, California

John M. Coffin, Ph.D.
Professor
Cancer Research Center
Tufts University School of Medicine
Boston, Massachusetts

Wayne L. Gerlach, Ph.D.
Principal Research Scientist
Division of Plant Industry
Commonwealth Science and Industrial Research
Canberra A.C.T.
Australia

Rosemarie W. Hammond, Ph.D.
Microbiology and Plant Pathology Laboratory
Agricultural Research Service
U.S. Department of Argiculture
Beltsville, Maryland

Arnold Hampel, Ph.D.
Professor
Department of Biololgical Science and Chemistry
Northern Illinois University
DeKalb, Illinois

T. C. Hodgman, Ph.D.
Research Assistant
MRC Laboratory of Molecular Biology
Cambridge, London

Thomas Hohn, Ph.D.
Friedrich Miescher Institut
Basel, Switzerland

Andrew M. Q. King, Ph.D.
Research Scientist
A.F.R.C. Institute for Animal Disease Research
Pirbright, Woking
Surrey, England

Christa Kuhn, Ph.D.
Research Scientist
Zentrum für Molekulare Biologie
Universität Heidelberg
Heidelberg, West Germany

Donald L. Nuss, Ph.D.
Associate Member
Department of Cell and Developmental Biology
Roche Institute of Molecular Biology
Nutley, New Jersey

Robert A. Owens, Ph.D.
Mirobiology and Plant Virology Laboratory
Agricultural Research Service
U.S. Department of Agriculture
Beltsville, Maryland

Pierre Pfeiffer, Ph.D.
Laboratoire de Virologie
IBMC, CNRS
Strasbourg, France

Heinz Schaller, Ph.D.
Professor
Zentrum für Molekulare Biologie
Universität Heidelberg
Heidelberg, West Germany

Sondra Schlesinger, Ph.D.
Professor
Department of Microbiology and Immunology
Washington University School of Medicine
St. Louis, Missouri

David Zimmern, Ph.D.
Staff Scientist
MRC Laboratory of Molecular Biology
Cambridge, England

RNA GENETICS

Volume I
RNA-DIRECTED VIRUS REPLICATION

RNA Replication of Positive Strand RNA Viruses
Kinetics of RNA Replication by Qβ Replicase
Replication of the Poliovirus Genome
RNA Replication in Comoviruses
Replication of the RNAs of Alphaviruses and Flaviviruses
RNA Replication of Brome Mosaic Virus and Related Viruses
Replication of Coronavirus RNA

RNA Replication of Negative Strand RNA Viruses
Replication of Nonsegmented Negative Strand RNA Viruses
Influenza Viral RNA Transcription and Replication

RNA Replication of Double-Stranded RNA Viruses
Replication of the Reoviridae: Information Derived from Gene Cloning and Expression
Replication of the dsRNA Mycoviruses

Volume II
RETROVIRUSES, VIROIDS, AND RNA RECOMBINATION

Reverse Transcribing Viruses and Retrotransposons
Replication of Retrovirus Genomes
Reverse Transcription in the Plant Virus, Cauliflower Mosaic Virus
Hepatitis B Virus Replication
Retrotransposons

Replication of Viroids and Satellites
Structure and Function Relationships in Plant Viroid RNAs
Replication of Small Satellite RNAs and Viroids: Possible Participation of Nonenzymic Reactions

Recombination in RNA Genomes
Genetic Recombination in Positive Strand RNA Viruses
The Generation and Amplification of Defective Interfering RNAs
Deletion Mutants of Double-Stranded RNA Genetic Elements Found in Plants and Fungi
Evolution of RNA Viruses

Volume III
VARIABILITY OF RNA GENOMES

Genetic Heterogeneity of RNA Genomes
High Error Rates, Population Equilibrium, and Evolution of RNA Replication Systems
Variability, Mutant Selection, and Mutant Stability in Plant RNA Viruses
Molecular Genetic Approaches to Replication and Gene Expression in Brome Mosaic and Other RNA Viruses
Sequence Variability in Plant Viroid RNAs

Gene Reassortment and Evolution in Segmented RNA Viruses
Genetic Diversity of Mammalian Reoviruses
Influenza Viruses: High Rate of Mutation and Evolution

Role of Genome Variation in Disease
Antigenic Variation in Influenza Virus Hemagglutinins
Variation of the HIV Genome: Implications for the Pathogenesis and Prevention of AIDS
Biological and Genomic Variability Among Arenaviruses
Modulation of Viral Plant Diseases by Secondary RNA Agents
Modulation of Viral Disease Processes by Defective Interfering Particles

Role of Genome Variation in Virus Evolution
Sequence Space and Quasispecies Distribution

TABLE OF CONTENTS

REVERSE TRANSCRIBING VIRUSES AND RETROTRANSPOSONS

Chapter 1
Replication of Retrovirus Genomes 3
John M. Coffin

Chapter 2
Reverse Transcription in the Plant Virus, Cauliflower Mosaic Virus 23
J. M. Bonneville, T. Hohn, and P. Pfeiffer

Chapter 3
Hepatitis B Virus Replication 43
Valerie Bosch, Christa Kuhn, and Heinz Schaller

Chapter 4
Retrotransposons 59
Jef D. Boeke

REPLICATION OF VIROIDS AND SATELLITES

Chapter 5
Structure and Function Relationships in Plant Viroid RNAs 107
Robert A. Owens and Rosemarie W. Hammond

Chapter 6
Replication of Small Satellite RNAs and Viroids: Possible Participation of Nonenzymic Reactions 127
George Bruening, Jamal M. Buzayan, Arnold Hampel, and Wayne L. Gerlach

RECOMBINATION IN RNA GENOMES

Chapter 7
Genetic Recombination in Positive Strand RNA Viruses 149
Andrew M. Q. King

Chapter 8
The Generation and Amplification of Defective Interfering RNAs 167
Sondra Schlesinger

Chapter 9
Deletion Mutants of Double-Stranded RNA Genetic Elements Found in Plants and Fungi 187
Donald L. Nuss

Chapter 10
Evolution of RNA Viruses 211
David Zimmern

INDEX 243

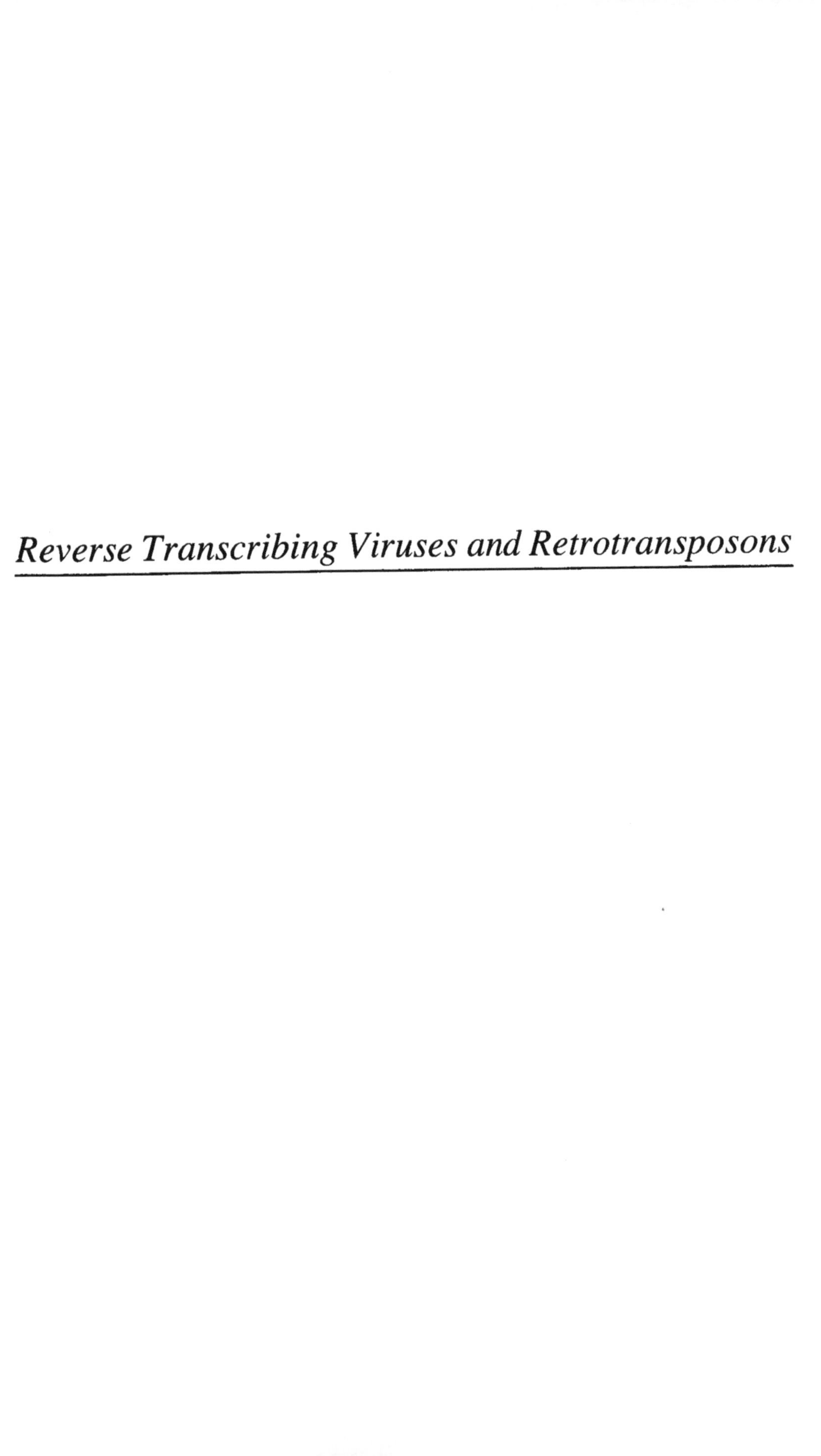

Reverse Transcribing Viruses and Retrotransposons

Chapter 1

REPLICATION OF RETROVIRUS GENOMES

John M. Coffin

TABLE OF CONTENTS

I. Introduction 4

II. Overview and Statement of the Problem 4

III. The Virion 6

IV. The Genome 8

V. Replication 10
- A. Initial Events 10
- B. Integration 12
- C. Expression 13
- D. RNA Processing 15
- E. Protein Synthesis and Processing 15
- F. Transactivation 16

VI. Consequences of Retrovirus Replication 17
- A. Transformation and Oncogenes 17
- B. Cytopathic Interactions 18

VII. Evolution and Genetics 19

References 20

I. INTRODUCTION

The study of retroviruses has proven to be a remarkably fruitful endeavor. Although regarded for many years as curiosities of little direct relevance, no one would presently dispute the importance of these agents to modern biology. Retroviruses are now universally recognized as important human and animal pathogens, as valuable models for carcinogenesis, and as paradigms for a mechanism of information transfer that has apparently been a strong force in modeling eukaryotic genomes. There is little reason to doubt that continued study of the basic molecular biology of these fascinating viruses will continue to yield large dividends in our understanding of fundamental processes of practical importance and intellectual appeal.

At first glance, retrovirus replication seems dauntingly complex, replete with jumps and other contortions involving a variety of enzymatic activities and unusual DNA and RNA structures. In this brief chapter, it is my intention to try to convince the reader that much of the complex phenomenology of these viruses follows in a straightforward way from a few basic principles, and that once these principles are understood, one can (with only a little guidance from prior knowledge) reconstruct much of the life cycle and genetics of the viruses.

Retroviruses are unique among vertebrate viruses in many respects. Of all viruses, they have the most intimate association with the host cell, using systems preexisting within the normal cell to accomplish all but the initial steps of the replication cycle, inserting their genetic information into the cell genome to form an irreversible lifetime association — one in which by far the most common outcome is that the cell becomes a producer of new virions, but is otherwise not significantly altered in its physiology. Indeed, alone among eukaryotic viruses, retrovirus infection can span generations, being passed on as endogenous proviruses following infection of the germline. The exceptions to this general rule, although dramatic, are rare. Human immunodeficiency virus (HIV) and the related lentiviruses cause disease as a consequence of the killing of specific target cells, yet the viruses can persist and replicate in the infected individuals for years without significant pathogenic effect. Another exception is found in the prototypic retrovirus isolates — the oncogene-containing viruses — which readily induce malignant transformation in appropriate infected target cells. Although the transforming viruses were the first and most intensively studied members of the family, and have been crucial to understanding the molecular basis of cancer, they are in fact the consequence of uncommon aberrations in the virus replication cycle. They thus represent evanescent phenomena which, if not provided with a good home in the laboratory, do not survive for long in nature.

Like other important biological aspects, the rich variety of retrovirus-host cell interactions — from benign to cytopathic to transforming — follows in a fairly direct way from their unique mode of replication. It is the intention of this chapter not only to describe the basic features of virus replication, but also to point out how these features lead into the diverse biologies and genetics of the various viruses. For more detailed information (and more complete references) the reader is encouraged to consult a number of recent reviews (References 1 to 4).

II. OVERVIEW AND STATEMENT OF THE PROBLEM

If one were to design a virus that follows the retrovirus lifestyle, i.e., to have a replication intermediate which sufficiently resembles a cellular gene that new genomes and virus can be synthesized using only the machinery of the normal host, one would have to consider a number of specific constraints.

1. Since the host cells do not have the means of making RNA copies from RNA templates or short DNA copies from DNA templates, the virus would have to use RNA polymerase for genome and mRNA synthesis. This condition necessitates a replicative intermediate of DNA and a genome of RNA. We will call the DNA replication intermediate the *provirus*.
2. The provirus should be associated with the cell in a stable, heritable fashion. While one could imagine episomal structures that might have this property (and have apparently been achieved by some DNA viruses), this can probably not be accomplished using host cell systems alone. More straightforward is to have the provirus become covalently associated with the cell genome so that it is replicated along with the cell DNA and regularly passed to progeny cells. Thus, the virus need only provide for the integration step, and the cell will take care of the rest automatically.
3. The integrated provirus must resemble a cellular gene. It must contain signals for initiation of RNA synthesis by cellular RNA polymerase II at appropriate times and rates; the transcripts must contain signals for their own processing into genome and mRNA; and the mRNAs must contain signals for translation into virion proteins. The necessity of transcriptional signals in the provirus creates the fundamental paradox of retrovirology: some of these signals (e.g., promoters) must reside in DNA *outside* of the region to be copied. Yet the region to be copied will become new genomes which must pass these signals to the next generation of infected cells. How can a virus provide replication signals outside its own genome? A solution is to arrange the process of proviral DNA synthesis so that sequences important for providing transcriptional control signals are present once in the genome, but duplicated in the provirus so that one copy can be used to direct the current round of replication while the unused copy provides genetic continuity into the next round. The duplicated sequence found at each end of the provirus and known as the long terminal repeat (LTR) is thus characteristic of all proviruses.
4. The RNA transcripts must serve multiple roles: as genomes and as mRNAs for internal structural and membrane proteins and enzymes. Except under exceptional circumstances, eukaryotic mRNAs apparently cannot be translated into multiple primary products. Special strategies therefore are required to obtain the variety of viral products needed. A variety of strategies are used by all retroviruses to obtain the necessary range of expression.
5. The infected cell must survive infection and remain in sufficiently good health to continue dividing and live out its normal life span. Thus, the virus infection must not be extremely virulent. There are several aspects to this constraint. First, the expression of the provirus must be limited to an extent which does not efficiently interfere with normal cell functions. Second, the number of proviruses must not be permitted to expand beyond the point that the cell can tolerate. Third, the exit of virus from the infected cell must not require irreparable damage to the cell membrane. The level expression of proviruses is often regulated in interesting (and poorly understood) ways; for example, it can be restricted to specific differentiation states or be made responsible to specific extracellular stimuli (such as hormonal signals), or to proteins encoded by the virus itself. Additionally, with most retroviruses, there seems to be a mechanism strictly limiting the accumulation of additional proviruses by the infected cell. Finally, all retroviruses are enveloped and released by budding — a mechanism that both provides the virus with a protective envelope derived from the cell membrane and permits release of virions without punching holes in the cell membrane.
6. An early step in virus replication must be the copying of the viral genome RNA into the DNA molecule which, after integration, will become the provirus. The uninfected cell, however, does not seem to contain the enzymatic machinery necessary for these

reactions. These must be provided by the virus and be available shortly after infection. In principle, this could be accomplished by translation of the incoming genome to yield the appropriate proteins, and then (after cleaning off the ribosomes) copying the same molecule into DNA. In practice, retroviruses have eschewed this strategy in favor of the simpler approach of bringing the enzymes into the cell in the same package as the genome. Thus, all retroviruses contain within the virion an RNA-directed DNA polymerase (or *reverse transcriptase*) as well as other enzymes necessary for DNA synthesis and integration.

7. The reverse transcriptase activity, like all DNA polymerases known, has a strict requirement for a primer — a preexisting polymer to which new nucleotides are added to form the growing DNA chain. This apparently absolute biochemical requirement raises two problems. First, an appropriate molecule must be provided and, second, some provision must be made for copying the ends of the genome into DNA. Primers for DNA synthesis in other systems include both protein and RNA molecules. Retroviruses have chosen the latter course. To initiate synthesis of the first strand (the *minus* strand, since it is complementary to the plus-stranded genome), a molecule of tRNA taken from the cell is used. Minus strand DNA synthesis, by contrast, is initiated on a nick introduced in the genome itself.

   The final problem of copying of the ends of the genome can be resolved in a clear way that also provides for synthesis of the LTR. Initiation of synthesis at an internal site on the genome can then be followed by elongation until the end is reached. A simple transfer (or ''jump'') of the nascent strand to the other end will generate a new template-primer combination which will allow completion of the strand through the primer site to the end of the template. A similar set of events, now using the plus strand DNA as template and nicked genome as primer leads to synthesis of the plus strand of DNA. Note that the ends of the final DNA molecule are derived from the internal sites marked by the locations of the primers. The region between the primer sites thus becomes the LTR.

8. To permit these events, the RNA genome must have specific signals that specify the correct initiation and jumping. These include a site for binding the tRNA primer, a recognition site for the cleavage that generates the plus strand primer site, and a short sequence directly repeated at either end of the genome to facilitate the end-to-end transfer of the growing chain. A final feature which must be provided is a site for recognition by the integration systems.

These constraints thus can be considered to define the retrovirus virion and replication cycle. It should be apparent that the overall cycle can be divided into two phases. In the early phase, the major events — the synthesis of LTR-containing proviral DNA and its integration into the cell genome — are catalyzed by enzymes brought into the cell with the virion. The late events — synthesis and processing of viral RNA and protein and assembly and release of virus — are carried out by cellular systems. Note that neither viral DNA synthesis, integration, nor expression require expression of viral genes. With most retroviruses, it is possible to substitute virtually all the coding region with other genes and (so long as virion proteins are supplied in *trans*) achieve all the major events in replication. It is this feature that has allowed retroviruses to serve as useful vectors — both of oncogenes, and of deliberately inserted sequences.

III. THE VIRION

Virions of retroviruses have a fairly simple protein composition, usually containing only eight or nine proteins. The internal structures are quite fragile and the details of the capsid

RETROVIRUS VIRION PROTEINS

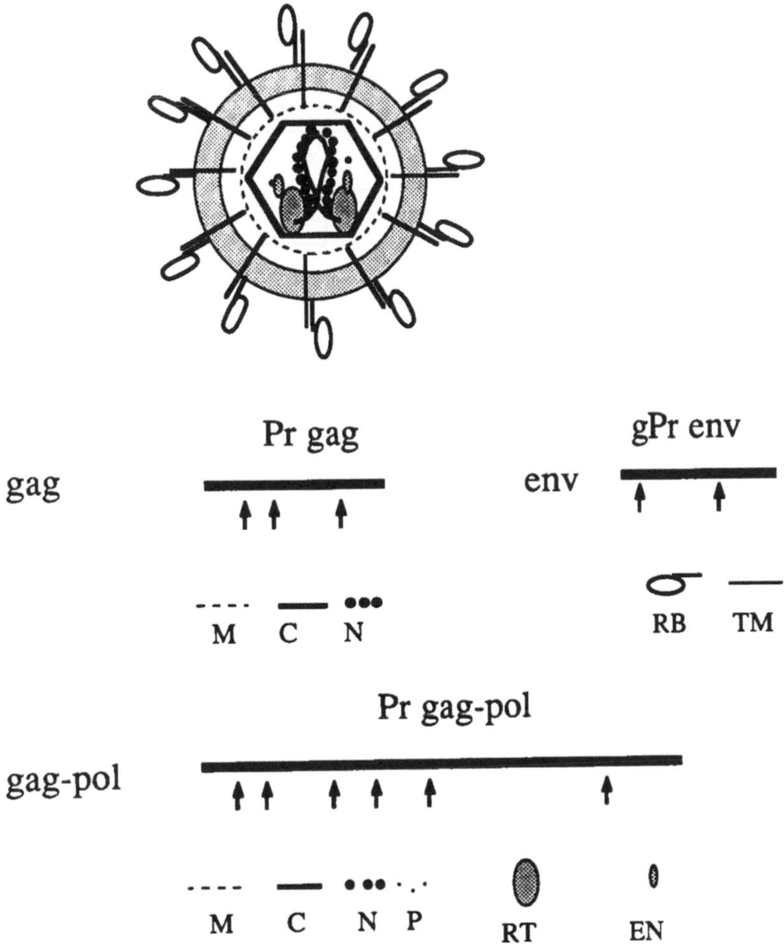

FIGURE 1. The retrovirus virion. This highly schematic sketch shows the approximate locations of the various virion components common to all retroviruses. Abbreviations: (gag), M, matrix protein; C, major core shell protein; N, nucleoprotein; p, protease; (pol), RT, reverse transcriptase; EN, endonuclease; (env), RB, receptor binding protein; TM, transmembrane protein.

organization are not well known. Figure 1 presents a diagram indicating, in schematic fashion, what is currently known about virion structure. The distribution on the genome of sequences coding for the various retrovirus proteins is shown in Figure 2.

The surface glycoprotein spike consists of two proteins, both products of the env gene. The larger one specifies binding to the specific cell receptor and is necessary for the initial interaction of the virus with the host cell. The smaller, transmembrane protein is joined to the larger by disulfide bonds and anchors the complex to the lipid envelope. The C terminal end of the smaller protein spans the membrane and must interact with an internal virus protein (probably the ''matrix'' protein just under the membrane). This interaction is not necessary for virion formation since mutants lacking the gene for these proteins still synthesize fairly normal virions.

The internal proteins of the virion are all products of two genes, gag and pol (Figure 1).

Coding Regions Of Retroviruses

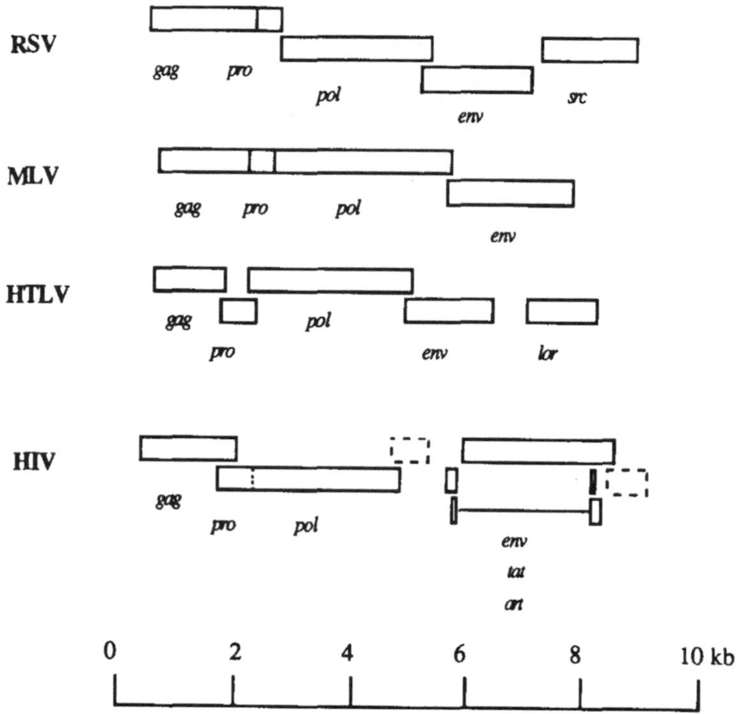

FIGURE 2. Reading frames. The various identified open reading frames are indicated by boxes. Dividing lines denote termination codons or cleavage sites. Dotted boxes indicate reading frames encoding unnamed proteins of unknown function. Note the different relative positions of the gag, protease (pro) and pol domains among the four groups of viruses. Note also the multiple use of the env region of HIV.

The gag gene (for *g*roup-specific *a*nti*g*en) contributes the bulk of structural proteins which vary in number from three to five, depending on the virus. The three proteins found in all retroviruses are (1) a "matrix" protein which lies adjacent to the membrane and probably interacts with it (sometimes via an added fatty acid group) to mediate budding; (2) the largest gag product which most likely forms the core shell visible in electron micrographs; and (3) a small, basic protein which is found in close association with the genome RNA and which behaves as a nonspecific nucleic acid binding protein. Some, but not all, retroviruses have a fourth gag protein of unknown function located on the precursor between the matrix and capsid proteins.

The remaining virion proteins serve enzymatic, rather than structural, roles. The first of these is a protease which is encoded C terminal to the nucleic acid binding protein, but in various reading frames relative to gag and pol (Figure 2). Finally, there are two or three polypeptides derived from the pol gene which carry out synthetic and nucleolytic events early in replication (synthesis of viral DNA and some steps necessary for integration).

IV. THE GENOME

Retrovirus genomes contain structures and sequences that reflect both their origin and

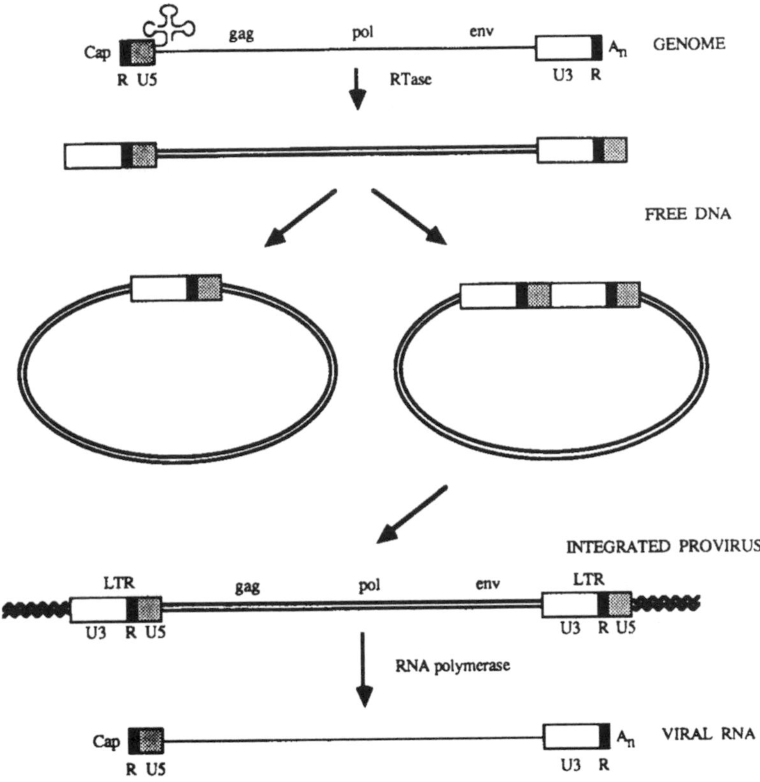

FIGURE 3. The virus replication cycle. Single lines indicate RNA. Double lines indicate viral DNA (with LTR sequences boxed). Wavy lines show cell DNA around the integration site. (Courtesy of S. A. Herman.)

function. Since they are synthesized and processed by the same mechanisms used for the synthesis of normal cell mRNA, they physically resemble cell mRNAs: they have a sequence of about 200 A residues at their 3′ ends, a capping group at their 5′ end, and are further modified by the presence of methylated A residues.[5] Retrovirus genomes are similar to one another in their overall organization the virion genes are always in the order gag-pol-env and differ only in mechanism of translation of pol and disposition of the protease reading frame (Figure 2). The function of additional reading frames is discussed below. In addition to the reading frames, all genomes contain a set of terminal regions essential for function as a template for reverse transcription and for proper function (integration and transcription) of viral DNA (Figure 3). From 5′ and 3′ these include:

- **R** — The directly repeated sequence (12 to 250 bases) also found adjacent to the poly(A) at the 3′ end. R serves the critical role during viral DNA synthesis of a "bridge" for transfer of the growing chain from one end of the genome to the other.
- **U5** — Unique sequence (80 to 200 bases) adjacent to R.
- **PB** — The binding site for the tRNA primer. The primer can be one of several different tRNAs and the primer binding site invariably consists of 18 bases perfectly complementary to the 3′ end of the tRNA.
- **Leader region** — The sequence between PB and the beginning of gag. It usually contains two well-defined functions: a splice donor site for the generation of subgenomic mRNAs and a signal (sometimes called "Ψ") that specifies assembly into virions, presumably by specifically interacting with a virion protein. These are usually arranged so that spliced mRNAs lack the signal and thus are not packaged.

Except for the splice acceptor and some additional donor sites, there are no known processing signals within the coding regions of the retrovirus genome, although specific sequences can have effects on RNA splicing (see later).

Defined genome regions near the 3′ end include:

- **PP** — (for polypurine tract) — A run of ten or more G and A residues which invariably mark the initiation site of plus-stranded DNA synthesis, by providing a specific cleavage and primer site when present in an RNA-DNA hybrid.
- **U3** — A unique region near the 3′ end varying in length from about 200 to 1200 bases. Because it forms the upstream end of the LTR, U3 consists mostly of sequences relevant to initiation and regulation of transcription (see below).
- **R** — Finally, the 3′ end of the genome contains a second copy of the repeated sequence identical to that of the 3′ end. Depending on the virus, either R or U3 contains a canonical signal (AAUAAA) for poly(A) addition.

In virions, the genome is invariably present in two copies joined to one another by base-pairing at multiple points, most strongly in a poorly defined region known as the "dimer linkage structure" near the 5′ end. The function of this unusual "diploid" arrangement is not known with certainty; it has the effect of causing a very high frequency of recombination,[6] and an attractive idea is that the presence of two copies permits repair of damage to the relatively fragile single-stranded RNA.

V. REPLICATION

A. Initial Events

Following penetration of the virion into the cytoplasm of the cell (an event probably accomplished by receptor-mediated endocytosis as for other viruses), the virion is partially uncoated to a structure which is poorly understood, but probably is very similar to the nucleocapsid, i.e., consists of the core shell surrounding the genome (and RNP protein) and reverse transcriptase. It is within this structure that DNA synthesis takes place. An outline of this process is presented in Figure 3. The overall process leads from a molecule of single-stranded RNA with the structure R–U5–genes–U3–R to a molecule of double-stranded DNA of the structure U3–R–U5–genes–U3–R–U5. The sequence "U3–R–U5" thus consitutues the LTR.

The first strand synthesized (the minus strand) is initiated at the 3′ end of the tRNA primer and proceeds toward the 5′ end of the genome (Figure 4). When the 5′ end of the genome is reached, the RNase H activity of reverse transcriptase removes the RNA just copied, and a new template primer pair is formed with the R sequence at the 3′ end of the genome. This event constitutes the first jump, and permits completion of the minus strand. While this is happening, a nick is made at the 3′ end of the PP sequence to create a primer for plus-strand synthesis using the U3–R portion of the minus-strand DNA as template. A second jump to the end of the minus strand then permits completion of both strands and formation of the LTRs.

While this process is usually depicted as involving only genome and reverse transcriptase, the capsid structure itself is probably of great importance to ensuring the specificity of a number of the steps — particularly those involving jumps of the growing chain. Although it is enzymatically active and capable of carrying out many of the relevant reactions on model templates, purified reverse transcriptase has not yet been used to accomplish all of the steps of viral DNA synthesis. In particular, the specificity of the jumps seems to be lost. Indeed, apparently complete viral DNA can be synthesized in vitro using detergent-disrupted virus; but the correctness of the reaction is exquisitely sensitive to detergent concentration.

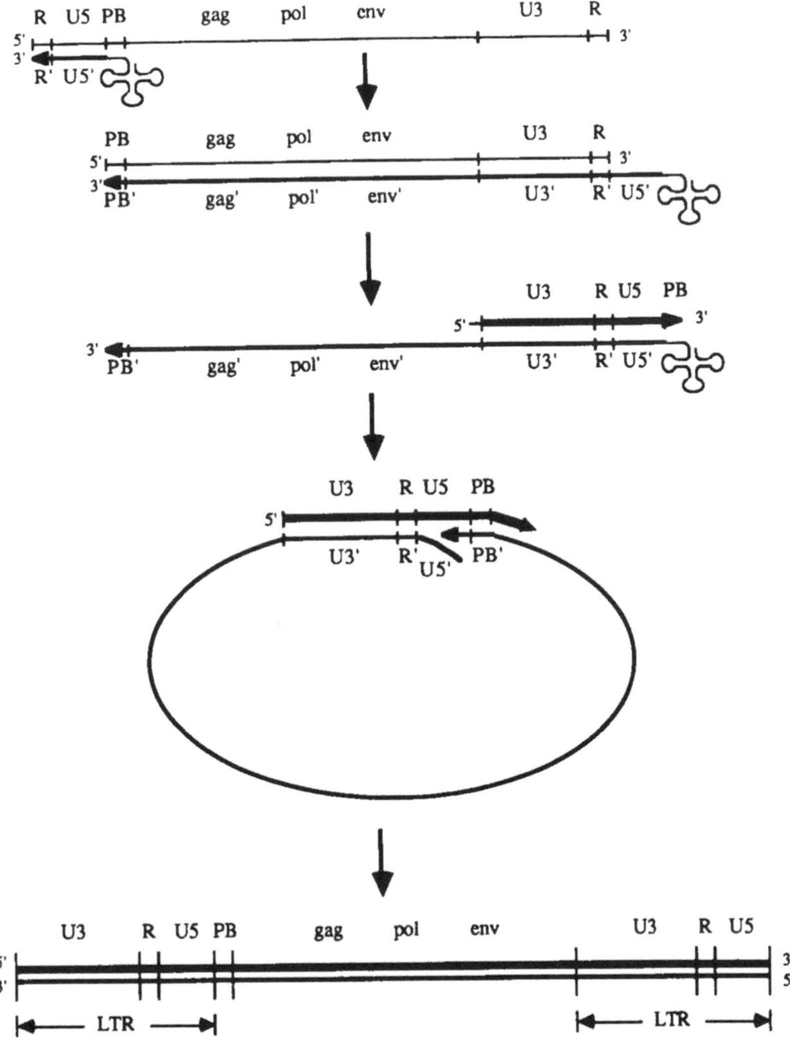

FIGURE 4. Synthesis of viral DNA. The direction of synthesis is shown by arrows. Note the successive "jumps" of reverse transcription; the first using the R sequence as a bridge, the second using a copy of the primer binding (PB) site. (Courtesy of S. A. Herman.)

If too much detergent is used, the major product is simply a copy of R–U5 called "strong stop DNA" and some aberrant elongation products in which synthesis is continued from random locations on the genome. It seems likely that reverse transcriptase, once it reaches the end of its template, can form new template-primer pairs more or less independent of base-pairing between the two. The major driving force seems to be affinity of enzyme for template and not homology between template and primer. (This phenomenon is well known to genetic engineers who use reverse transcriptase to make cDNA copies of RNA, since it is evidenced in this case by the tendency of the enzyme to double back at the end of the template and continue copying the newly synthesized DNA strand, leading to hairpin structures.) Thus, the specificity for the first jump is probably not dictated so much by homology between R and its DNA copy as by the capsid structure which probably holds the ends in correct alignment.

Another subtlety concerning the jumping mechanism of DNA synthesis relates to the use of the two genome subunits. Are the jumps specifically from one end of one template genome to the same molecule or are they specifically to the other molecule or are they random? Considering that the two genomes are probably intertwined in a very complex structure, it is difficult to imagine how the ends could be readily distinguished. Consistent with this, the progeny of virus heterozygous for markers at each end of the genome (but separated by only a few bases in the DNA) has been found to contain random combinations of the markers, indicating that jumping occurs at random, and is nonselective for the ends.[7]

Finally, it should be noted that although there are two copies of the genome in each virion, they do not seem to yield two viral DNA molecules, but rather one provirus seems to be derived from each infecting virion. This may be simply a consequence of an overall inefficiency in the process (only one virion in a hundred or so is infectious anyway), but it may also reflect the use of both RNA molecules to make a single molecule of DNA, by a copy choice repair mechanism (see later).

B. Integration

The linear DNA molecules (probably still in association with capsid proteins) are transported to the nucleus where they are circularized either by direct joining of the ends or by homologous recombination between two LTRs. The next event is integration of the viral DNA into that of the host. The integration event is a regular and indispensable part of the replication cycle and distinguishes retroviruses from all other virus families. In contrast to DNA synthesis, integration has not yet been duplicated in vitro, and the mechanism is therefore much more poorly understood. Integration is a highly specific process, since the provirus is always joined in the same way to cellular DNA to provide a structure flanked by the LTRs. Invariably, there is a duplication of a short (4 to 6 bases) sequence of cell DNA and a loss of two bases at the exposed end of each LTR. Figure 5 presents a hypothetical scheme which could lead to the structure found, involving staggered cuts in viral and cellular DNA which are subsequently repaired. Specificity for the cleavage probably resides in the virion endonuclease, since this protein has been shown to be dispensable for DNA synthesis, but required for integration.[8] The length of cell DNA duplicated varies with the type of virus, and the enzyme can recognize the surrounding viral sequences with at least moderate specificity.[9] In addition to endonucleases, this provirus must also use a ligase to join the ends and a DNA polymerase to accomplish the repair synthesis. It is likely that there are host enzymes involved, but it must be emphasized that none has been identified.

The viral DNA intermediate to integration remained unknown until recently. It now seems highly probable that the relevant intermediate is the two-LTR circle, and that the integration system (presumably the viral endonuclease) specifically recognizes the sequence formed by the junction of the LTRs.[10,11] This combination of sequences is found nowhere else in the viral DNA and is destroyed by the integration process, rendering the event unique and irreversible.

Again, it should be emphasized that, like DNA synthesis, integration requires the mediation of at least one enzymatic activity which enters the cell in the virion. It is also likely that the event is accomplished by a structure considerably more complex than free DNA plus enzyme. Presumably such a structure would also be descended from the viral nucleocapsid. One might thus predict that introduction of naked viral DNA, even with the appropriate target sequence, would not lead to its integration, even in the presence of the viral "integrase", and indeed, such experiments have not yet yielded successful results. Naked viral DNA introduced into cells becomes associated with cell DNA in the same sort of inefficient and nonspecific way that any other DNA does. To date, efficient, viral-like integration at the ends of the LTRs has been accomplished only as part of the usual infection cycle.

The other remaining issue regarding integration concerns its specificity. If host cell sequences flanking integration sites are compared, no obvious specificity emerges.[12] One might imagine, however, that it would be appropriate for the provirus to be inserted into tran-

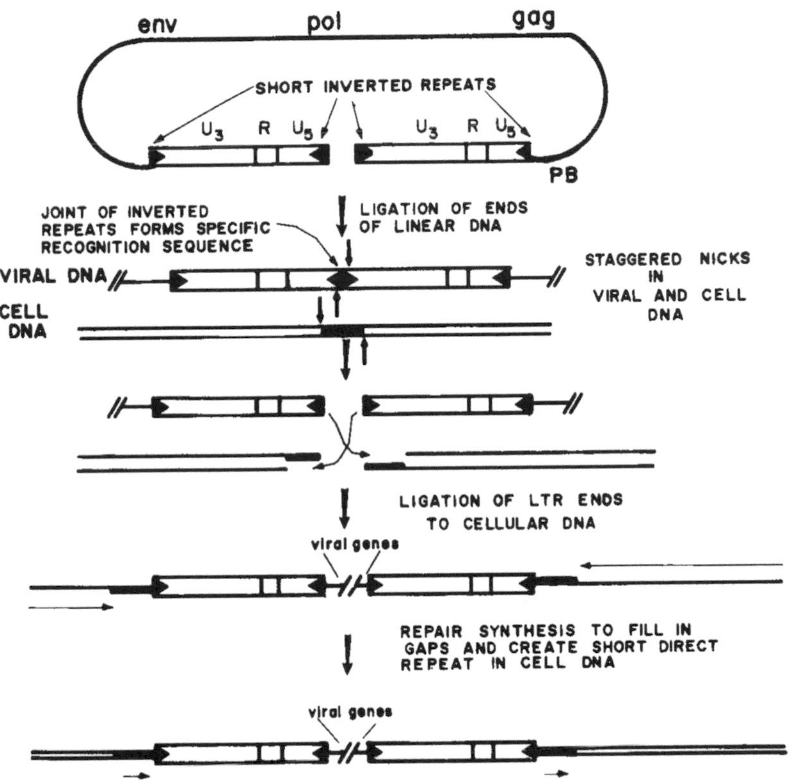

FIGURE 5. A possible scheme for integration of viral DNA. This outline is based only on structural analysis. None of the enzymatic reactions suggested has been definitively characterized.

scriptionally active regions of the genome. Indeed, it has recently been suggested, based on analysis of a few integration targets, that integration tends to occur near chromatin sites characterized by high sensitivity to DNase,[13] a probable hallmark of transcriptionally active regions.

Once integrated, the provirus is almost completely stable and is replicated regularly along with the cell DNA. There is no excision or direct transposition process, and loss of provirus (when it occurs) seems to be a consequence of random deletion events.

C. Expression

The integrated provirus behaves as though it were a cellular gene: it serves as an efficient template for mRNA synthesis by RNA polymerase II. For this purpose, much of the U_3 region of the LTR consists of sequences which resemble signals used by the normal cell (Figure 5). 5′ of the initiation site for RNA synthesis (or "cap" site) are standard consensus signals for eukaryotic transcription: a "TATA" box about 24 bases upstream, and a "CAAT" box about 80 bases upstream. In addition, there are more or less well-defined enhancer elements which can stimulate synthesis from any promoter in relatively distance- and orientation-independent fashion. These are often present in multiple copies, as, for example, in the MLV LTR which can contain a perfectly repeated sequence of about 70 base-pairs in length and HIV which contains repeated binding sites for the Sp1 transcription factor.[14] The enhancer sequences can provide considerable biological specificity by affecting both

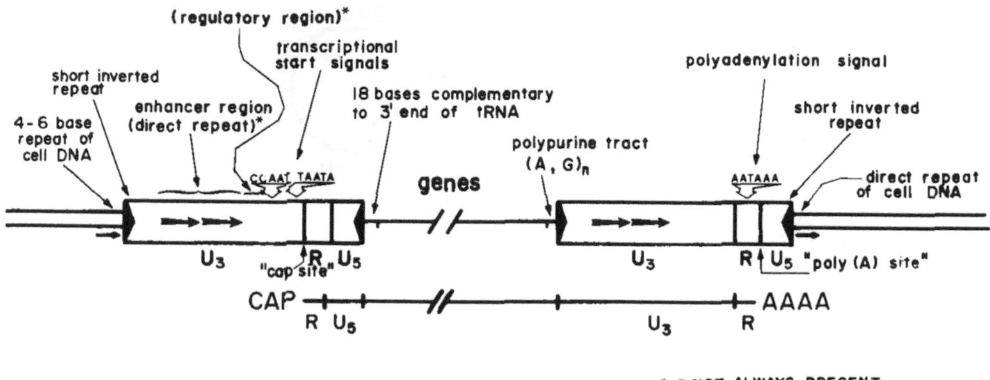

FIGURE 6. Structural features of LTRs. Although the LTRs are identical in sequence, the signals are only shown for the LTR in which they are important. Thus, initiation signals are shown only in the 5′ LTR, and 3′ processing signals only in the 3′ LTR.

the rate of virus replication (presumably by regulating the rate of initiation of transcription) as well as the cell specificity of replication and pathogenesis. For example, the differential ability of various MLV strains to infect different cell types resides in relatively small sequence differences in the enhancer region of the LTR.[15-17] In addition to specifying replication rate and tissue specificity, the LTR can also provide regulatory sequences. For example, the mammary tumor virus (MTV) U3 region contains sequences which respond to the action of glucocorticoid hormones, by binding hormone-receptor complex and stimulating transcription. Similarly, the HTLV-1 and -2 LTRs contain sequences which respond directly or indirectly to the ‘‘transactivating’’ influence of a virus-coded protein (see below).

In addition to specifying the initiation of transcription and hence the 5′ end of the viral transcript, the LTR also specifies the 3′ end of the RNA. As with most eukaryotic RNAs, the primary transcript seems to extend well beyond the final 3′ end. Thus, synthesis of viral RNA almost certainly proceeds through the LTR into adjacent cell DNA, and the primary transcript is then rapidly cleaved and polyadenylated. A consensus addition signal (AAUAAA) is invariably found within either U3 or R 16 to 20 bases upstream of the poly(A) addition site. Interestingly, cleavage and polyadenylation are not absolutely specific for the viral poly(A) site. A significant fraction of genomes can be derived from RNA polyadenylated at sites derived from adjoining host DNA.[18]

As shown in Figure 6, the LTR thus contains a complex collection of signals used at various stages of virus replication. Despite the wide diversity of viruses and the virtually complete lack of sequence homology among different groups, all LTRs contain these signals and they can often be recognized this way.

Since the two LTRs are identical in sequence, but different in function, the question arises as to whether and how they are functionally distinguished. That there is a functional distinction is strongly suggested by the relative rarity (less than 1 to 2%) of stable transcripts initiated at the downstream LTR in the usual infected cell.[18] Several possibilities have been suggested to account for such a functional distinction. For example, it has been proposed that initiation at the 5′ LTR is interfered with by transcription through it,[19] leading to a sort of ‘‘dominance’’ of the upstream LTR. Second, there are recent suggestions (Beemon, K., personal communication) that there may be one or more regions near the 5′ end, but within the coding portion of the provirus with enhancer-like activity. These could tend to favor transcription from the nearby LTR at the expense of the 3′ LTR. Finally, there is a suggestion

(Norton, P. and Coffin, J., in preparation) that some signal within the leader region may specify the nearby LTR for initiation of transcription. Alternatively, the presence of this sequence in the transcript may specifically protect it against degradation. These considerations have important consequences for understanding pathogenesis by some retroviruses (see later).

D. RNA Processing

The viral RNA transcripts have three distinct fates: they can become new genomes, they can serve as unspliced mRNA for the gag and gag-pol precursors, or they can be spliced to form env or other mRNAs. Clearly there must be a balance among these uses; too much or too little splicing, for example, would be deleterious to the virus replication. What determines this balance is, at present, not well understood. The inefficient use of the splicing signals is almost certainly not due to the nature of the signals themselves or to any kind of direct regulation. Rather, it seems more likely that some structural feature of the RNA in the gag-pol region determines the efficiency of utilization of the env splice acceptor, since altering the sequence within this region can drastically affect the relative amounts of spliced and unspliced mRNA.[20]

E. Protein Synthesis and Processing

The gag and pol proteins are both translated from genome-length RNA and their synthesis and processing are fairly well understood. In all retroviruses, the pol gene is expressed as a C terminal extension of about 5 to 10% of the gag precursor molecules. Both gene products are cleaved via the activity of the virion protease. This scheme gives rise to the problem of how the protease itself is generated. It now seems probable that the protease is active in the precursor form, but only slightly so, so that a critical mass of precursor has to assemble before cleavage begins. Once the initial cleavages take place, releasing active protease, the remaining cleavage would be relatively rapid. By this model, assembly and budding would be of gag and gag-pol precursors, ensuring both an appropriate balance of the virus proteins in virions as well as the absence of active reverse transcription complexes within the infected cell (see later). This scheme would agree with electron micrograph evidence which shows that there is a significant structural rearrangement (from an open to a condensed core) shortly after budding.

The assembly and release of virions by budding is an unusual process in at least two respects. First, with most retroviruses, assembly and budding occur simultaneously. Second, analysis of mutants shows that only a relatively small fraction of the virion proteins is required. Mutants lacking pol, env, and genome RNA still yield normal-appearing virions (except for the absence of surface projections). In fact, not even all of gag is required since RSV mutants lacking the C terminal third (including the protease and about half the RNA-binding region) yield virus, albeit aberrant in appearance.[21] Thus, all the ability to form capsids and bud out of a cell seems to reside in just the core and matrix proteins.

Until recently, the mechanisms by which pol was translated remained unclear. In all cases (except HTLV-1 and -2, see below), nucleotide sequencing reveals that the gag and pol genes are directly adjoining and separated by a single termination codon, usually also with a change in reading frame. It was originally proposed that the bypassing of the terminator and shift of reading frame was a consequence of splicing of a small percent of the mRNAs to allow synthesis of a complete gag-pol precursor. Such a spliced RNA could not be detected, however, and it is now clear that novel (for eukaryotic cells) mechanisms of translation suppression account for the expression of pol in a fixed ratio to gag.[22,23] Apparently, under certain conditions which are not fully understood, ribosomes can either skip a terminator (in the case of MLV, a glutamine residue is inserted in its place) or change reading frames. The frequency with which this aberration occurs determines the ratio of gag to gag-pol precursor synthesized. At the moment, this mechanism seems limited to retroviruses; no normal cell or other virus process is known to use it.

The envelope glycoproteins are the products of the env gene, and their synthesis and processing are virtually identical among all retroviruses. The env gene products are translated from a spliced, subgenomic RNA and, like cellular glyoproteins, are synthesized on membrane-bound polyribosomes, followed by removal of an N terminal signal peptide. Glycosylation and transport occur via the usual cellular pathways through the endoplasmic reticulum and Golgi apparatus where cleavage (by a cellular enzyme) into the two subunits occurs, invariably following a sequence of 3/4 or 4/4 basic amino acids. A structure of about 24 uncharged amino acid residues in the smaller protein serves as a membrane anchor.

F. Transactivation

Until recently, it seemed that expression of all the proviruses was accomplished entirely by factors and systems already existent in the cell prior to infection, for which the provirus and its transcript simply provided appropriate signals. Indeed, this property of the retrovirus infection cycle is crucial to the ability of some viruses to acquire oncogenes by exchanging host cell information for some or all of the coding sequence of the virus. This process would not be so readily feasible if these viruses required a viral gene product for expression. More recently it has been found that some of the "newer" retroviruses, including HTLV and HIV, do not follow this formerly general role. Rather, full expression of these viruses apparently requires the presence of one or more protein products of the virus genome itself. This phenomenon was first described with HTLV-I when it was found that reconstructed DNAs containing an HTLV-I LTR joined to a gene with a readily assayable product (chloramphenicol acetyl transferase or CAT) were expressed much more efficiently in HTLV-I-infected than in uninfected cells.[24] This effect was found to be due to the protein encoded by the lor (also called pX or tat-1) sequence.[25-27] This nuclear protein causes the level of viral transcripts to be greatly increased, presumably by affecting the rate of initiation of transcription via interaction (possibly indirectly) with specific sequences in the U_3 region of the LTR.[28,29]

When the same sort of experiment is done with HIV, a similar result is obtained: the expression of HIV-like DNA constructs is greatly stimulated by prior infection of the cell.[30] This effect was initially also attributed to transcriptional activation. However, it now seems to be due to some posttranscriptional event, possibly the rate of translation.[31] HIV, in fact, encodes two such activities. The first one described (tat or tat-III) has a target sequence within U_5 and thus presumably effects all viral mRNAs; the second (art) seems to be specific for gag and env expression.[32] Both of these proteins are encoded by use of different alternative reading frames in the same region of env (Figure 2).[33-36]

It is apparent that the tat gene is required for replication of HIV;[37] however, the role of transactivation in the life of these retroviruses remains to be clairified. The phenomena observed may reveal only a portion of a more complex regulatory network in which (for example) the activity of the transactivating proteins themselves might be somehow regulated by interaction with a cell component. Indeed, both types of viruses seem to have quiescent phases in infected individuals in which viral expression is very difficult to detect. Alternatively, transactivation may allow higher levels of expression or a more rapid infection cycle than would be available with unmodified cell systems. It should be emphasized that transactivation has been studied almost solely using the artificial constructs and transfection methodology, which (although quite revealing) may only disclose a fraction of the overall picture. Experimentation using more complete infection cycles is needed to clarify the issue further.

VI. CONSEQUENCES OF RETROVIRUS REPLICATION

As noted above, retrovirus replication is usually without effect on the infected cell. It becomes permanantly virus producing, but is otherwise unaltered. Aberrations in this process that lead to pathogenic virus are rare, but highly instructive. The two basic types of aberrant interactions that have been seen are, first, malignant transformation of the infected host cell; and, second, interactions which lead to death of the infected cell. These are due to fundamentally different sorts of mechanisms. The first case usually involves the mediation of oncogenes, either genetically acquired by the virus, or created by insertion of the provirus near or within the related cellular sequence (a c-*onc* or protooncogene). The second involves either slight variation in the overall replication cycle to permit overreplication, or a more direct toxic effect of some virus gene product.

A. Transformation and Oncogenes

The activation event which converts a protooncogene into an oncogene can be one of three fundamentally different types. First, all or part of the nucleotide sequence can be inserted into the viral genome, usually in place of viral genes. This rare set of events gives rise to rapidly transforming viruses such as Rous sarcoma virus, Abelson murine leukemia virus, and simian sarcoma virus.

The second type of activation can occur following integration of a provirus into or near protooncogenes. This type of event, although it probably occurs less than once per million infected cells, can occur at a high frequency in certain circumstances, as in chickens infected with avian leukosis virus (ALV) or mice infected with murine leukemia virus (MLV) or mammary tumor virus (MTV). For example, integration with the c-*myc* gene is found in virtually all B-cell lymphomas induced by ALV, and many T-lymphomas induced by MLV.

At least four different mechanisms seem to be responsible for activation of different protooncogenes in different diseases. The first mechanism is usually seen in the case of ALV-induced lymphoma, where the 3′ LTR can act as a newly inserted promoter to drive transcription of the downstream portion of c-*myc*. There are three direct consequences of this, all of which may be important to inducing the disease. For one, the level of transcription (and hence of the protein product) is greatly elevated. Second, expression of c-*myc* is separated from the usual regulation influences. Third, in the usual case, a 5′ noncoding portion of the transcript is absent because of the insertion point of the provirus between the first two exons. Thus, although the product is identical, its level of expression and transcriptional and translational regulation are likely to be quite different. Considering the prior discussion regarding differential function of the two LTRs, it is noteworthy that the inserted provirus is almost always deleted in a region adjacent to the 5′ LTR[38] (Hayward, W., personal communication).

The second mechanism is generally referred to as "enhancer insertion" and is usually characterized by integration of proviruses completely outside the transcribed region and in the opposite transcriptional orientation. In these cases, transcription seems to be initiated at the "correct" promoter, but again, with regulation and levels determined by the LTR enhancer rather than the usual signals.

The third case also involves integration within and in the same transcriptional orientation as the oncogene, but the 5′ LTR provides the initiation site for transcription. This sort of activation is found in activation of c-*erb*B in ALV-induced erythroleukemia.[39] A truncated form of the c-*erb*B protein (missing the domains normally expressed in exons upstream of the integration site) is translated from a complex transcript in which env and leader sequences are spliced onto the 3′ exons of c-*erb*B. This results in both aberrant expression and altered structure of the protein product.

Finally, a recently described case implies what might be called "terminator insertion", in which a provirus is integrated downstream of the *pim-1* gene in MLV-induced lymphoma in the same transcriptional orientation. In this instance, the importance of the insertion seems to be the provision of the poly(A) site derived from the 5′ LTR and consequent truncation of the transcript without altering the product. This truncation has the effect of removing a signal which specifies instability of the RNA. Thus, although the structure of the protein and the regulation and level of transcription may be substantially unaltered, increased stability of the mRNA may lead to a substantially greater level of product.

Similar or identical oncogenes can also apparently be activated by nonviral events. For example, chromosome breakage and rejoining may accomplish activation of c-*myc* in certain lymphoid tumors in mouse and man, and point mutations activate *ras* in both spontaneous (or chemically induced) tumors and certain transforming viruses.

In exception to the other viruses, carcinogenesis by HTLV and BLV seems not to involve the action of typical oncogenes; these viruses do not contain cell-derived sequences, and there seems to be no similarity of integration regions from one tumor to the next, as would be expected for insertional activation. Rather, it has been suggested that carcinogenesis might be due to transactivation of cellular protooncogenes mediated by the lor gene product. A recently suggested pair of targets for such transactivation is the T-cell growth factor IL-2 as well as the cell receptor it uses (Rosen, C., personal communication). It could be imagined that overexpression of this protein could cause cells to be stimulated to divide in the complete absence of the usual IL-2 signal, but direct evidence for this mechanism remains scanty.

B. Cytopathic Interactions

Infection with a number of retroviruses — including HIV and even some oncoviruses — can lead to cell death in certain cell culture models. Two fundamentally different mechanisms have been proposed, both of which may be important in HIV infection. The first mechanism involves apparent "overreplication" of the virus. It has been repeatedly noted that cells infected with cytopathic retroviruses have excessive amounts of unintegrated DNA — hundreds of copies as compared to one or a few copies in noncytopathic infections.[40-43] Since retroviral DNA apparently does not contain signals for its own replication, the excess copies most probably arise by repeated cycles of transcription — reverse transcription either via complete infection cycles involving reinfection of a cell with virus produced by it, or by a "short-circuit" cycle involving reverse transcription of newly synthesized RNA. Two features of the "usual" replication cycle serve to prevent this from happening. First, cleavage of the gag and gag-pol precursor proteins into an active "mature" form does not take place until the virion is released (or about to be released) by the infected cell. Thus, an active reverse transcriptase-template complex seems never to be present within the cell. Second, expression of the env gene product at the cell surface confers a very high level of resistance to super-infection, preventing nascent virions from reentering the same cell. Indeed, cytopathic variants of normally noncytopathic ALV strains differ specifically in the particular receptor utilized for infection,[44,45] indicating that it may be more difficult to establish superinfection resistance with some retroviral receptors than others. In the case of HIV and most other cytopathic retroviruses, the mechanism of accumulation of viral DNA is not known.

Also not known is why excess viral DNA should be associated with cell killing. It may simply be that the excess of viral signals, such as binding sites for transcriptional or translational factors, renders them unavailable to the cell for its own purposes. Alternatively, the DNA itself may be toxic or there may be a consequent high level of expression of some more toxic viral gene products.

An alternative cytopathic mechanism is based on the observation that HIV-infected cultures contain numerous multicellular syncytia comprised of large numbers of cells fused together. Such cells can apparently form by interaction of the env protein on the surface of an infected cell with T4 protein on a neighboring cell, followed by membrane fusion analogous to that used for virus infection. This event has been observed directly on mixtures of T4-positive cells with cells expressing env protein.[46,47] It has been proposed that this mechanism would permit a relatively small number of infected cells to do considerable damage by incapacitating a much larger number of T4-positive uninfected cells.

VII. EVOLUTION AND GENETICS

Retroviruses have been around for at least a substantial fraction of vertebrate evolution. Unlike most infectious agents, retroviruses have left "fossil" traces of prior association with their host in the form of endogenous proviruses in the germ line. Although that class of endogenous proviruses which can be readily expressed as infectious virus seems to be relatively recently acquired (at least postspeciation), recent experimentation has revealed additional, new classes of endogenous viruses in the DNA of several species including humans. These are distantly related to the modern oncoviruses (such as murine leukemia virus), but are clearly of great antiquity since they have suffered many point mutations which would render their coding regions unusable, and even the LTRs (which should have been identical at the time of insertion) have diverged in sequence.[48] Furthermore, the similarity of integration sites of some elements in humans and chimpanzees[49] implies that their presence predates divergence of the species. Thus, the association of retroviruses with human ancestors would seem to be quite longstanding and involve repeated waves of "endogenization" of proviruses. Unfortunately, other groups of retroviruses, such as lentiviruses, do not seem to have left such distinct traces in the germ line. Although clearly very ancient, retroviruses are also capable of very rapid variation and adaptation to new niches as evidenced by two recent examples. The first is the relatively frequent appearance of viruses containing oncogenes. These are presumably the consequence of a very rare series of events involving integration at specific sites and illegitimate recombination. Although the probability of any retrovirus-infected cell giving rise to an oncogene-containing virus could be substantially less than 10^{-10}, the consequences are readily visible and a very large number of animals can be easily screened by visiting a slaughterhouse or a veterinarian. Because the horizontal transmission of retroviruses is very poor, viruses as virulent as the transforming viruses almost certainly die with the animal in which they arose, unless they are given a good home by the curious virologist. Thus, the laboratory can be considered the niche to which these viruses have adapted.

The spread of HIV into the human population represents a second example of recent retrovirus evolution. In this case, the virus is most likely a variant of much older viruses, which probably existed in Africa in association with either a monkey or possibly an isolated human population.[50-52] Its appearance internationally probably reflects changes in living conditions including increased international travel.

It should be clear from the above discussion that retroviruses have adapted to a wide variety of niches, and have done so despite their very low rate of transmission and the lability of the virion. The adaptability of these viruses is intimately related to special features of the replication cycle. Retrovirus replication involves a fairly high frequency of both point mutations and rearrangements and an extraordinarily high frequency of recombination. The former presumably reflects infidelity of the replication systems (probably both RNA polymerase and reverse transcriptase) and gives rise to about one error per replication cycle.[53] The recombination frequency seems to be a consequence of the diploid genome of retroviruses and the ability of the reverse transcriptase system to change templates during replication.[6]

The result of this "genetic flexibility" is a significant amount of genetic variation within even a closely related population of viruses. Different HIV or ALV isolates, for example, can differ by 5 to 10% from one another. The ability to generate this sort of variation may be important not only in allowing the viruses to adapt to new environments, but also to aid survival, as in the apparently rapid sequence and possibly antigenic variation in HIV and some other viruses. It should be apparent that the rapid variability of these viruses must be borne in mind when developing strategies to deal with them.

REFERENCES

1. **Weiss, R., Teich, N., Varmus, H. E., and Coffin, J. M., Eds.,** *Molecular Biology of Tumor Viruses: RNA Tumor Viruses,* Vol. 1, Cold Spring Harbor Press, Cold Spring Harbor, N.Y., 1982.
2. **Weiss, R., Teich, N., Varmus, H. E., and Coffin, J. M., Eds.,** *Molecular Biology of Tumor Viruses: RNA Tumor Viruses,* Vol. 2, 2nd ed., Cold Spring Harbor Press, Cold Spring Harbor, N.Y., 1985.
3. **Varmus, H. E.,** Form and function of retroviral proviruses, *Science,* 216, 812, 1982.
4. **Baltimore, D.,** Retroviruses and retrotransposons: the role of reverse transcription in shapping the eukaryotic genome, *Cell,* 40, 481, 1985.
5. **Kane, S. E. and Beemon, K.,** Precise localization of m^6A in Rous Sarcoma virus RNA reveals clustering of methylation sites: implications for RNA processing, *Mol. Cell. Biochem.,* 5, 22998, 1985.
6. **Coffin, J. M.,** Structure, replication and recombination of retrovirus genomes: some unifying hypotheses, *J. Gen. Virol.,* 42, 1, 1979.
7. **Tsichlis, P. N. and Coffin, J. M.,** Role of the C region in relative growth rates of endogenous and exogenous avian oncoviruses, *Cold Spring Harbor Symp. Quant. Biol.,* 54, 1123, 1980.
8. **Donehower, L. A. and Varmus, H. E.** A mutant murine leukemia virus with a single missense codon in pol is defective in a function affecting integration, *Proc. Natl. Acad. Sci. U.S.A.,* 81, 6461, 1984.
9. **Duyk, G., Longiaru, M., Corinik, D., Kowal, R., deHaseth, P., Skalka, A. M., and Leis, J.,** Circles with two tandem long terminal repeats are specifically cleaved by pol gene-associated endonuclease from avian sarcoma and leukosis viruses: nucleotide sequences required for site-specific cleavage, *J. Virol.,* 56, 589, 1985.
10. **Panganiban, A. T. and Temin, H. M.,** The terminal nucleotides of retrovirus DNA are required for integration but not virus production, *Nature (London),* 306, 155, 1983.
11. **Panganiban, A. T. and Temin, H. M.,** Circles with two tandem LTRs are precursors to integrated retrovirus DNAs, *Cell,* 36, 673, 1983.
12. **Varmus, H. E.,** Retroviruses, in *Mobile Genetic Elements,* Shapiro, J.A., Ed., Academic Press, New York, 1983, 411.
13. **Vijaya, S., Steffen, D. L., and Robinson, H. L.,** Acceptor sites for retroviral integrations map near DNase-1 hypersensitive sites in chromatin, *J. Virol.,* 60, 683, 1986.
14. **Jones, K. A., Kadonaga, J. T., Luciw, P. A., and Tjian, R.,** Activation of the AIDS retrovirus promoter by the cellular transcription factor Sp1, *Science,* 232, 755, 1986.
15. **Chatis, P. A., Holland, C. A., Silver, J. E., Fredrickson, T. N., Hopkins, N., and Hartley, J. W.,** A 3′ end fragment encompassing the transcriptional enhancers of nondefective Friend virus confers erythrolukemogenicity on Moloney leukemia virus, *J. Virol.,* 52, 248, 1984.
16. **Rosen, C. A., Haseltine, W. A., Lenz, J., Ruprecht, R., and Cloyd, M. W.,** Tissue selectivity of murine leukemia virus infection is determined by long terminal repeat sequences, *J. Virol.,* 55, 862, 1985.
17. **Stocking, C., Kollek, R., Berholz, U., and Ostertag, W.,** Point mutations in the U3 region of the long terminal repeat of Moloney murine leukemia virus determine disease specificity of the myeloproliferative sarcoma virus, *Virology,* 153, 145, 1986.
18. **Herman, S. A. and Coffin, J. M.,** Differential transcription from the long terminal repeats of integrated avian leukosis virus DNA, *J. Virol.,* 60, 497, 1986.
19. **Cullen, B. R., Lomedico, P. T., and Ju, G.,** Transcriptional interference in avian retroviruses — implications for the promoter insertion model of leukemogenesis, *Nature (London),* 307, 241, 1984.
20. **Miller, C. K. and Temin, H. M.,** Insertion of several different DNAs in reticuloendotheliosis virus strain T suppresses transformation by reducing the amount of subgenomic DNA, *J. Virol.,* 58, 75, 1986.
21. **Voynow, S. L. and Coffin, J. M.,** Truncated gag-related proteins are produced by large deletion mutants of Rous sarcoma virus and form virus particles, *J. Virol.,* 55, 79, 1985.
22. **Yoshinaka, Y., Katoh, M. I., Copeland, T. D., and Oroszlan, S.,** Translational readthrough of an amber termination codon during synthesis of feline leukemia virus protease, *J. Virol.,* 55, 870, 1985.
23. **Jacks, T. and Varmus, H. E.,** Expression of the Rous sarcoma virus pol gene by ribosomal frameshifting, *Science,* 230, 1237, 1985.

24. **Sodroski, J. G., Rosen, C. A., and Haseltine, W. A.,** Trans-acting transcriptional activation of the long terminal repeat of human T lymphotropic viruses in infected cells, *Science*, 225, 381, 1984.
25. **Sodroski, J., Rosen, C., Goh, W. C., and Haseltine, W.,** A transcriptional activator protein encoded by the x-lor region of the human T-cell leukemia virus, *Science*, 228, 1430, 1985.
26. **Sodroski, J. G., Goh, W. C., Rosen, C. A., Salahuddin, S. Z., Aldovini, A., Franchini, G., Wong-Staal, F., Gallo, R. C., Sugamura, K., Hinuma, Y., and Haseltine, W. A.,** Trans-activation of the human T-cell leukemia virus long terminal repeat correlates with expression of the x-lor protein, *J. Virol.*, 55, 831, 1985.
27. **Seiki, M., Hikioshi, A., Taniguchi, T., and Yoshida, M.,** Expression of the pX gene of HTLV-I: general splicing mechanism in the HTLV family, *Science*, 228, 1532, 1985.
28. **Rosen, C. A., Sodroski, J. G., and Haseltine, W. A.,** Location of cis-acting regulatory sequences in the human T-cell leukemia virus type I long terminal repeat, *Proc. Natl. Acad. Sci. U.S.A.*, 82, 6502, 1985.
29. **Rosen, C. A., Sodroski, J. G., Kettman, R., and Haseltine, W. A.,** Activation of enhancer sequences in Type II human T-cell leukemia virus and bovine leukemia virus long terminal repeats to virus-associated trans-acting regulator factors, *J. Virol.*, 57, 738, 1986.
30. **Sodroski, J., Rosen, C., Wong-Staal, F., Salahuddin, S. Z., Popovic, M., Arya, S., Gallo, R., and Haseltine, W. A,** Trans-acting transcriptional regulation of human T-cell leukemia virus type III long terminal repeat., *Science*, 227, 171, 1985.
31. **Rosen, C. A., Sodroski, J. G., Goh, W. C., Dayton, A. I., Lippke, J., and Haseltine, W. A.,** Post-transcriptional regulation accounts for the trans-activation of the human T-lymphotropic virus type III, *Nature (London)*, 319, 555, 1986.
32. **Sodroski, J., Goh, W. C., Rosen, C., Dayton, A., Terwilliger, E., and Haseltine, W.,** A second post-transcriptional trans-activator gene required for HTLV-III replication, *Nature (London)*, 321, 412, 1986.
33. **Arya, S. K., Guo, C., Josephs, S. F., and Wong-Staal, F.,** Trans-activator gene of human T-lymphotropic virus type III (HTLV-III), *Science*, 229, 69, 1985.
34. **Sodroski, J., Patarca, R., Rosen, C., Wong-Staal, F., and Haseltine, W.,** Location of the trans-activating region on the genome of human T-cell lymphotropic virus type III, *Science*, 229, 74, 1985.
35. **Goh, W. C., Rosen, C., Sodroski, J., Ho, D. D., and Haseltine, W. A.,** Identification of a protein encoded by the trans activator gene tatIII of human T-cell lymphotropic retrovirus type III, *J. Virol.*, 59, 181, 1986.
36. **Seigel, L. J., Ratner, L., Josephs, S. F., Derse, D., Feinberg, M. B., Reyes, G. R., O'Brien, S. J., and Wong-Staal, F.,** Transactivation induced by human T-lymphotropic virus type III (HTLV-III) maps to a viral sequence encoding 58 amino acids and lacks tissue specificity, *Virology*, 148, 226, 1986.
37. **Dayton, A. I., Sodroski, J. G., Rosen, C. A., Goh, W. C., and Haseltine, W. A.,** The trans-activator gene of the human T cell lymphotropic virus type III is required for replication, *Cell*, 44, 941, 1986.
38. **Robinson, H. L. and Gagnon, G.,** Patterns of proviral insertion and deletion in avian leukosis virus-induced lymphomas, *J. Virol.*, 57, 28, 1986.
39. **Nitsen, T. W., Maroney, P. A., Goodwin, R. G., Reitman, F. M., Crittenden, L. B., Raines, M. A., and Kung, H-J.,** C-erbB activation in ALV-induced erythroblastosis: novel RNA processing and promoter insertion result in expression of an amino-truncated EGF receptor, *Cell*, 41, 719, 1985.
40. **Keshet, E. and Temin, H. M.,** Cell killing by spleen necrosis virus DNA, *J. Virol.*, 31, 376, 1979.
41. **Mullins, J. I., Chen, C. S., and Hoover, E. A.,** Disease-specific and tissue-specific production of unintegrated feline leukemia virus variant DNA in feline AIDS, *Nature (London)*, 319, 333, 1986.
42. **Shaw, G. M., Hahn, B. H., Arya, S. K., Groopman, J. E., Gallo, R. C., and Wong-Staal, F.,** Molecular characterization of human T-cell leukemia (lymphotropic) virus type III in the acquired immune deficiency syndrome, *Science*, 226, 1165, 1984.
43. **Weller, S. K., Joy, A. E., and Tamin, H. M.,** Correlation between cell killing by members of some subgroups of avian leukosis virus and the transient accumulation of unintegrated linear viral DNA, *J. Virol.*, 33, 494, 1980.
44. **Weller, S. K. and Temin, H. M.,** Cell killing by avian leukosis viruses, *J. Virol.*, 39, 713, 1981.
45. **Dorner, A. J. and Coffin, J. M.,** Determinants for receptor interacting and cell killing on the avian retrovirus glycoprotein gp85, *Cell*, 45, 365, 1986.
46. **Lifson, J. D., Reyes, G. R., McGrath, M. S., Stein, B. S., and Engelman, E. G.,** AIDS retrovirus induced cytopathology: giant cell formation and involvement of CD4 antigen, *Science*, 232, 1123, 1986.
47. **Sodroski, J., Goh, W. C., Rosen, C., Campbell, K., and Haseltine, W. A.,** Role of the HTLV-III/LAV envelope in syncytium formation and cytopathicity, *Nature (London)*, 322, 470, 1986.
48. **Repaske, R., Steele, P. E., O'Neill, R. R., Rabson, A. B., and Martin, M. A.,** Nucleotide sequence of a full-length human endogenous retroviral segment, *J. Virol.*, 54, 764, 1985.
49. **Steele, P. E., Martin, M. A., Rabson, A. B., Bryan, T., and O'Brien, S. J.,** Amplification and chromosomal dispersion of human endogenous retroviral sequences, *J. Virol.*, 59, 545, 1986.
50. **Curran, J. W., Morgan, W. M., Hardy, A. M., Jaffe, H. W., Darrow, W. W., and Dowdle, W. R.,** The epidemiology of AIDS: current status and future prospects, *Science*, 229, 1352, 1985.

51. **Kanki, P. J., Barin, F., M'Boup, S., Allan, J. S., Romet-Lemonne, J. L., Marlink, R., McLane, M. F., Lee, T-H., Arbeille, B.,Denis, F., and Essex, M.,** New human T-lymphotropic retrovirus related to simian T-lymphotropic virus type III (STLV-IIIAGM), *Science,* 232, 238, 1986.
52. **Kanki, P. J., Alroy, J., and Essex, M.,** Isolation of T-lymphotropic retrovirus related to HTLV-III/LAV from wild-caught African green monkeys, *Science,* 230, 951, 1985.
53. **Coffin, J. M., Tsichlis, P. N., Barker, C. S., Voynow, S., and Robinson, H. L.,** Variation in avian retrovirus genomes, *Ann. N.Y. Acad. Sci.,* 354, 410, 1980.

Chapter 2

REVERSE TRANSCRIPTION IN THE PLANT VIRUS, CAULIFLOWER MOSAIC VIRUS

J. M. Bonneville, T. Hohn, and P. Pfeiffer

TABLE OF CONTENTS

I. Introduction 24

II. CaMV up to the Replication Model 24
- A. The Virion Particle 24
- B. The Viral Genome: Structure and Genetic Organization 24
- C. CaMV Proteins 26
- D. Transcription 27
- E. Viral DNA Synthesis: Early Experiments and Replication Model 27

III. Confirmation of the Involvement of Reverse Transcription 28
- A. Genetic Evidence for the Replication Model 28
  - 1. Mutants and Conserved Sequences at the Discontinuity Sites 28
  - 2. Recombination Events Mapping in the R Region of the Viral DNA 30
  - 3. Deletions Explained by Splicing Events 31
- B. Towards Direct Evidence for Reverse Transcription 32
  - 1. In Vivo Replication Intermediates 32
  - 2. A Nuclear Phase and a Cytoplasmic Phase: Transport Problems and Intracellular Sites of DNA Synthesis 32
  - 3. In Vitro CaMV DNA Synthesis Directed by Viral Replicative Complexes (VRCs) 33
  - 4. Enzymatic Questions: What is the Functional CaMV Reverse Transcriptase? 34
    - a. Reverse Transcriptase Activity in Infected Plants and a Viral ORF as a Strong Candidate for its Gene 35
    - b. ORF V Products in Infected Plants and in Heterologous Systems 36

IV. Implications of the CaMV Replication Mode for Gene Expression 36

Acknowledgments 38

References 38

I. INTRODUCTION

Cauliflower mosaic virus (CaMV) is the type member of the caulimoviruses, a group of isometric plant viruses whose virions contain dsDNA. Not many years ago, the discussion of these DNA-containing viruses in a series devoted to RNA genetics would have been an unexpected and apparently inappropriate inclusion. However, the CaMV circular DNA molecule found in mature virions orignates from reverse transcription of a long RNA transcript that represents the entire viral genome and features a direct terminal repeat. As with retroviruses and retrotransposons, the CaMV genome passes through an RNA phase, and as such may be susceptible to the genetic variation events specific to RNA.

Literature in the field of CaMV has been reviewed in the last years, with emphasis on viral DNA manipulation,[1] close comparison to hepadna- and retroviruses,[2,22] or replication in protoplasts.[3] In this chapter, after a brief description of the virus, we shall present the experiments leading to, and those confirming, the reverse transcription model for the replication of the viral genome. Recent data on CaMV will be used to draw a parallel with retroviruses and other retroid elements. We will also consider mechanisms by which the 35 S RNA, which is used as an intermediate of replication, may also serve as a messenger RNA for the expression of CaMV genes.

II. CaMV UP TO THE REPLICATION MODEL

A. The Virion Particle

Caulimovirus particles appear under the electron microscope as isometric spheres about 50 nm in diameter. Virions are located in the cytoplasm of infected cells and accumulate in large numbers (sometimes in paracrystalline arrays) within amorphous proteinaceous inclusion bodies, also called viroplasms, which are characteristic of infection by caulimoviruses. There is no membrane surrounding these inclusion bodies, nor are the virus particles enveloped. CaMV particles are very resistant to dissociating agents, and, indeed, extraction of the viral DNA requires protease treatment prior to phenol extraction.

Polyacrylamide gel analysis of the viral proteins reveals a complex and somewhat variable polypeptide pattern, and analogies with some animal viruses suggested that some of these polypeptides may be internal proteins that interact with the viral nucleic acid to form a nucleoprotein core.[4] Al Ani et al.[5] showed, nevertheless, that the apparent multiplicity of the protein bands observed could be accounted for by proteolysis and aggregation. Furthermore, all polypeptides are equally accessible to proteolytic attack, indicating that the virus does not contain an internal protein layer. Neutron-scattering studies confirmed that CaMV virions are essentially hollow spheres in which the DNA interacts with the inner side of the protein subunits.[6] The shell of the virus is assembled from 57-kdalton protein subunits that undergo a series of proteolytic cleavages. Whether the responsible protease is host or virus encoded remains unknown.

B. The Viral Genome: Structure and Genetic Organization

Caulimovirus particles contain a circular, double-stranded DNA molecule; these DNA circles are relaxed and migrate in agarose gels as several bands that nevertheless share the same restriction map.[7] These distinct topological forms have been identified as DNA circles with an increasing number of knots.[8] The circular DNA molecules are relaxed due to the presence of single-stranded interruptions at specific positions in both strands. The number of these S1 nuclease-sensitive sites varies among caulimoviruses, but invariably, one strand (the transcribed strand in the case of CaMV) possesses a single interruption called Δ_1, while the other strand bears one to three interruptions called Δ_2, Δ_3, etc. (Figure 1).[7,9,10] Chemical sequencing of CaMV reveals that these so-called "gaps" have in fact a triple-stranded

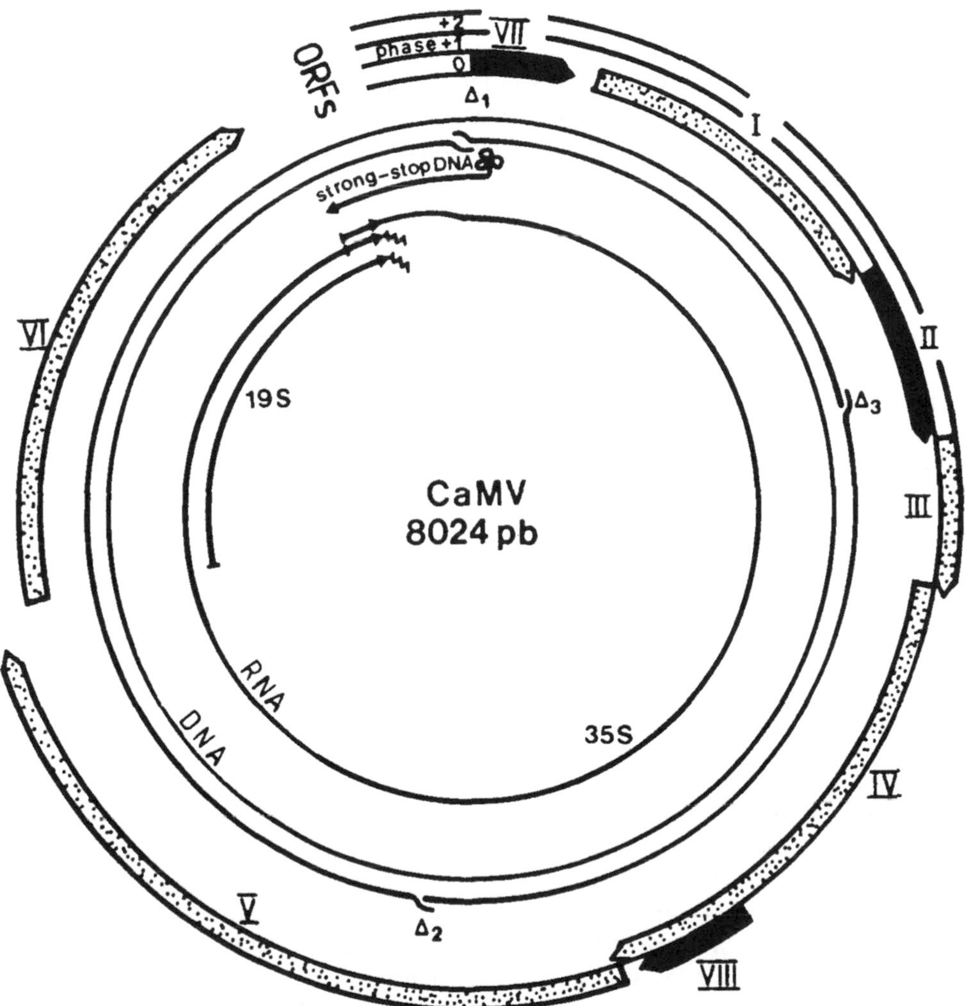

FIGURE 1. CaMV nucleic acids (Cabb-S isolate). From inside to outside: 19 S and 35 S RNA (arrow: terminal repeat; wavy line: poly(A) tail), α_1 "strong-stop" DNA, CaMV minus and plus DNA strands with the single strand interruptions ($\Delta_{1,2,3}$), and open reading frames. Dotted and solid arrows represent essential and dispensible cistrons, respectively (ORF VIII status is not known).

structure in which a short stretch of DNA is redundant, with the 3′ and 5′ ends of the interrupted chains overlapping by 8 to 43 nucleotides.[11] The eight redundant nucleotides at Δ_1 include only A and T residues, which makes possible a three strand helix configuration.[12]

These bizarre structures were first suspected to be either encapsidation signals or to be related to the replication process, by a rolling circle mechanism, for instance. These speculations were supported by the fact that CaMV DNA cloned into a bacterial plasmid loses its gaps. When the cloned DNA is precisely excised from its prokaryotic vector, it recovers the ability of the naked virion DNA to infect turnip plants.[13-15] The progeny virus again contains the gaps at their original position: these structural features are therefore not required for infectivity, but arise during the replication cycle of the virus. It was later discovered that ribonucleotides are often covalently bound 5′ to the first deoxynucleotide of the DNA chains, suggesting RNA priming and providing further evidence that the gaps are not merely postreplicative decorations.[16]

At the same time, CaMV transcripts were found to span over the minus strand gap,[17,18] making the encapsidated form of the DNA an unlikely template for transcription. This paradoxal result led to the discovery of nonencapsidated, covalently closed, supercoiled CaMV DNA[19,20] associated with histones to form transcribing minichromosomes[21] with an average of 41 nucleosomes.[8] So far, CaMV DNA has not been found integrated into the genome of its hosts, which include plants of the crucifer family and, for some virus isolates, a few additional solanaceous species.[22]

The complete DNA sequence (8 kilobases) is available for four CaMV isolates: the Strasbourg isolate Cabb-S,[23] the Californian strain CM 1841,[24] the Hungarian strain DH,[25] and the Xinjing isolate from China.[25a] They differ by an average of 5%, mainly in the wobble base in coding DNA. All three isolates exhibit the same asymmetry in their genetic organization: seven open reading frames (ORF) can be read from a single DNA strand (the plus strand) in the same respective positions and phases (Figure 1). A striking feature of the potential CaMV genes is their tight packing: all together, they use up to 90% of the genome. A single nucleotide separates ORF I and II. The same is true for ORF II and III in three strains, while they overlap by five nucleotides in the fourth one. ORFs III and IV overlap by 13 nucleotides, and ORFs IV and V by 34. ORF VI, the most variable CaMV gene, is by contrast flanked by noncoding regions, a short one (≃100 base-pairs) upstream and a longer one (≃700 base pairs) downstream. This last intergenic sequence is the most conserved part of the genome.

The capsid proteins of the caulimovirus group members show immunological cross-reactions (reviewed in Reference 26). The insect transmission factors somehow cross-complement, since a CaMV ORF II mutant normally deficient for insect transmission can be transmitted by aphids prefed with plants infected with carnation etched ring virus (CERV) or figwort mosaic virus (FMV).[26a] Restriction maps are available for dahlia[10] and mirabilis mosaic virus,[27] as well as for soybean chlorotic mottle virus[27a] and horseradish latent virus.[27b] CERV and FMV have recently been sequenced.[27c,d] Both sequences display homologies with CaMV for their overall organization and their deduced coding capacity: homologues to the CaMV ORFs I, II, III, IV, V, and VI are found at conserved positions. The ORFs I and V have the most conserved coding potential. The ORFs I-II-III-IV-V are also tightly packed, with small overlaps at all of the ORF interfaces; with respect to CaMV, the ORF junctions of CERV are simpler, with the stop and start codon entangled in the sequence ATGA. The strongly divergent ORF VI is again flanked by a short noncoding region in 5′ and a long one in 3′. A small ORF is present upstream of ORF I for both FMV and CERV, but differs in sequence from CaMV ORF VII. For 3 CaMV isolates and for CERV, a small reading frame is contained within the 3′ region of ORF IV in a different phase; this overlapping ORF VIII is, however, absent in CAMV-Xinjing and in FMV. Another small, nonconserved (and therefore probably fortuitous) reading frame, ORF IX, overlaps the ORF III-IV junction in FMV. FMV shows less sequence homologies to CaMV than CERV does; the conserved sequences are therefore likely to point out conserved biological functions: among these are the central part of ORF VI and a 35-base-pair sequence in the long intercistronic region.

C. CaMV Proteins

In vitro mutagenesis of cloned CaMV DNA shows that alterations in the CaMV ORFs generally give rise to defective genomes: in-frame[28] and out-of-frame linker insertions[29,30] in ORFs I, III, IV, V, and VI are lethal to virus spread, while the small ORFs VII and II can accept minor insertions[30,31] or even complete deletion[32,33] without loss of infectivity. Mutants in ORF II lose the ability to be transmitted by aphids, the natural caulimovirus vector. These results indicate that potential coding regions are in fact expressed. Indeed, translation products of most of the viral ORFs have been identified.

The ORF I protein has been recently detected in infected plants:[33a,b,c] its function remains unelucidated, but it exhibits structural homologies both to the ATP binding domain of animal kinases and to the 30- to 34-kdalton protein involved in the cell-to-cell spread of RNA plant viruses.[26c,33a,34] Genetic analysis has assigned to the dispensible ORF II an 18-kdalton protein;[35,36] this polypeptide is a component of the proteinaceous viroplasm matrix.[37] Definitive evidence for the expression of this ORF comes from its substitution by the coding region of a dihydrofolate reductase gene: plants infected by a virus modified as such express the novel enzyme.[33] ORF III is expressed as a 15-kdalton protein representing the entire coding capacity of the reading frame.[38] This gene product exhibits DNA binding properties,[38a] and a related polypeptide is virion-associated (Reference 38a and Gordon et al., submitted). ORF IV encodes the coat protein precursor.[24,39] The 57-kdalton primary translation product, not unlike the gag gene of retroviruses, is phosphorylated and undergoes specific cleavages to give rise to the main components of the virion shell.[40,41] Coat protein phoshorylation can be reproduced in vitro since the responsible kinase, of unknown origin, is tightly associated with the virion particle.[41a,41b] ORF V expression products detected in infected plants also show limited proteolysis.[42,42a] One of the ORF V protein products is most probably the viral polymerase, as discussed in Section II.B.4.

ORF VI is the gene encoding the main structural component of the viroplasm matrix; its 66-kdalton product is implicated in symptom severity;[43] it is also a host range determinant.[43a] No protein has been assigned as yet to ORF VII, but certain frameshift mutations in this small gene are lethal,[32] and a mutation in its initiation codon reverts in vivo to wild type,[44] which suggests that it is indeed translated. No evidence for translation of ORF VIII exists so far.

D. Transcription

Two major capped and polyadenylated transcripts can be detected in CaMV-infected plants: the 19 S RNA, which is the messenger for gene VI,[17,18,45] and a 35 S RNA that covers the whole viral genome plus a terminal direct repeat of 180 nucleotides (Figure 1). These two RNAs are synthesized by the host RNA polymerase II and have separate 5′ ends and promoters (which have both been widely and successfully used in plant genetic engineering), but they share the same 3′ terminus.[46] How 3′ end RNA maturation is by-passed at the upstream polyadenylation site on the 35 S RNA remains unknown.

Despite extensive search, no convincing evidence exists for subgenomic messengers other than the 19 S RNA. On the basis of S1 nuclease mapping experiments, Meagher and Condit proposed the existence of several genome-length transcripts with different endpoints on the DNA circle.[47] However, to date this suggestion has not been confirmed by primer extension analysis or any other data. Plant et al.[48] proposed the existence of a subgenomic mRNA for ORF V from hybrid selection experiments followed by in vitro translation, but this molecule has never been seen on a Northern blot. It therefore seems that the 35 S RNA, despite its polycistronic nature, may serve as a messenger for the expression of all the viral genes except for ORF VI. Such an uncommon expression strategy is supported by indirect genetic evidence and can be considered as a consequence of the mode of replication; this will be discussed in Section III. Viral transcripts have not yet been mapped for other caulimoviruses. The FMV sequence, however, reinforces the assumption that again, a subgenomic messenger will be found for ORF VI in addition to a terminally redundant genomic RNA: two putative TATA boxes and a possible polyadenylation signal exist at the same respective positions as in CaMV.[27d]

E. Viral DNA Synthesis: Early Experiments and Replication Model

The existence of a transcribing CaMV minichromosome suggested that this molecule, like the minichromosome of the better characterized animal DNA virus, SV40, might also be used as a template for DNA replication. Pfeiffer and Hohn[49] decided to examine CaMV-infected plants using procedures that had led to the isolation of replicating SV40 minichromosomes.[50] Fractions containing nuclei and viroplasms were leached in hypotonic buffer in order to recover diffusible transcription and replication complexes without loss of salt-sensitive proteins. Sucrose gradient centrifugation of the solubilized extract separated the components engaged in the synthesis of either viral RNA or DNA: while CaMV transcription was directed by heavy fractions containing supercoiled DNA organized into minichromosomes as already described,[19] CaMV DNA synthesis was confined to the lighter fractions of the gradient, in a zone where clearly no minichromosomes were sedimenting. Analysis of the DNA content of these fractions, either by electron microscopy or by gel electrophoresis, revealed the presence of heterodisperse material with genome-size viral DNA in both relaxed circular and linear forms, but no common intermediates of circular DNA replication such as θ-shaped molecules or rolling circles (Pfeiffer, P., unpublished). In addition, electrophoresis revealed a smear of nucleic acids, some of which were found to be RNA.

This observation, together with a survey of the literature of CaMV and a comparison with the life cycle of duck hepatitis B virus,[51] led these authors to propose that CaMV replicates by reverse transcription of its longer transcript.[49] The replication model was proposed simultaneously and independently by two other groups, based on independent observations. Hull and Covey proposed the same strategy after analyzing unencapsidated and incomplete CaMV DNA molecules, which appear to be replicated counterclockwise starting from the Δ_1 discontinuity.[52] Guilley et al.[16] came to the same conclusion upon discovering α_1 DNA, a single-stranded molecule of negative polarity which resembles strongly the strong-stop DNA of animal retroviruses (see Figure 1). The 3′ end of $\alpha_1$is very close (two nucleotides away) from the 5′ end of the 35 S RNA, while its 5′ end is the same as that of the full-length, minus strand DNA. Immediately upstream of the 5′ end of α_1 DNA in the CaMV sequence is a stretch of 14 nucleotides complementary to the 3′ end of plant initiator $tRNA^{met}$, which therefore could be used as a primer for α_1 DNA synthesis.

A model for CaMV nucleic acid replication based on these observations is illustrated in Figure 2. It is closely related to that of retroviruses: a tRNA primer is used to start the synthesis of the first DNA strand, polypurine tracts (PPTs) are used to prime second-strand synthesis, and the RNA terminal direct repeat allows the first template switch to occur, while the second one takes place at the minus strand priming site. A key feature of the model is the emergence of a covalently closed circular DNA only by the repair in the nucleus of gapped molecules originating from virions. Minichromosomes are thus indeed replicative intermediates, but only as templates for transcription and not as a source of θ-shaped or rolling circles molecules. The model postulates a reverse transcriptase activity for minus strand synthesis, an RNase H activity for RNA template degradation, and a DNA polymerase activity for plus strand synthesis. RTases are known to copy DNA as well as RNA templates, and to carry RNase H activity, so that only one enzyme is finally required.

The initial data supporting this model are strong, but certainly neither definitive nor direct. The model accounts for some of the puzzling observations made about the virus, and has received a wide acceptance by most groups working in the field. Even more important, the model has quite precise implications and therefore was useful in the development of independent and crucial tests needed to resolve the molecular biology of CaMV replication.

III. CONFIRMATION OF THE INVOLVEMENT OF REVERSE TRANSCRIPTION

A. Genetic Evidence for the Replication Model

1. Mutants and Conserved Sequences at the Discontinuity Sites

According to the replication model, second-strand priming takes place at the level of polypurine tracts spared by the RNase H activity and remaining base-paired to the minus strand DNA. The sequence responsible for the formation of the Δ_3 interruption has been

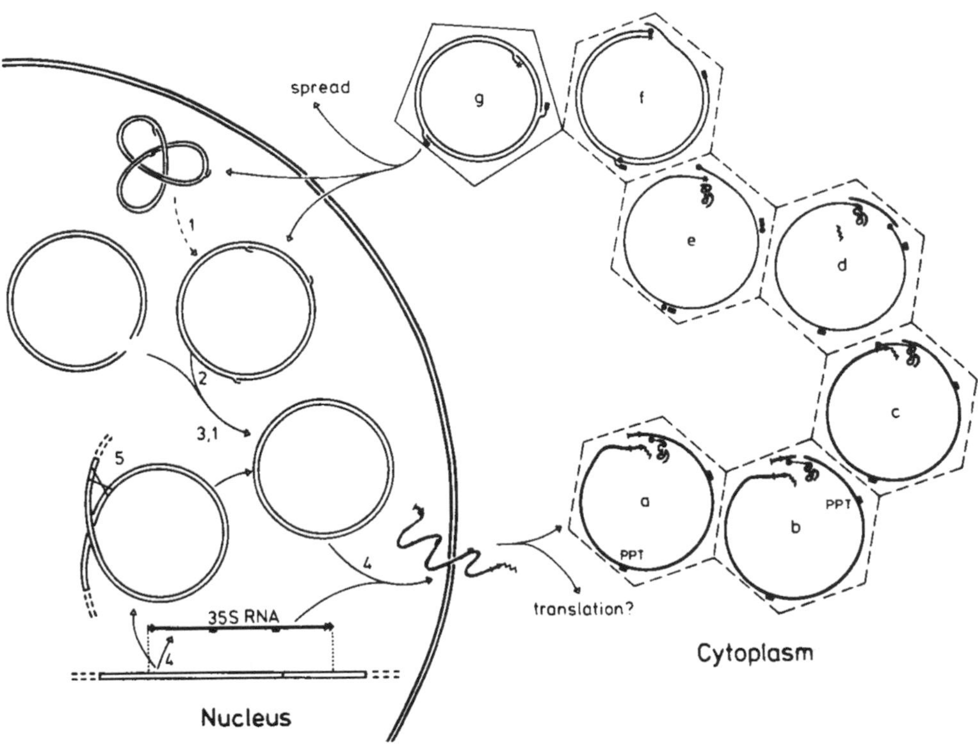

FIGURE 2. Likely interconversion pathways between CaMV nucleic acids. Nuclear events involve topoisomerase (1), DNA repair (2), and ligase (3) activities to give rise to minichromosomes from encapsidated or linearized (cloned) DNA. Minichromosomes direct viral transcription (4), and 19 S and 35 S RNAs are capped, polyadenylated, and exported to the cytoplasm. CaMV concatemers inserted, for example, between Ti-plasmid T-DNA borders, give rise to 35 S RNA, mainly by direct transcription, but also through generation of minichromosomes by recombination at the DNA level.[5] Cytoplasmic events: viral RNAs are translated, and 35 S RNA is transcribed into DNA, probably within virion-like structures (hexagonal boxes, a through f, see Section II.B.2). (a) Postulated early packaging of 35 S RNA with its tRNA primer and reverse transcriptase (●), (b) DNA synthesis procedes up to the 5′ end of the template which is subsequently degraded by RNAse H; the α_1 "strong-stop" DNA generated anneals to the 3′ repeat of the 35 S RNA, allowing the first template switch (c); reverse transcription resumes and completes the minus strand (d,e); the primer tRNA is cleaved, leaving its first 3′ nucleotide on the DNA (*); RNAse H digests the copied RNA, except the polypurine tracts (solid boxes), used in turn as primers for plus strand synthesis (f). Completion of plus strand implies a second template jump, at the Δ_1 discontinuity (the minus strand may use its 3′ end to invade the double helix at its 5′ end; reverse transcriptase would jump from this three-strand helix configuration), and plus strand is completed (g). Possible coat protein modifications from previrion to virion particle (pentagonal box) occur here or before. Some replication steps (e.g., d through f) may occur simultaneously. Note that the overlaps at the minus and plus strand discontinuities do not have the same origin: DNA repeat at Δ_1 implies that the first-copied nucleotides of 35 S RNA are not immediately digested by RNAse H and can be copied a second time at the end of minus strand synthesis, while it is not known what limits the extent of plus strand displacement at Δ_2 and Δ_3.

cloned as a 119-base-pair fragment overlapping the gap; in good agreement with the model, this fragment bears the polypurine tract and does not give rise to the DNA discontinuity when reinserted in the viral DNA in reverse orientation.[53] Polypurine tracts are also present in the sequence of CERV and FMV at the vicinity of the plus strand gaps. As far as the minus strand is concerned, CERV and FMV also show homology to the 3′ end of plant $tRNA_i^{met}$, although at a different position (immediately upstream of ORF I and not of ORF VII).[26c,d] In CaMV, removal of the tRNA binding site destroys the virus infectivity: counterclockwise deletions starting from the adjacent ORF VII can remove up to the first base of the initiation codon without loss of viability, but a deletion larger by 17-base pairs, removing the immediately contiguous 13 nucleotides complementary to the initiator tRNA plus 4 noncoding nucleotides, renders the virus nonviable.[32] This last deletion mutant could be used to change the tRNA homology to another amino acid acceptor, or to move the tRNA binding site elsewhere in the CaMV genome.

2. Recombination Events Mapping in the R Region of the Viral DNA

Since the replication model points to the terminal repeats of the 35 S RNA as the site for template switching, one can expect the corresponding area of the viral DNA (R region) to be a hot spot for recombination events, due to switching by the reverse transcriptase from a first RNA template used to synthesize the minus strand, strong-stop DNA, to a second (and possibly distinct) template copied into the complementary part of the minus strand. Some CaMV DNAs, coming from either a natural or a provoked recombination situation, can indeed be interpreted as the fossils of replicative recombination events.

A natural case is CM 4-184, a CaMV strain considered as a single-deletion derivative of the CM 1841 strain until Dixon et al.[54] sequenced a portion of its genome and found some sequence deviations, chiefly single base substitutions, between the supposed mother and daughter genomes. All of these deviations fall in the area covered by the α_1 "strong-stop DNA" molecule. Comparison to the Cabb-S virus sequence reveals a complementary deviation pattern, with all but one of the sequence deviations now falling outside this area. The authors proposed that in the genetic history of CM 4-184, a recombination event occurred in a plant infected simultaneously by two CaMV strains, with reverse transcription being primed first on 35 S RNA from a Cabb-S-type virus, then switching by an *inter*molecular jump to a CM 1841-derived RNA template after completion of the strong-stop molecule, and thus giving rise to the minus strand DNA of the ancestor of the CM 4-184 isolate. This implies that the first template switch allows CaMV replication to make use of two different and successive RNA templates. In this case, genome circularization would occur later in the replication cycle, during the second strand switch for instance. Whether CaMV replication can occur on a single 35 S RNA template or needs a second one (as may be the case with retroviruses, the genomic RNA molecules of which are associated pairwise),[55] remains unknown.

To provoke an experimental recombination situation, CaMV DNAs have been engineered in partial tandem arrays and inserted between the T-DNA borders of the Ti plasmid of *Agrobacterium tumefaciens*. Inoculation of the resulting bacteria on a wounded turnip generates unit-length viral DNA circles and thereby fully infectious viruses that spread and give rise to the systemic chlorotic lesions characteristic to CaMV.[56] These "agroinfection" experiments indicate two alternative ways of genome escape. The first one, homologous recombination events at the DNA level, was proposed to account for the earlier data of infectivity of a naked and uncut plasmid DNA bearing a larger-than-unit-length CaMV genome.[15] Recombination at the DNA level between CaMV genomes has indeed been further documented and can occur at different places on the viral genome, at least for intermolecular events.[29,57,58] The alternative possibility is that the CaMV partial dimers are transcribed like a retroviral provirus to yield genomic RNA, which can then be copied into DNA by reverse transcription. It is known that the CaMV RNA expression signals are recognized on an integrated viral template, since transgenic plants harboring a full-length CaMV genome (monomeric DNA without terminal redundancy) produce transcripts equivalent to the 19 and 35 S RNAs.[59]

The partial CaMV dimer used for agroinfection was constructed from two distinct parent viruses, CM 4-184 and Cabb-S, and their distinct restriction maps enabled Grimsley et al.[60] to map the crossover point of the escaping (hybrid) viruses. The provirus hypothesis predicts the synthesis of a hybrid 35 S RNA from this DNA in a plant cell, with a CM 4-184 5′ part and a Cabb-S 3′ complement; the junction between these two genomes after reverse transcription of this RNA is then expected to occur within the 180-base-pair R sequence. On the other hand, circularization by DNA crossing-over can occur elsewhere within the much longer (3.4 kilobases) repeated DNA sequence. It turned out that the bulk of the progeny viruses arising from the T-DNA bearing the partial dimer have the restriction map expected from reverse transcription of the hybrid 35 S RNA transcribed from the unrecombined, chromosomally integrated CaMV DNA copy.[60] Reciprocal DNA crossing over or gene conversion events account for the generation of a minor fraction of the viral progeny. Sequence analysis of cloned viruses from the first and prevalent population reveals furthermore that the DNA information in the R sequence is imported mostly, or perhaps exclusively, from the CM 4-184 parent, i.e., from the 5′ repeat of the 35 S RNA. From this observation, and from the fact that the strong-stop DNA extends up to the 5′ end of the 35 S template,[16] it is likely that the template switch during the minus strand DNA synthesis occurs at the 5′ end of the template, and not further upstream within the RNA repeat. This is indeed compatible with the copied RNA being completely digested by RNase H upon strong-stop DNA synthesis, and DNA synthesis resuming upon annealing of the strong-stop DNA with the 3′ end of the 35 S RNA.

3. Deletions Explained by Splicing Events

Hohn et al.[60a] have inserted an intervening sequence from a soybean leghemoglobin gene within the infectivity-dispensable CaMV ORF II. Upon inoculation, this intervening sequence was precisely deleted, albeit at a surprisingly slow rate, and the spliced form of the 35 S RNA could be detected. Hirochika et al.[61] presented parallel evidence for a splicing event by analyzing a natural CaMV isolate confusedly named CaMV-S, but distinct from the Cabb-S strain. In plants infected with either CaMV-S virus or a full-length clone of CaMV-S DNA, they detect a discrete, noninfectious and deleted DNA molecule in addition to the full-length viral DNA. The 5′ and 3′ borders of this reproducible 856-base-pair deletion that spans part of ORF I and are homologous with the consensus donor and acceptor sites for RNA splicing, respectively, site-directed mutagenesis of the donor site strongly backs the interpretation of deletion by splicing of an RNA intermediate: the signature deletion does not occur any longer from the inactivated donor site, and, furthermore, smaller deletions are found that still end up at the same acceptor site, but start from less canonic splicing donor sites located downstream of the mutated startpoint. Another noninfectious deleted clone of CaMV, pUM 124, may have arisen by a similar pathway. This spontaneous-deletion derivative, that lacks most of ORF I, was picked up by cloning viral DNA from Cabb-S infected plants;[62] here again, the deletion donor and acceptor sites fit approximately their splicing counterparts (Melcher, V., personal communication).

It has been stressed that both donor and acceptor sites present in the CaMV-S sequence also exist in other CaMV strains for which this deletion event has not been be detected.[61] This suggests that some or most CaMV strains may possess a mechanism for suppressing splicing of the 35 S RNA, accounting for the slow rate of splicing observed by Hohn et al.[60a] If so, CaMV-S may be a mutant defective in this suppression, which might involve the action in *trans* of a virus-encoded product. The hypothesis of a *trans*-acting inhibitor predicts suppression of the appearance of the deleted satellite in a plant coinfected by CaMV-S and another CaMV strain. In any case, unique properties of given CaMV strains can be mapped on the viral genome by exchanging restriction fragments between two parent viruses of different biological behavior and assaying the phenotype of the hybrid virus. Such an approach has been successfully used to assign functions to ORF II and VI,[35,43] and could also be used here to test whether substitution of sequences from other CaMV strains would correct the putative deficiency in splicing control of the CaMV-S strain.

B. Towards Direct Evidence for Reverse Transcription

1. In Vivo Replication Intermediates

Several classes of minor CaMV nucleic acids have been observed in vivo. First, additional small molecules that map in the large intergenic region have been found. Turner and Covey demonstrated that the virion associated sa-DNA, which corresponds to the α_1 "strong-stop" DNA, can be found with a 75-nucleotide RNA primer covalently linked to its 5′ end.[63] Identity of this primer with $tRNA^{Met}$ was supported by the presence of the modified base, m^7G, at the expected location on the RNA moiety of this RNA-DNA hybrid molecule. Marco and Howell described a hairpin double strand DNA corresponding to the α_1 DNA:[64] this immediately evokes the well-known tendency of RTases to perform snap-back synthesis. Longer hairpin dsDNAs have been described later[64a]

A second class includes full-length, encapsidated molecules with a nonclassical plus strand structure, either with two missing or with an additional gap.[65] The sites for these extra gaps also present PPTs. These results further suggest that initiation of plus strand synthesis requires at least one, but can make use of several, polypurine primer(s) and that their location is not critical.

The third category includes nonencapsidated, supercoiled, and quite puzzling viral DNA circles. Olzewski and Guilfoyle reported less than 8-kilobase-long type I DNA molecules,[66] in a range varying continuously from almost full length to 10% of the CaMV genome size; they may correspond to illegitimate template switch events following strong-stop DNA synthesis, but it is not clear whether these deleted circles have gapped counterparts. Rollo and Covey described the fate of viral DNA upon callus inception from infected leaves:[67] the proportion of viral DNA to total DNA first drops and then apparently stabilizes, and the bulk of the persistent CaMV DNA remains supercoiled, either full length or with discrete deletions. Furthermore, a minor proportion of calli-derived protoplasts from CaMV-infected turnips still contains infective particles, and here again, CaMV DNA is mainly supercoiled.[68] By contrast, the bulk of viral DNA from infected leaves and from *Brassica* protoplasts shortly after infection is encapsidated and either open circular or linear,[69] suggesting that packaging or some other mechanism may limit the transport of newly synthesized viral DNA into the nucleus of already-infected cells. The contrasting situation in calli may reflect a failure of this mechanism in proliferating tissue.

2. A Nuclear Phase and a Cytoplasmic Phase: Transport Problems and Intracellular Sites of DNA Synthesis

Exporting the viral genome from the nucleus to the cytoplasm as a polyadenylated transcript is a logical hypothesis, but which form of the CaMV genome does the nucleus import? Two lines of evidence argue against repair of the gapped DNA before entry into the nucleus: (1) the infectivity of linearized and, hence, inactive cloned viral DNA suggests that the uptake of viral DNA by nuclei is a passive phenomenon; (2) the existence of minor transcripts ending at gap Δ_1, suggestive of transcription of the gapped DNA form.[46] We do not know how the encapsidated DNA enters the nucleus, except that upon in vitro infection of protoplasts with CaMV virions, gapped DNA is released from the nucleocapsid very quickly.[69] The puzzling observation has been made, however, that virion particles of a CaMV argentinian isolate can be found in the nucleus, as well as in cytoplasmic inclusion bodies, of infected *Nicotiana clevelandii* cells.[70] The infected cells of this plant (at the limit of CaMV host range) with virus in their nucleus contained no inclusion bodies, and there is no evidence that these cells produced the observed viral particles, rather than simply accumulated virus replicated by neighboring cells.

Results concerning the site of viral replication have been controversial, with proposals for viroplasm involvement[71,72] and claims of nuclear replication.[73] The replication model predicts that DNA synthesis is cytoplasmic. Experimental data confirm and refine this point:

inclusion bodies support DNA synthesis in a cell-free system,[74] and upon further purification, support viral DNA synthesis only.[75]

Furthermore, the knots and gaps present on the packaged genome indicate that DNA synthesis occurs in a compartment devoid of topoisomerase, DNA repair, and ligase activities; such a DNA structure is quite compatible with a very simple replication complex containing RTase as the sole enzyme involved in DNA metabolism. Several additional pieces of evidence point to the replication compartment being the virion itself or a direct precursor of it: first, the sheer presence within virions of DNA molecules synthesized very early in the replication cycle, such as the minus strand strong-stop DNA;[16,63,76] second, the nuclease resistance of at least some CaMV replication complexes, associated with fast-sedimenting particles;[75,76,78] and, third, the presence of a "cysteine motif" (a feature conserved among the gag polyproteins and implicated in RNA binding) in the C terminal part of ORF IV translation products.[79] Finally, growing minus strand DNA chains, base-paired with RNA, are revealed only after proteinase K digestion; these molecules, smearing from 8000 down to 650 nucleotides (size of α_1), copurify extensively with virion particles and can be immunoprecipitated by antibodies raised to the main capsid protein.[79a] Clearly enough, some CaMV virion or previrion particles contain replicative intermediates encapsidated before completion of the first (minus) strand synthesis; these particles, like those of retroviruses, are able to resume DNA synthesis in vitro when supplied with the necessary ingredients.

Thus, a common trait of the retroid replication strategy involves the packaging of the genomic RNA together with reverse transcriptase into virion or virion-like structures. The whole DNA synthesis is then suspected to occur "in virion". Unlike retrotransposons, however, infectious retroid viruses need a transient interruption in their life cycle corresponding to the step of cell-to-cell movement and reinfection. Very different tactics are adopted here by proper retroviruses, hepadnaviruses, and caulimoviruses: retroviruses do not start DNA synthesis before reinfection, hepadnaviruses export themselves as a semifinished DNA product (with a nicked minus strand and a gap of variable length in the plus strand), and the traveling form of the caulimovirus genome, a double strand DNA circle with limited strand displacement, is ready for repair and ligation. Very little is known about the events triggering the entry into and exit from this kind of dormancy step, but at least for hepadnaviruses and CaMV it is clear that the blocking step implies a stop in DNA elongation, this may be blocking the access of deoxynucleotides to the packaged replication complexes (see also the chapters by Coffin and Bosch et al. on retro- and hepadnavirus replication, respectively).

3. *In Vitro CaMV DNA Synthesis Directed by Viral Replicative Complexes (VRCs)*

Several groups have studied in vitro the synthesis of CaMV DNA directed by viral replication complexes (VRCs) preinitiated in vivo. The conflicting results reported arise chiefly from the differences in the isolation procedures used and in the type of VRCs studied. For instance, in their attempts to assign the CaMV DNA synthesis to a defined cellular compartment, Ansa et al.[73] used high salt treatment and CsCl cushion centrifugation to obtain highly purified viroplasms. In retrospect, their failure to detect any viroplasm-associated CaMV DNA synthesis appears to have resulted from the solubilization of the CaMV-specific DNA polymerase from the VRCs upon ammonium sulfate precipitation of the viroplasms. Indeed, this free enzyme could subsequently be purified by phosphocellulose chromatography.[80] Pfeiffer and co-workers[49,81] took care to avoid exposure of the VRCs to high salt concentrations and dealt exclusively with "soluble VRCs" that could be released by suspensions of organelles (chiefly nuclei and viroplasms) in a hypotonic buffer. Such "soluble VRCs" have a rather low sedimentation coefficient (in the 25 to 30 S range), and their RNA moiety is highly exposed and susceptible to the RNAse and RNAse H activities present in this hypotonic extract (Pfeiffer, P., unpublished). In vitro, these VRCs synthesize CaMV

DNA of both strand polarities, and the relative ratio of plus and minus strand DNA synthesized remained essentially constant in response to changes in the incubation conditions (divalent cation, dideoxynucleotides),[81] suggesting that a single enzyme performs both steps of the CaMV DNA synthesis. Surprisingly, actinomycin D treatment reduced DNA synthesis on both strands in the soluble VRCs, while the synthesis of minus strand DNA from an RNA template should have remained unaffected. A similar observation was made by Marsh et al.[77] on the soluble VRCs present in "cleared nuclear lysates" obtained by the salt- and EDTA-induced lysis of organelle suspensions and fractionated on sucrose gradients: here again, DNA of both strands is being synthesized in vitro, and actinomycin D depletes the synthesis of both species to the same extent. These unexpected results suggest that the minus strand DNA made by the soluble complexes is the copy of a DNA template. Minus strand DNA, however, has been found associated with RNA.[77]

In addition to these soluble complexes, fast-sedimenting VRCs can be isolated from infected leaves.[65,75,78] In "microsomal pellets", Marsh et al.[77] found DNA synthesis of both viral strands, and minus strand synthesis is resistant to actinomycin D. The viroplasms prepared by Mazzolini et al.[75] essentially direct the synthesis of minus strand DNA from circular and linear full-length templates; the in vitro products hybridize to all CaMV restriction fragments without regional preference. In addition to the genomic DNA, a small DNA molecule originating from the intergenic region, but of plus strand polarity, was found heavily labeled (Mazzolini, L., personal communication); this may indicate the presence of some remnants of the RNA template along the α_1 molecule that can be used in vitro as primer.

Finally, the ability of the enzyme engaged in such VRCs to switch to another externally provided template depends very much on the type of VRC considered and on its history: reverse transcriptase remains associated with viroplasms along their purification,[75] with most of the activity probably remaining engaged on viral endogenous templates. Thomas et al.[78] mentioned that the reverse transcriptase present in their microsomal fraction (i.e., in fact, within virus particles) would copy a specific template only after having undergone one cycle of freeze-thaw.[78] We have made a similar observation with "soluble VRCs" (Pfeiffer, P., unpublished).

A critical assessement of the conditions used to isolate VRCs is therefore required in any study prior to drawing any conclusions as to in vitro CaMV synthesis. Questions which must be addressed include the nature of the VRCs studies (virus-like particles or naked, soluble VRCs), the salt conditions used for purification (dissociation of soluble VRCs), whether nucleases were used for any purfications steps (hydrolysis of exposed nucleic acids), and finally, whether the preparation was submitted to any freeze-thaw cycle.

4. *Enzymatic Questions: What is the Functional CaMV Reverse Transcriptase?*

At the time the involvement of an RNA step in replication of the CaMV genome was proposed, a question concerning the better-studied RNA plant viruses was under active discussion: do these viruses use a host RNA-dependent RNA polymerase to replicate, or do they encode their own enzyme? Similarly, the original model for CaMV predicted a reverse transcriptase activity in infected plants, but left open several possible scenarios for its origin: (1) CaMV uses for its replication a host activity capable of copying both RNA and DNA templates — such as the γ-type plant DNA polymerase which is able to copy an RNA homopolymer, like poly(A), into DNA; (2) this polymerase could be used as such, or in association with a virus-encoded subunit which would confer to the host enzyme the ability to copy heteropolymeric RNA; or (3) CaMV codes for its own reverse transcriptase as animal retroviruses do. The latter hypothesis very quickly turned out to be the most attractive one. In any case, the plant DNA polymerase α, the enzyme responsible for replication of the nuclear DNA, is not involved in CaMV DNA replication. This enzyme can be completely and specifically inhibited by aphidicolin,[82] a dCTP competitor, and in vitro CaMV-directed DNA synthesis, either from soluble extracts[81] or from purified viroplasms,[83] is not only fully resistant, but even stimulated by this drug, thus indicating that CaMV DNA is not replicated by this host enzyme.

a. Reverse Transcriptase Activity in Infected Plants and a Viral ORF as a Strong Candidate for its Gene

Computer-assisted search reveals significant protein homology between the translation product of ORF V and that of the pol gene of animal retroviruses, especially of Moloney murine leukemia virus (MoMLV).[80,84] Furthermore, extracts from infected plants exhibit an RNA-dependent DNA polymerase activity absent from their healthy counterparts.[80] This novel enzyme, unlike the host γ-like activity, accepts poly(C)-oligo(dG) as a template-primer, the most stringent homopolymer for assaying reverse transcriptase. It is also able to copy natural (heteropolymeric) messenger RNA, albeit with lower apparent efficiency,[78] and, finally, it could be clearly separated from the γ-like polymerase by phosphocellulose chromatography.

Detection of DNA polymerase activities in acrylamide gels of protein mixtures renatured after denaturing electrophoresis allowed Hohn et al.[85] to assign a molecular weight of 75 to 80 kdaltons to the poly(C)-oligo(dG) accepting activity.

A key prediction of the CaMV replication model — the existence of a reverse transcriptase activity in infected plants — was thus demonstrated, and the existence of sequence homologies places ORF V as a strong candidate for being the corresponding gene. During this time, the list of retroid elements was growing, mainly from the retrotransposon side, and DNA sequences revealed the conservation of the arrangement of ORF, with the ''gag'' ORF always located upstream, and very often shortly overlapping, the ''pol'' ORF. In all cases known, the function of this conserved gene arrangement is the unequal production of two precursor polypeptides, with the gag polyprotein overproduced with respect to a gag-pol polyprotein. Translation of the fusion polypeptide requires a frameshift in the case of Rous sarcoma virus (RSV) and Ty-1 element, or the suppression of a termination codon for MoMLV. In retroviruses, typically MoMLV or RSV, this gag-pol alignment corresponds to polyproteins that are cleaved to give rise to the following polypeptides (in some cases separated by smaller cleavage products), reading from the 5′ end of gag to the 3′ end of pol: the structural component of the virion shell, a viral RNA binding protein, a protease responsible for the processing of the gag polyprotein, the reverse transcriptase polypeptide, and finally, the viral integrase responsible of the integration of the proviral DNA into the host chromosome (see the review by Coffin in this volume).

The structural organization of CaMV ORF IV and V also follows this gag-pol scheme[79] and may correspond to a similar strategy of gene expression. The CaMV sequence was evaluated in view of this conserved gene arrangement,[86] and a computer search reveals that in addition to the initially described protein homology thought to be part of the catalytic domain of reverse transcriptase, a shorter stretch of amino acid residues in the N terminal part of the ORF V translation product is homologous to the protease domain of animal retroviruses. No homology is found between the pol gene of either HBV or CaMV on one side, and the integrase domain of proper retroviruses on the other side, an observation that correlates well with the fact that the former viruses, contrary to the latter ones, undergo no obligatory step as integrated provirus.

It is also worth pointing out that sequence homologies are often stronger between CaMV and proper retroviruses (e.g., MoMLV) than between HBV and proper retroviruses. In other words, the phylogenetic tree of retroid elements does not match that of their hosts. If, as these homologies suggest, all these viruses once had a common ancestor, the implication is then that there has been some horizontal information transfer between animal and plant kingdom, a story that might involve the insect vector of CaMV.

b. ORF V Products in Infected Plants and in Heterologous Systems

Ziegler et al.[42] raised antibodies against a synthetic peptide covering the 25 amino acids translated from the 3′ end of ORF V. They used these anti-C-terminal antibodies to reveal Western blots (i.e., nitrocellulose transfers of protein electrophoregrams) of infected plant extracts, and detected an 80-kdalton band, i.e., an apparent full-size translation product of ORF V, in addition to a 33- and a probable 56-kdalton species and some host protein that interfered with the reaction. The ORF V polypeptides present in infected leaves were further mapped using antibodies raised against either the full-length translation product or its N-terminal part.[42a] A full length product and trimmed versions thereof were detected: a 58- (corresponding to the 56- above) and a 18-kdalton species map at the C terminus, while a 62- and a 22-kdalton species contain the N-terminal epitopes.

We also know from immunological data that the ORF V product is associated with virus particles: antibodies raised against CaMV virions, dissociated after cesium chloride gradient centrifugation, immunoprecipitate the ORF V protein translated in vitro, and antibodies raised against the ORF V protein produced in *E. coli* recognize antigenic proteins in Western blots of virus preparations (Pfeiffer, P. and Pietzrack, M., unpublished observations). A DNA polymerase activity copurifies with crude virus preparations,[87] and preliminary evidence that the reverse transcriptase activity can be partially inhibited by the C terminal antibodies has been published.[88]

Attempts have been made to express a DNA polymerase activity from ORF V protein expressed in heterologous systems. The viral reading frame was put under the control of bacterial inducible transcription units, resulting in translation of ORF V protein fused at its N terminus to a few residues of bacterial proteins.[42,42a] These fusion proteins are clearly detected on electrophoregrams of induced extracts, but show no catalytic properties on DNA polymerase activity gels. Trivial inhibition or irreversible inactivation of the activity by the bacterial extract are unlikely, because activity gel analysis reveals the bacterial polymerase; furthermore, the absence of activity is not a consequence of the amino terminal fusion: ORF V has been translated in vitro, using either the wheat germ or the reticulocyte lysate systems, from artificial mRNAs covering the complete reading frame from its first, or even second, AUG (messengers obtained in vitro using the SP6 bacteriophage RNA polymerase). Here again, the abundant proteins obtained failed in all cases to exert any DNA polymerase activity (Gordon et al., submitted).

This leads us to conclude that the CaMV reverse transcriptase is *not* the primary translation product of ORF V. Because the active form of the enzyme migrates slightly faster than full-length ORF V product translated in vitro (Gordon et al., submitted), we argue that the functional CaMV reverse transcriptase may be a trimmed form of this protein. Alternatively or in addition, the CaMV reverse transcriptase activation may require another posttranslational modification event, such as phophorylation, as in RSV. The reverse transcriptase of this virus is highly stimulated by a protein kinase.[89]

Takatsuji et al.[90] reported a successful expression of reverse transcriptase activity from ORF V cloned in yeast under the control of an inducible promoter. They also reported the lack of such activity from *E. coli* expressing ORF V, thus showing the need for choosing a more homologous expression host. Unfortunately, they did not present immunological data nor showed a protein gel of the yeast-produced enzyme they purified 4000 times from the crude, induced extract. However, HTLV-III gag protein processing has been shown to occur in yeast,[91] and it is indeed possible that a correct processing also occurs with the yeast expressing ORF V.

IV. IMPLICATIONS OF THE CaMV REPLICATION MODE FOR GENE EXPRESSION

The mere fact of using RNA as a replicative intermediate has some obvious and drastic consequence on the transcription and translation strategies of retroid elements in general and of CaMV in particular. First, all the DNA has to be transcribed, but on one strand only, since anti-sense RNA interferences must be avoided; as a consequence, all ORFs map on the same strand. A second implication, either from a linear, terminally redundant transcriptional template or from a circular minichromosome, is that a terminally redundant RNA has to be made. This, in turn, implies that the site for RNA 3′ end cleavage and polyadenylation has to be by-passed when it is located upstream, but recognized when it is downstream of the bulk of the retroid element genome. A third implication would be that only one polyadenylation site is allowed on a retroid DNA, which, in turn, implies that all transcripts are coterminal in 3′, a feature that severely limits the generation of distinct RNA species. On the other hand, studies of eukaryotic gene expression have established the rule that translation initiates on the first AUG read from the 5′ end of the mRNA (yeast), providing this AUG is located in a favorable nucleotide context (mammals): this is the essential point of the scanning hypothesis (see Reference 92 for a review). Retroid elements apparently need to encode several proteins in order to ensure their replication, and furthermore the retroid viruses require additional proteins to spread from cell to cell and from host to host. Retroid elements developed strategies that overcome these limitations and allow regulation of protein synthesis with an impressive economy of RNA expression signals; these strategies involve RNA synthesis, RNA processing, translation, and protein processing.

Synthesis of distinct RNAs with a common 3′ end can be achieved either by making use of different promoters or through splicing of the pregenome. The first solution has been adopted by HBV, and by CaMV for ORF VI expression. Splicing is used by retroviruses to produce their env mRNA. Splicing events must be regulated in order to avoid the complete disappearing of the genomic RNA, and a mechanism must exist which avoids the reverse transcription of all types of subgenomic RNA. Proteolytic processing is an obvious possibility to make several polypeptides from a single reading frame. The ORF IV and V proteins of CaMV undergo specific cleavages, and the N terminal part of ORF V is suspected to encode the responsible protease. Cleavage of a precursor polypeptide, however, does not allow the unbalanced production of different proteins, and retroid elements make use of additional tricks to direct and tune the translation of more than one polypeptide from one mRNA molecule. Despite the structural similitudes between CaMV and retroviruses at the gag-pol interface, there is no evidence for ribosome frameshifting at the CaMV ORF IV-V junction;[42a] in fact, indications for translation reinitiation at the first AUG of ORF V exist (M. Schultze, personal communication).

The CaMV 35 S RNA is so far the only demonstrated transcript covering ORFs VII, I, II, III, IV, and V. It is thus tempting to propose that the genomic RNA is also a polycistronic messenger, despite the absence of messenger activity of 35 S RNA in the classical but heterologous cell-free translation systems obtained from reticulocyte lysate or wheat germ extracts. Two models have been advanced to explain CaMV translation. By analogy with an earlier proposal for RSV,[93] Hull proposed the following scheme:[94] (1) ribosomes bind to a sequence *a* located within ORF VII; (2) close and 3′ to *a* exists a sequence *b*; *b*′ sequences that are imperfectly complementary to *b* can be found close and 5′ to the different viral frames; these *b*′ sequences base-pair with *b*, thus generating a set of small-stem + huge-loop alternative configurations; and (3) ribosomes start scanning and by-pass the loop, find the corresponding AUG, and start translation. This model is ruled out by the infectivity of ORF VII deletion derivatives lacking both *a* and *b* sequences.[32] In addition, translation of alternate 35 S configurations cannot explain the polarity of certain frameshift mutations in the dispensible ORF II and VII: when the frame of these genes is shifted, resulting in an overlap between the mutated reading frame and the following, essential ORF,[31,32,95] the virus infectivity is either suppressed or delayed. A second cycle of infection, using as inoculum leaves that showed delayed symptoms, results in a much faster symptom appearance. This kinetics suggests that back mutations occurred in the primary infected plants. Indeed, sequence analysis of the fast-propagating progeny viruses reveals deletions of variable extent that all restore the original reading frame. These results suggest that ribosomes must translate to the end of ORFs VII and II to allow translation of the essential gene which follows 3′ in each case (ORFs I and II, respectively). The authors proposed a polar "relay race" model for CaMV translation that also accounts for the compact genetic organization of the virus. According to this hypothesis, ribosomes start translation on the first, 5′ ORF of 35 S RNA (i.e., ORF VII). They stop protein synthesis at the corresponding stop codon, but remain on the RNA, scan it, and resume protein synthesis when finding the next and closely located downstream AUG; thus, translation would procede sequentially until the stop codon of ORF III is reached, and may continue further down if ribosomes can backtrack the ORF III-IV and IV-V small overlaps.

Two RNA molecules are candidates for being such a polycistronic messenger: the genomic RNA or major 35 S transcript, that bears a 700-nucleotide leader of untranslated RNA, and the minor 35 S transcript whose 5′ end is at Δ_1,[46] only 14 nucleotides away of the AUG of ORF VII.

The possibility then exists that CaMV makes use of the same template for both reverse transcription and translation. The fascinating point is that this generates a decision problem, since both events cannot take place simultaneously on the same molecule. At least two regulation pathways could resolve this potential dilemma, each involving the possible requirement for a viral translation product to mediate the shift from 35 S RNA translation to reverse transcription. First, the decision between protein and DNA synthesis may be determined by the mode of interaction of 35 S RNA and the $tRNA^{met}$ which initiates both processes. The choice might be governed by the effective concentration of reverse transcriptase, which is required to melt the primer and anneal it to the template in the case of avian myeloblastosis virus,[96] and/or by the degree of aminoacylation of the $tRNA^{met}$, which is required to start translation, but would inhibit the start of reverse transcription. Secondly, since reverse transcription may take place in virion particles, early encapsidation would transfer 35 S RNA from a pool available for translation to a pool available for reverse transcription.

ACKNOWLEDGMENTS

We wish to thank Drs. K. Gordon, E. Hiebert, and H. Sanfacon for critical reading of the manuscript. J. M. Bonneville was supported by an EMBO long-term fellowship.

REFERENCES

1. **Dixon, L. and Hohn, T.,** Cloning and manipulating cauliflower mosaic virus, in *Recombinant DNA Research and Viruses,* Becker, Y., Ed., Martinus Nijhoff, Netherlands, 1985, 247.
2. **Mason, W. S., Taylor, J. M., and Hull, R.,** Retroid virus genome replication, *Adv. Virus Res.,* 1986.
2a. **Fuetterer, J. and Hohn, T.,** Involvement of nucleocapsids in reverse transcription: a general phenomenon?, *TIBS,* 12, 92, 1987.
3. **Maule, A. J.,** Replication of caulimoviruses in plants and protoplats, in *Molecular Plant Virology,* Vol. II, Davies, J. W., Ed., CRC Press, Boca Raton, Fla., 1985, 161.
4. **Hull, R. and Shepherd, R. J.,** The coat protein of cauliflower mosaic virus, *Virology,* 70, 217, 1976.
5. **Al Ani, R., Pfeiffer, P., and Lebeurier, G.,** The structure of cauliflower mosaic virus. II. Identity and location of the viral polypeptides, *Virology,* 93, 188, 1979.
6. **Chauvin, C., Jacrot, B., Leubeurier, G., and Hirth, L.,** The structure of cauliflower mosaic virus. A neutron diffraction study, *Virology,* 96, 640, 1979.
7. **Volovitch, M., Drugeon, G., and Yot, P.,** Studies on the single-stranded discontinuities of the cauliflower mosaic virus genome, *Nucleic Acids Res.,* 5, 2913, 1978.

8. **Menissier, J., De Murcia, G., Lebeurier, G., and Hirth, L.,** Electron microscopic studies of the different topological forms of the cauliflower mosaic virus DNA: knotted encapsidated DNA and nuclear minichromosome, *EMBO J.*, 2, 1067, 1983.
9. **Hull, R. and Howell, S. H.,** Structure of the cauliflower mosaic virus genome. II. Variation in DNA structure and sequence between isolates, *Virology*, 86, 482, 1978.
10. **Hull, R. and Donson, J.,** Physical mapping of the DNAs of carnation etched ring and figwort mosaic viruses, *J. Gen. Virol.*, 60, 125, 1982.
11. **Richards, K. E., Guilley, H., and Jonard, G.,** Further characterization of the discontinuities in cauliflower mosaic virus DNA, *FEBS Lett.*, 134, 67, 1981.
12. **Hohn, T., Richards, K., and Lebeurier, G.,** Cauliflower mosaic virus on its way to becoming a useful plant vector, *Curr. Top. Microbiol. Immunol.*, 96, 1982.
13. **Lebeurier, G., Hirth, L., Hohn, T., and Hohn, B.,** Infectivities of native and cloned DNA of cauliflower mosaic virus, *Gene*, 12, 139, 1980.
14. **Howell, S. H., Walker, L. L., and Dudley, R. K.,** Cloned cauliflower mosaic virus DNA infects turnips (Brassica rapa), *Science*, 208, 1265, 1980.
15. **Lebeurier, G., Hirth, L., Hohn, B., and Hohn, T.,** In vivo recombination of cauliflower mosaic virus DNA, *Proc. Natl. Acad. Sci. USA*, 79, 2932, 1982.
16. **Guilley, H., Richards, K. E., and Jonard, G.,** Observations concerning the discontinuous DNAs of cauliflower mosaic virus, *EMBO J.*, 77, 282, 1983.
17. **Covey, S. N. and Hull, R.,** Transcription of cauliflower mosaic virus DNA. Detection of transcripts, properties, and location of the gene encoding the virus inclusion body protein, *Virology*, 111, 463, 1981.
18. **Dudley, R. K., Odell, J. T., and Howell, S. H.,** Structure and 5′-termini of the large and 19 S RNA transcripts encoded by the cauliflower mosaic virus genome, *Virology*, 117, 19, 1982.
19. **Olszewski, N. and Guilfoyle, T.,** Nuclei purified from cauliflower mosaic virus-infected turnip leaves contain subgenomic, covalently closed circular cauliflower mosaic virus DNAs, *Nucleic Acid Res.*, 11, 8901, 1983.
20. **Menissier, J., Lebeurier, G., and Hirth, L.,** Free cauliflower mosaic virus supercoiled DNA in infected plants, *Virology*, 117, 322, 1982.
21. **Olszewski, N., Hagen, G., and Guilfoyle, T. J.,** A transcriptionally active, covalently closed minichromosome of cauliflower mosaic virus DNA isolated from infected turnip leaves, *Cell*, 29, 395, 1982.
22. **Schoelz, J. E., Shepherd, R. J., and Richins, R. D.,** Properties of an unusual strain of cauliflower mosaic virus, *Phytopathology*, 76, 451, 1986.
23. **Franck, A., Guilley, H., Jonard, G., Richards, K., and Hirth, L.,** Nucleotide sequence of cauliflower mosaic virus DNA, *Cell*, 21, 285, 1980.
24. **Gardner, R. C., Howarth, A. J., Hahn, P., Brown-Leudi, M., Shepherd, R. J., and Messing, J.,** The complete nucleotide sequence of an infectious clone of cauliflower mosaic virus by m13mp7 shotgun sequencing, *Nucleic Acids Res.*, 9, 2871, 1981.
25. **Balazs, E., Guilley, H., Jonard, G., and Richards, K.,** Nucleotide sequence of DNA from an altered-virulence isolate D/H of the cauliflower mosaic virus, *Gene*, 19, 239, 1982.

25a. **Rongxiang, F., Xiaojun, W., Ming, B., Yingchuan, T., Faxing, C., and Keqiang, M.,** Complete sequence of cauliflower mosaic virus (Xinjing isolate) genomic DNA, *Chin. J. Virol.*, 1, 247, 1985.

26. **Shepherd, R. J.,** DNA plant viruses, *Adv. Virus Res.*, 20, 405, 1976.

26a. **Markham, P. G. and Hull, R.,** Cauliflower mosaic virus aphid transmission facilitated by transmission factors from other caulimoviruses, *J. Gen. Virol.*, 66, 921, 1985.

27. **Richins, R. D. and Shepherd, R. J.,** Physical maps of the genomes of dahlia mosaic virus and mirabilis mosaic virus — two members of the caulimovirus group, *Virology*, 124, 208, 1983.

27a. **Verver, J., Schijns, P., Hibi, T., and Goldbach, R.,** Characterization of the genome of soybean chlorotic mottle virus, *J. Gen. Virol.*, 68, 159, 1987.

27b. **Richins, R. D. and Shepherd, R. J.,** Horseradish latent virus, a new member of the caulimoviruse group, *Phytopatology*, 76, 749, 1986.

27c. **Hull, R., Sadler, J., and Longstaff, M.,** The sequence of carnation etched ring virus DNA: comparison with cauliflower mosaic virus and retroviruses, *EMBO J.*, 5, 3083, 1986.

27d. **Richins, R. D., Schlolthof, H. B., and Shepherd, R. J.,** Sequence of figwort mosaic virus DNA (caulimovirus group), *Nucleic Acids Res.*, 15, 8451, 1987.

28. **Daubert, S., Shepherd, R. J., and Gardner, R. C.,** Insertional mutagenesis of the cauliflower mosaic virus genome, *Gene*, 25, 201, 1983.
29. **Howell, S. H., Walker, L. L., and Walden, R. M.,** Rescue of in vitro generated mutants of cloned cauliflower mosaic virus genome in infected plants, *Nature (London)*, 293, 483, 1981.
30. **Dixon, L. K., Koenig, I., and Hohn, T.,** Mutagenesis of cauliflower mosaic virus, *Gene*, 25, 189, 1983.
31. **Gronenborn, B., Gardner, R. C., Schaefer, S., and Shepherd, R. J.,** Propagation of foreign DNA in plants using cauliflower mosaic virus as a vector, *Nature (London)*, 294, 773, 1981.

32. **Dixon, L. and Hohn, T.,** Initiation of translation of the cauliflower mosaic virus genome from a polycistronic mRNA: evidence from deletion mutagenesis, *EMBO J.*, 3, 2731, 1984.
33. **Brisson, N., Paszkowski, J., Penswick, J. R., Gronenborn, B., Potrykus, I., and Hohn, T.,** Expression of a bacterial gene in plants by using a viral vector, *Nature (London)*, 310, 511, 1984.
33a. **Martinez-Izquierdo, J. A., Fuetterer, J., and Hohn, T.,** Protein encoded by ORF I of cauliflower mosaic virus is part of the viral inclusion body, *Virology*, 160, 527, 1987.
33b. **Young, M. J., Daubert, S. D., and Shepherd, R. J.,** Gene I products of cauliflower mosaic virus detected in extracts of infected tissue, *Virology*, 158, 444, 1987.
33c. **Harker, C. L., Mullineaux, P. M., Bryant, J. A., and Maule, A. J.,** Detection of CaMV gene I and VI protein products in vivo using antisera raised to COOH-terminal betagalactosidase fusion proteins, *Plant Mol. Biol.*, 8, 275, 1987.
34. **Hull, R. and Covey, S. N.,** Cauliflower mosaic virus: pathways of infection, *Bioessays*, 3, 160, 1985.
35. **Woolston, C. J., Covey, S. N., Penswick, J. R., and Davies, J. W.,** Aphid transmission and a polypeptide are specified by a defined region of the cauliflower mosaic virus genome, *Gene*, 23, 15, 1983.
36. **Armour, S. L., Melcher, U., Pirone, T. P., Lyttle, D. J., and Essenberg, R. C.,** Helper component for aphid transmission encoded by region II of cauliflower mosaic virus DNA, *Virology*, 129, 25, 1983.
37. **Givord, L., Xiong, C., Giband, M., Koenig, I., Hohn, T., Lebeurier, G., and Hirth, L.,** A second cauliflower mosaic virus gene product influences the structure of the viral inclusion body, *EMBO J.*, 3, 1423, 1984.
38. **Xiong, C., Lebeurier, G., and Hirth, L.,** Detection of a new gene product (geneIII) of cauliflower mosaic virus, *Proc. Natl. Acad. Sci. U.S.A.*, 81, 6608, 1984.
38a. **Giband, M., Mesnard, J. M., and Lebeurier, G.,** The gene III product (P15) of cauliflower mosaic virus is a DNA-binding protein while an immunologically related P11 polypeptide is associated with virions, *EMBO J.*, 5, 2433, 1986.
39. **Daubert, S., Richins, R., Shepherd, R. J., and Gardner, C. R.,** Mapping of the coat protein gene of cauliflower mosaic virus by its expression in a prokaryotic system, *Virology*, 122, 444, 1982.
40. **Hahn, P. and Shepherd, R. J.,** Phosphorylated proteins in cauliflower mosaic virus, *Virology*, 107, 295, 1980.
41. **Hahn, P. and Shepherd, R. J.,** Evidence for a 58-kilodalton polypeptide as precursor of the coat protein of cauliflower mosaic virus, *Virology*, 116, 480, 1982.
41a. **Menissier de Murcia, J., Geldreich, A. and Lebeurier, G.,** *J. Gen. Virol.*, 67, 1885, 1986.
41b. **Martinez-Izquierdo, J. A. and Hohn, T.,** Cauliflower mosaic virus coat protein is phosphorylated in vitro by a virion-associated protein kinase, *Proc. Natl. Acad. Sci. U.S.A.*, 84, 1824, 1987.
42. **Ziegler, V., Laquel, P., Guilley, H., Richards, K., and Jonard, G.,** Immunological detection of CaMV gene V protein produced in engineered bacteria or infected plants, *Gene*, 36, 271, 1985.
42a. **Pietzrack, M. and Hohn, T.,** Translation products of cauliflower mosaic virus ORF V, the coding region corresponding to the retrovirus pol gene, *Virus Genes*, 1, 83, 1987.
43. **Daubert, S. D., Schoelz, J., Debao, L., and Shepherd, R. J.,** Expression of disease symptoms in CaMV genomic hybrids, *J. Mol. Appl. Genet.*, 2, 537, 1984.
43a. **Schoeltz, J., Shepherd, R. J., and Daubert, S.,** Region VI of cauliflower mosaic virus encodes a host range determinant, *Mol. Cell. Biol.*, 6, 2632, 1986.
44. **Dixon, L. K., Jiricny, J., and Hohn, T.,** Oligonucleotide directed mutagenesis of cauliflower mosaic virus DNA using a repair-resistant nucleoside analogue: identification of an agnogene initiation codon, *Gene*, 41, 225, 1986.
45. **Xiong, C., Muller, S., Lebeurier, G., and Hirth, L.,** Identification by immunoprecipitation of cauliflower mosaic virus in vitro major translation product with a specific serum against viroplasm protein, *EMBO J.*, 1, 971, 1982.
46. **Guilley, H., Dudley, R. K., Jonard, G., Balazs, E., and Richards, K.,** Transcription of cauliflower mosaic virus DNA: detection of promoter sequences, and characterization of transcripts, *Cell*, 30, 763, 1982.
47. **Condit, C. and Meagher, R. B.,** Multiple, discrete 35S transcripts of cauliflower mosaic virus, *J. Mol. Appl. Genet.*, 2, 301, 1983.
48. **Plant, A. L., Covey, S. N., and Grierson, D.,** Detection of a subgenomic mRNA for gene V, the putative reverse transcriptase gene of cauliflower mosaic virus, *Nucleic Acids Res.*, 13, 8305, 1985.
49. **Pfeiffer, P. and Hohn, T.,** Involvement of reverse transcription in the replication of cauliflower mosaic virus: a detailed model and test of some aspects, *Cell*, 33, 781, 1983.
50. **Edenberg, K.,** Extraction of transcribing SV40 minichromosomes at very low salt, *Nucleic Acids Res.*, 8, 473, 1980.
51. **Summers, J. and Mason, W. S.,** Replication of the genome of an hepatitis-B like virus by reverse transcription of an RNA intermediate, *Cell*, 29, 403, 1982.
52. **Hull, R. and Covey, S. N.,** Does cauliflower mosaic virus replicate by reverse transcription?, *TIBS*, 8, 119, 1983.

53. **Pietrzak, M. and Hohn, T.,** Replication of the cauliflower mosaic virus role and stability of the cloned delta-3 discontinuity sequence, *Gene,* 33, 169, 1985.
54. **Dixon, L. K., Nyffenegger, T., Delley, G., Martinez-Izquierdo, J., and Hohn, T.,** Evidence for replicative recombination in cauliflower mosaic virus, *Virology,* 150, 463, 1985.
55. **Coffin, J. M.,** Structure of the retroviral genomes, in *Molecular Biology of Tumorviruses. RNA Tumor Viruses,* Weiss, R. A., Teich, N., Varmus, H. E., and Coffin, J. M., Eds., Cold Spring Harbor Laboratory, Cold Spring Harbor, N.Y., 1982.
56. **Grimsley, N., Hohn, B., Hohn, T., and Walden, R.,** Agroinfection, a novel route for plant viral infection using Ti plasmid, *Proc. Natl. Acad. Sci. U.S.A.,* 83, 3282, 1986.
57. **Walden, R. M. and Howell, S. H.,** Intergenomic recombination events among pairs of defective cauliflower mosaic virus genomes in plants, *J. Mol. Appl. Genet.,* 1, 447, 1982.
58. **Walden, R. M. and Howell, S. H.,** Uncut recombinant plasmids bearing nested cauliflower mosaic virus genomes infect plants by intragenomic recombination, *Plant Mol. Biol.,* 2, 27, 1983.
59. **Shewmaker, C. K., Caton, J. R., Houk, C. M., and Gardner, R. C.,** Transcription of cauliflower mosaic virus integrated into plant genomes, *Virology,* 140, 281, 1985.
60. **Grimsley, N., Hohn, T., and Hohn, B.,** Recombination in a plant virus, *EMBO J.,* 5, 641, 1986.
60a. **Hohn, B., Balazs, E., Ruegg, D., and Hohn, T.,** Splicing of an intervening sequence from hybrid cauliflower mosaic viral RNA, *EMBO J.,* 5, 2759, 1986.
61. **Hirochika, H., Takatsuji, H., Ubasawa, A., and Ikeda, J. E.,** Site specific deletion in CaMV DNA: possible involvement of RNA splicing and reverse transcription, *EMBO J.,* 4, 1673, 1985.
62. **Choe, S., Melcher, U., Richards, K., Lebeurier, G., and Essenberg, R. C.,** Recombination between mutant CaMV DNAs, *Plant Mol. Biol.,* 5, 281, 1985.
63. **Turner, D. S. and Covey, S. N.,** A putative primer for the replication of CaMV by reverse transcription is virion associated, *FEBS Lett.,* 165, 285, 1984.
64. **Marco, Y. and Howell, S. H.,** Intracellular forms of viral DNA consistent with a model of reverse transcriptional replication of the cauliflower mosaic virus, *Nucleic Acids Res.,* 12, 1517, 1984.
64a. **Covey, S. N. and Turner, D. S.,** Hairpin DNAs of cauliflower mosaic virus generated by reverse transcription in vivo, *EMBO J.,* 5, 2763, 1986.
65. **Maule, A. J. and Thomas, C. M.,** Evidence from CaMV virion DNA for additional discontinuities in the plus strand, *Nucleic Acid Res.,* 13, 7359, 1985.
66. **Olszewski, N. E. and Guilfoyle, T. J.,** Nuclei purified from cauliflower mosaic virus-infected turnip leaves contain subgenomic, covalently closed circular cauliflower mosaic virus DNAs, *Nucleic Acid Res.,* 11, 8901, 1983.
67. **Rollo, F. and Covey, S. N.,** CaMV DNA persists as supercoiled forms in cultured turnip cells, *J. Gen. Virol.,* 66, 603, 1985.
68. **Paszkowski, J., Shinshi, H., Koenig, I., Lazar, G. B., Hohn, T., Mandak, V., and Potrykus, I.,** Proliferation of cauliflower mosaic virus in protoplast-derived clones of turnip (Brassica rapa), in *Protoplasts 1983 Poster Proceedings,* 6th International Protoplast Symposium, Birrkaauser Verlag, Basel, 1983.
69. **Maule, A. J.,** Partial characterization of different classes of viral DNA, and kinetics of DNA synthesis in turnip protoplasts infected with CaMV, *Plant Mol. Biol.,* 5, 25, 1985.
70. **Gracia, O. and Shepherd, R. J.,** CaMV in the nucleus of Nicotiana, *Virology,* 146, 141, 1985.
71. **Kamei, T., Rubio-Huertos, M., and Matsui, C.,** Thymidine-3H uptake by X-bodies associated with cauliflower mosaic virus infection, *Virology,* 37, 506, 1969.
72. **Favali, M. A., Basi, M., and Conti, G. G.,** A quantitative autoradiographic study of intracellular sites for replication of cauliflower mosaic virus, *Virology,* 53, 115, 1973.
73. **Ansa, O. A., Bowyer, J. W., and Shepherd, R. J.,** Evidence for replication of cauliflower mosaic virus DNA in plant nuclei, *Virology,* 121, 147, 1982.
74. **Modjtahedi, N., Volovitch, M., Sossountzov, L., Habricot, Y., Bonneville, J. M., and Yot, P.,** CaMV induced viroplasms support viral DNA synthesis in a cell free system, *Virology,* 133, 289, 1984.
75. **Mazzolini, L., Bonneville, J. M., Volovitch, M., Magazin, M., and Yot, P.,** Strand-specific viral DNA synthesis in purified viroplasm isolated from turnip leaves infected with CaMV, *Virology,* 145, 293, 1985.
76. **Covey, S. N., Turner, D., and Mulder, G.,** A small DNA molecule containing covalently-linked ribonucleotides originates from the large intergenic region of the cauliflower mosaic virus genome, *Nucleic Acids Res.,* 11, 251, 1983.
77. **Marsh, L., Kuzj, A., and Guilfoyle, T.,** Identification and characterization of CaMV replication complexes — analogy to HBV, *Virology,* 143, 212, 1985.
78. **Thomas, C. M., Hull, R., and Maule, A. J.,** Isolation of a fraction of CaMV infected protoplasts which is active in the synthesis of (+) and (−) viral DNA and reverse transcription of primed RNA templates, *Nucleic Acids Res.,* 12, 4557, 1985.
79. **Bonneville, J. M., Fuetterer, J., Gordon, K., Hohn, T., Martinez-Izquierdo, J., Pfeiffer, P., and Pietrzak, M.,** The replication cycle of CaMV in relation to other retroid elements; current perspectives, *UCLA Symp. New Ser.,* 48, 267, 1986.

79a. **Marsh, L. E. and Guilfoyle, T. J.,** Cauliflower mosaic virus replication intermediates are encapsidated into virion-like particles, *Virology,* 161, 129, 1987

80. **Volovitch, M., Modjtahedi, N., Yot, P., and Brun, G.,** RNA-dependent DNA polymerase activity in cauliflower mosaic virus infected plant leaves, *EMBO J.,* 3, 309, 1984.

81. **Pfeiffer, P., Laquel, P., and Hohn, T.,** Cauliflower mosaic virus replication complexes: characterization of the associated enzymes and of the polarity of the DNA synthesized in vitro, *Plant Mol. Biol.,* 3, 261, 1984.

82. **Huberman, J. A.** New view of the biochemistry of eucaryotic DNA replication revealed by aphidicolin, an unusual inhibitor of DNA polymerase α, *Cell,* 23, 647, 1981.

83. **Bonneville, J. M., Volovitch, M., Modjtahedi, N., Demery, D., and Yot, P.,** In virto synthesis of cauliflower mosaic virus DNA in viroplasms, in *Proteins Involved in DNA Replication,* Huebscher, U. and Spadari, S., Eds., Plenum Press, New York, 1984, 113.

84. **Toh, H., Hayashida, H., and Miyata, T.,** Sequence homology between retroviral reverse transcriptase and putative polymerases of hepatitis B virus and cauliflower mosaic virus, *Nature (London),* 305, 827, 1983.

85. **Hohn, T., Hohn, B., and Pfeiffer, P.,** Reverse transcription in a plant virus, *TIBS,* 10, 205, 1985.

86. **Toh, H., Kikuno, R., Hayashida, H., Miyata, T., Kugimiya, W., Inouye, S., Yuki, S., and Saigo, K.,** Close structural resemblance between putative polymerase of a Drosophila transposable element 17.6 and pol gene product of MoMuLV, *EMBO J.,* 4, 1267, 1985.

87. **Menissier, J., Laquel, P., Lebeurier, G., and Hirth, L.,** A DNA polymerase activity is associated with CaMV, *Nucleic Acids Res.,* 12, 8769, 1984.

88. **Laquel, P., Ziegler, V., and Hirth, L.,** The 80 K polypeptide associated with the replication complex of CaMV is recognized by antibodies to gene V translation product, *J. Gen. Virol.,* 67, 197, 1986.

89. **Lee, X. G., Miceli, M. V., Jungman, R. A., and Hung, P. P.,** Protein kinase and its regulatory effects on reverse transcriptase activity of Rous sarcoma virus, *Proc. Natl. Acad. Sci. U.S.A.,* 72, 2945, 1975.

90. **Takatsuji, H., Hirochika, H., Fukushi, T., and Ikeda, J. E.,** Expression of CaMV reverse transcriptase in yeast, *Nature (London),* 319, 241, 1986.

91. **Kramer, R. A., Schaber, M. D., Skalka, A. M., Ganguly, K., Woong-staal, F., and Reddy, E. P.,** HTLV-III gag protein is processed in yeast cells by the virus pol-protease, *Science,* 231, 1580, 1986.

92. **Kozak, M.,** Evaluation of the "scanning model" for initiation of protein synthesis in eukaryotes, *Cell,* 22, 7, 1980.

93. **Darlix, J. L., Zuker, M., and Spahr, P. F.,** Structure-function relationship of Rous sarcoma virus leader RNA, *Nucleic Acids Res.,* 10, 5183, 1982.

94. **Hull, R.,** A model for expression of CaMV nucleic acid, *Plant Mol. Biol.,* 3, 121, 1984.

95. **Sieg, K. and Gronenborn, B.,** Evidence for polycistronic messenger RNA encoded by cauliflower mosaic virus, Abstr. NATO Advanced Studies Inst., Advanced course 1982, 1982, 154.

96. **Garret, M., Romby, P., Giege, R., and Litvak, S.,** Interactions between avian myeloblastosis reverse transcriptase and $tRNA^{Trp}$. Mapping of complexed tRNA with chemicals and nucleases, *Nucleic Acids Res.,* 12, 2259, 1984.

Chapter 3

HEPATITIS B VIRUS REPLICATION

Valerie Bosch, Christa Kuhn, and Heinz Schaller

TABLE OF CONTENTS

I. Introduction 44
A. General 44
B. Virus Structure and Viral Proteins 44

II. Viral Replication 45
A. Viral Nucleic Acid in Infected Liver Cells and Virions 45
1. DNA 45
2. RNA 46
3. Mapping of the Ends of Minus and Plus Strand DNA 48
B. Control of Gene Expression 49
1. Transcriptional Control 49
2. Translational Control 51
C. Model for HBV Replication 51
1. Production of Pregenome 51
2. Minus Strand Synthesis 51
3. Plus Strand Synthesis 52
4. Proteins Involved in the Replication Cycle 52

III. Comparison of the Life Cycle of Hepatitis B Virus with Other Retroviruses and Retrotransposons 53

References 55

I. INTRODUCTION

A. General

Hepatitis B virus (HBV) is a major world health problem due to the high percentage of chronic HBV carriers, worldwide now approximately 200 million people, and the association of HBV with hepatocellular carcinoma, one of the most common human cancers. HBV has a very limited host range and infects only humans and chimpanzees, but in the last few years, related viruses have been characterized from woodchucks (WHV), ground squirrels (GSHV), ducks (DHBV), and others (for reviews on hepatitis viruses, see References 1 through 3a). Although mature HBV contains a DNA genome, its replication involves reverse transcription of an RNA pregenome. In this regard, HBV replication resembles that of cauliflower mosaic virus (described in the previous chapter) as well as the transposition of some eukaryotic mobile elements such as yeast Ty elements and copia-like elements of *Drosophila* (described in the following chapter). We should like to focus here on the processes involved in HBV replication via an RNA intermediate and to make a brief comparison of this replication strategy with those employed by other retroviruses and retrotransposons.

Hepatitis B virus is essentially hepatotropic in the infected animal, but cells of other tissues are infected to a minor extent. Until recently there was no in vitro cell culture system available for the efficient propagation of HBV, and this fact has greatly hampered the elucidation of all aspects of the virus life cycle. Recently, however, it has been possible to achieve HBV expression in cell culture by transfection with HBV-DNA.[4,5] This now allows the analysis of HBV genomes, specifically mutagenized in vitro, so that it can be anticipated that many of the questions, still open with regard to the replication cycle, may be solved in these systems. However, most of the information on the replication of HBV which is available up to now has been derived from an analysis of the viral gene products in infected liver and sera using gene technological methods. The discovery of HBV-related viruses in animals greatly accelerated the elucidation of the virus replication cycle. These animal systems will undoubtedly continue to be important in the future. In addition to their use as models for an acute hepatitis B infection, the animal systems, especially the WHV system, are important models for chronic hepatitis B infections and hepatocellular carcinomas.

The nucleotide sequence of several human HBV isolates, two woodchuck isolates, two ground squirrel isolates, and two duck isolates have been elucidated to date,[6-17] and a comparison of these sequences gives information as to putative viral gene products (open reading frames, ORFs) as well as to regions of regulatory importance (see Section II.B). The HBV genome, with only 3.2 kilobases, is one of the smallest known animal DNA virus genomes. Presumably as a result of its small size, alternative, very efficient modes of usage of the available ORFs have been developed.

B. Virus Structure and Viral Proteins

Figure 1 shows a schematic representation of an HBV particle (Dane particle) and in Figure 3, its genome organization is shown. HBV is a spherical, enveloped virus with a diameter of approximately 42 nm which appears to derive its lipid bilayer by budding from the endoplasmic reticulum of the infected cell.[18,19] Embedded in the lipid bilayer are three related proteins, *Pre*-S1, *Pre*-S2, and HBsAg, which are products of the *Pre*-S/S open-reading frame[20-23] and are involved in host cell recognition and uptake.[24] The virus capsid lies within this envelope and consists of 180 core proteins[25] (products of the C-gene) which have extremely arginine-rich C terminal sequences which presumably interact with the viral nucleic acid. The capsid encloses the viral genome which consists of a partially double-stranded, circular DNA molecule. One strand (the minus, coding strand) covers the whole genome and carries a protein covalently attached to its 5′ end.[26,27] The complementary, plus strand bridges the break in the minus strand and results in the formation of circular DNA

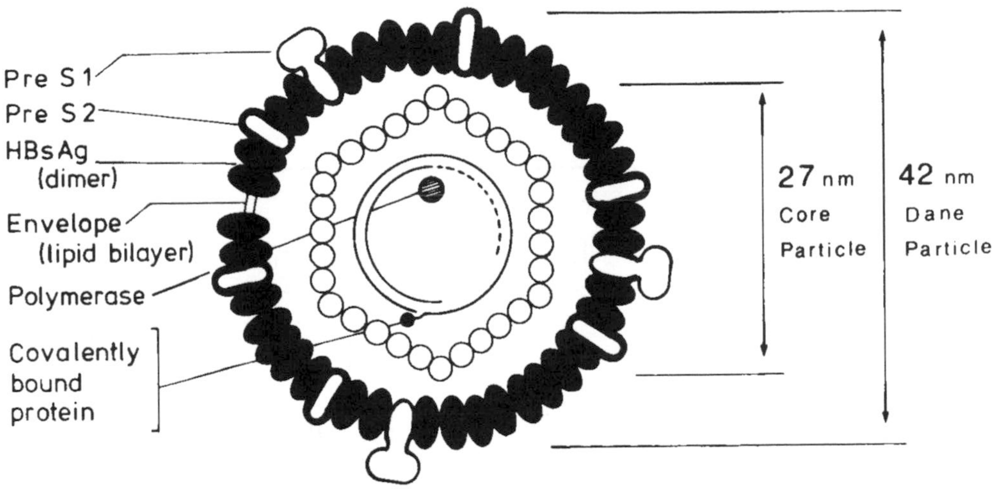

FIGURE 1. Schematic representation of a Dane particle. See text for description.

molecules. The position of the 3′ end of this plus strand is variable, so that in HBV, 30 to 50% of the genome is single stranded. A polymerase activity is present in the capsid and, on provision of deoxynucleoside triphosphates, this polymerase can complete plus strand synthesis in vitro.[28] It is assumed that the long P-ORF, which overlaps with all three other ORFs, codes for the viral polymerase (discussed in detail in Section II.C.4). In some infected individuals, antibodies, directed against protein sequences of the P-ORF, can be detected, demonstrating that P-protein expression has occurred in vivo (C. Schroeder, Heidelberg, personal communication). A further viral protein, the e-antigen, which is not a component of the virion, but is secreted from infected hepatocytes into blood,[29] most likely arises by translation of the entire *Pre*-C/C gene, resulting in a product which is initially membrane bound and is further processed to soluble e-antigen lacking most of the *Pre*-C sequence at the N terminus and the basic region at the C terminus.[30,31] Recent experiments with DHBV, however, have shown that the synthesis and secretion of e-antigen is not necessary for virus production.[31a] The X-ORF codes for a putative product of 154 amino acids which has not yet been unambiguously identified in infected cells. The fact that some HBV-infected individuals produce antibodies against X-sequences points to the protein being expressed in vivo.[32-35] In vitro experiments indicate that a possible function of the X gene product is that of a transactivator of transcription.[35a] However, the fact that DHBV lacks an X-region raises the question as to its essential role in the normal replication cycle.

II. VIRAL REPLICATION

Figure 2 shows the major steps in the replication of HBV. In the following sections (Sections II.A.1 and 2), the initial evidence leading to the proposal of this scheme will briefly be described, and in Sections II.A.3 and II.C, more precise details will be presented.

A. Viral Nucleic Acid in Infected Liver Cells and Virions

1. DNA

In the nucleus of infected liver cells, free, covalently closed, circular (ccc), supercoiled, viral DNA can be detected, and evidence points to this being produced shortly after infection.[36-39] This cccDNA is produced from input virion DNA by completion of the plus strand and further modifications (see Sections II.A.3. and II.C). It is the template for the production of viral mRNA. In addition to cccDNA, viral minus and plus strands of varying

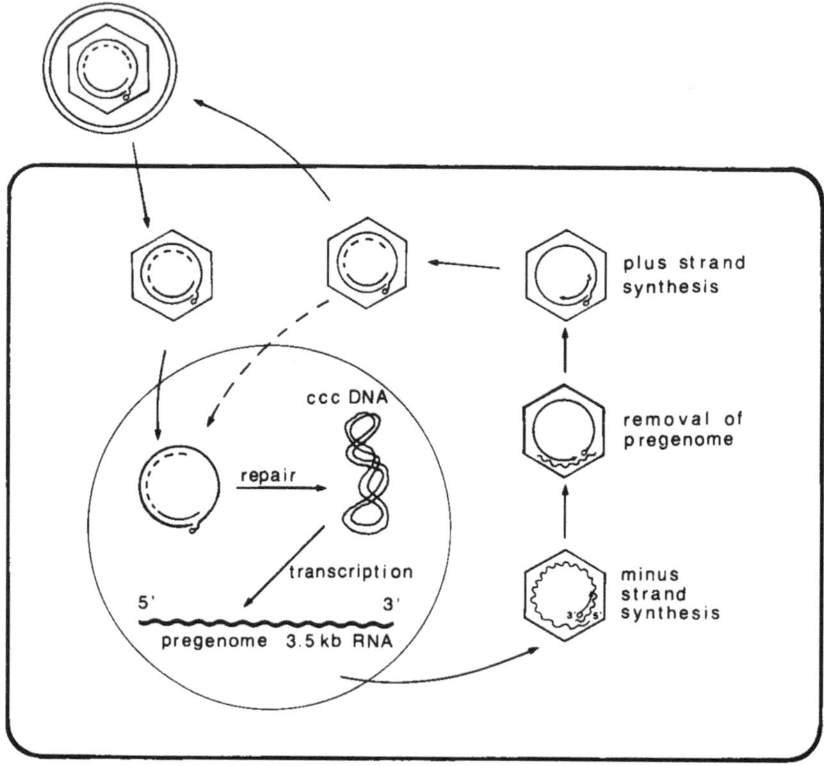

FIGURE 2. Replication cycle of hepatitis B viruses. See text for details.

length, which are intermediates in the replication cycle, are found intracellularly in immature viral cores.

Since hepatitis B virus is a DNA virus, it was not, *a priori,* to be expected that its replication would involve an obligatory RNA intermediate. However, in the case of the duck hepatitis B virus (DHBV), Summers and Mason[40] could show that the ongoing replication in cytoplasmic viral cores was partially resistant to actinomycin D. This, together with an analysis of the replicative intermediates found in immature cytoplasmic viral cores, allowed them to propose in 1982 that the viral minus DNA strand arises by reverse transcription of an RNA "pregenome", a concept which since then could be confirmed for HBV (for review see Reference 1). After its completion, the minus strand is used as a template for the synthesis of the viral plus strand. This latter strand is, however, not completed and virions containing circular genomes with single-stranded gaps of varying lengths are released from the cell (Figure 2). It could be recently shown[39] that, at least in in vitro cell culture, the partially double-stranded DNA genome which has been produced from the RNA pregenome, may also be used directly to amplify the pool of nuclear cccDNA (Figure 2).

2. *RNA*

Viral RNA transcripts in infected chimpanzee,[41,42] human,[43] duck,[44] woodchuck,[45] and ground squirrel[46] livers have been examined by Northern blot analysis, S1 mapping, and primer extension. In Figure 3, the transcriptional maps for HBV and DHBV are shown. In the case of DHBV infected liver, two subgenomic transcripts and one overlength transcript, present in approximately equimolar amounts, can be detected. The subgenomic transcripts presumably function as mRNAs for the surface proteins *Pre*-S/S and HBsAg.[44] In the case of HBV, the same transcripts are found. However, the larger subgenomic transcript (*Pre-*

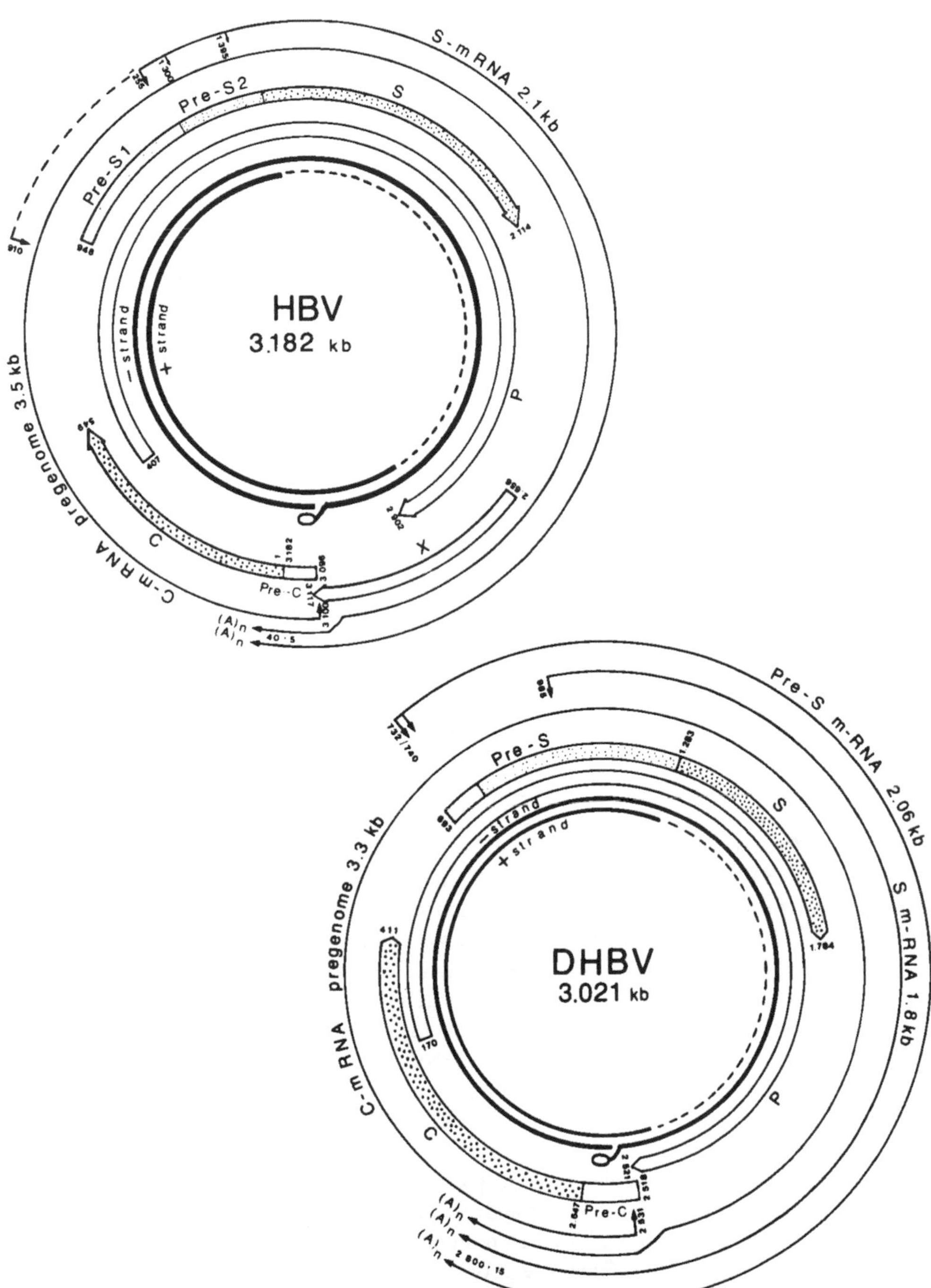

FIGURE 3. Gene organization and transcriptional maps of hepatitis B virus (upper) and duck hepatitis B virus (lower). For HBV, the numbering system of Pasek et al.[7] has been employed. The inner double ring shows the partially single-stranded genome with the protein primer on the 5′ end of the minus strand shown as a small circle. The open reading frames, ORFs, *Pre*-S/S, *Pre*-C/C, P, and X (HBV), are depicted as bars with the numbers referring to the nucleotide positions at the beginning and end of each ORF. The different patterns used to fill in the bars are intended to demarcate the alternative products which can be produced from the *Pre*-S/S and *Pre*-C/C ORFs. The transcripts (mRNAs) which have been detected in infected liver are shown as unbroken lines exterior to the ORFs. With HBV, the broken line at the 5′ end of the S-mRNA indicates the very probable existence of a minor *Pre*-SI mRNA.[43] The S-mRNA of HBV has a major start at position 1255 and minor starts at positions 1300 and 1395.

S-mRNA) cannot be detected in vivo by Northern analysis, but only by mapping the 5′ ends of HBV RNA in the infected liver. This transcript with a 5′ end at approximately position 910[43] must be a minor species in the infected liver and could serve as mRNA for HBV *Pre*-S1/S. In addition, in a recently published in vitro cell culture system for HBV, a transcript with the appropriate size for a *Pre*-S1 mRNA could be detected, but was not further characterized.[4] The major subgenomic transcript, S-mRNA, in HBV-infected liver actually has multiple starts. The most frequent start is at position 1255, but there are additional starts at approximately positions 1300 and 1395.

The overlength transcript, in both HBV and DHBV, initiates within the *Pre*-C region, continues over the full length of the viral genome and coterminates with the other transcripts. Thus, this transcript, which has failed to terminate on the first round of transcription, is terminally redundant. Its functions are to serve as mRNA for the major capsid protein, C, as well as possibly for the product(s) of the P-gene (see Section II.C.4). In addition, this transcript is presumably the "pregenomic" RNA which serves as template for reverse transcription (Figure 2 and Sections II.A.3. and II.C). In WHV and GSHV, two major transcripts representing the S-mRNA and the terminally redundant C-mRNA/pregenome, respectively, are also found.[45,46] No transcripts initiating directly upstream of the *Pre*-S region could be detected, but these may be present as very minor species as is the case with HBV.[43,44] The terminally redundant overlength mRNAs from GSHV and WHV have multiple 5′ ends so that species initiating within the *Pre*-C region can be detected (as with HBV and DHBV) as well as species initiating upstream from the *Pre*-C AUG.[45,46] These latter species may serve as mRNAs for the translation of *Pre*-C/C proteins which would be processed to HBeAg secreted into the serum.[29] Although in HBV and DHBV, such *Pre*-C MRNA transcripts have not yet been detected, these are presumably present as minor species.

3. *Mapping of the Ends of Minus and Plus Strand DNA*

In order to gain further insight into the HBV replication mechanism, the exact 5′ ends of both viral minus and plus DNA strands and the 3′ end of the minus DNA strand have been mapped for DHBV,[47,48] HBV,[43] and GSHV.[49] The 3′ end of the plus strand is variable and does not give further insight into the replication mechanism. In the transcription map of HBV shown in Figure 3, the positions of the minus and plus strand ends are shown relative to the 5′ end of the pregenomic RNA. Figure 5 now shows the exact sequence of the overlap region between minus and plus DNA strands for HBV. The salient features of the overlap region appear to be common for all the hepatitis viruses and have been elucidated initially for DHBV,[47,48] and recently for GSHV[49] and HBV.[43] An important characteristic, whose function was initially not understood, is the presence of a direct repeat of 11 nucleotides near the 5′ end of the minus strand (direct repeat, DR1) and the 5′ end of the plus strand (direct repeat, DR2).

The 5′ end of the minus strand maps to position 3108 within DR1.[43] A protein which presumably acts as a primer for minus strand synthesis is covalently attached to this 5′ end. The 3′ end of the minus strand maps at position 3100 so that this strand is terminally redundant by nine nucleotides. This 3′ end of the minus strand coincides exactly with the 5′ end of the overlength C-mRNA, making it very likely that it has arisen by reverse transcription of this RNA species. Interestingly, also in GSHV, a discrete minus strand 3′ end is detected at a position corresponding to the 5′ end of the C-mRNA which has initiated within the Pre-C region, and not to the 5′ end of the C-mRNA which has initiated upstream of *Pre*-C.[49]

As mentioned above, the protein covalently attached to the 5′ end of the minus strand is presumed to act as primer for reverse transcription. However, until recently, the primer for plus strand synthesis was not known. The 5′ end of the DNA plus strand maps to position 2883 (i.e., exactly 3′ to the DR2 border[43]), and it could be demonstrated, initially for

DHBV[48] and later for GSHV[49] and HBV,[43] that a ribooligonucleotide, with a capped 5′ end, is still attached to the 5′ end of the plus strand. This is the primer for plus strand synthesis and has arisen from the 5′ end of the pregenomic RNA, i.e., the sequence corresponding to DR1 has been translocated to DR2 to serve as primer.[48] The fact that the sequence has arisen from DR1 could be elegantly confirmed by in vitro mutagenesis of GSHV-DNA to yield a single base change in DR1 only. Transfection into ground squirrels resulted in the production of viable virus with the mutated DR1 oligoribonucleotide attached to the 5′ end of the plus strand.[49]

B. Control of Gene Expression

Because of the small size of the HBV genome, the available sequence information has to be used very economically in order to synthesize the many different proteins which are required during the viral life cycle. As mentioned previously, the open reading frames (genes) are extensively overlapping (Figure 3), so that the genome is actually used one and a half times. The viral regulatory sequences which control the transcription and replication of the viral nucleic acid are themselves coding sequences, a fact which imposes unusual restrictions on their sequence.

In the case of hepatitis B virus, several gene products are coded for by a single ORF, and this is achieved by regulating gene expression both at the transcriptional and translational level. The *Pre*-S/S ORF codes for three different protein moieties, namely, *Pre*-S1/S (initiating at the AUG at position 948), *Pre*-S2/S (initiating at the AUG at position 1272), and S (initiating at the AUG at position 1437) (see Figures 1 and 3). Furthermore, these individual proteins are differently glycosylated so that at least six different products are synthesized from this single ORF.[20] The *Pre*-C/C ORF can code for two primary gene products, namely, *Pre*-C/C and C. The C protein is the major capsid protein detected in vivo. *Pre*-C/C is not detected in vivo, but evidence, recently available,[30,31a] points to *Pre*-C/C being the precursor to the HBeAg detected in serum in vivo.[29]

The replication scheme, described in detail in Section II.C, requires several enzyme activities which are presumably products of the P-ORF. This will be discussed separately in Section II.C.4.

1. Transcriptional Control

None of the major or minor transcripts found either in infected liver cells[41,43] or in tissue culture cells of various kinds[50] are spliced. Thus, the production of the various transcripts described above requires that several promotors are active in vivo. Transfection experiments with all or parts of the HBV genome clearly show that the signals controlling the expression of the S-gene are largely tissue independent. Thus, the S gene can be expressed in a number of tissue culture systems under control of its own[51-53] or heterologous promotors.[54] The situation with the C-mRNA/pregenome is quite different, and it has been rather difficult to express significant amounts of C-mRNA or C-protein in tissue culture using either the authentic promotor region[55] or heterologous promotors.[56] It would appear that, on the one hand, the control regions operate in a defined tissue and differentiation specific manner and, on the other hand, even when a functional (heterologous) promotor is present, the production of C-protein is in some way down regulated. In Figure 4, putative in vivo transcriptional regulatory regions on the HBV genome are shown and will be described here briefly.

Considering first of all the region upstream of the major subgenomic transcript, the S-mRNA, no TATA box can be identified at the appropriate distance from the 5′ end. However, as indicated in Figure 4, the 40 nucleotides directly upstream of the 5′ end of the S-mRNA show a strong sequence homology to SV40 sequences; in particular, the −20 to −30 region shows a strong homology to the SV40 promotor region. In addition, at approximately −50 to −70 (relative to the 5′ end of the S-mRNA), there is a strong sequence homology to the

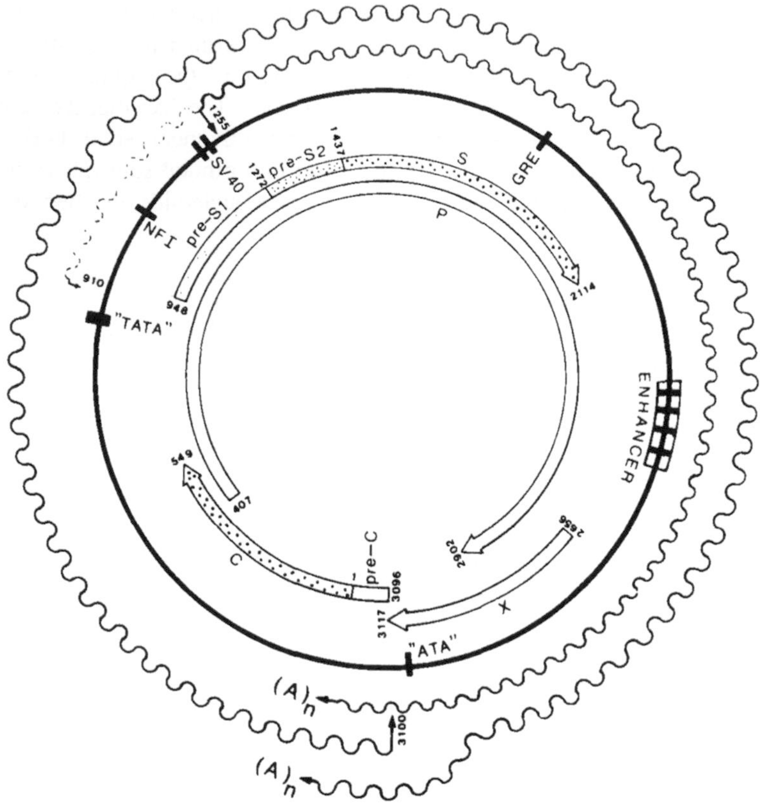

FIGURE 4. Putative transcriptional control elements on the HBV genome. The open reading frames are shown on the inside of the circular HBV genome and the RNA transcripts outside. The positions of putative control elements are shown as black bars and are abbreviated as follows. "TATA" indicates the position of a TATA box at position 876-882 which may be a component of the promotor for the minor *Pre*-SI mRNA.[58,59] NFI shows the position of a high-affinity nuclear factor I binding site at position 1068-1082.[57] SV-40 shows the regions of homology to the SV40 origin (1188-1202) and to the SV40 late promotor (1222-1234).[42] GRE shows the region of homology to the concensus sequence for glucocorticoid responsive elements.[63] The region (2361-2519) within which the putative enhancer lies is shown.[60-62] "ATA" indicates the position (3071-3077) of the so-called "ATA-box" presumably a component of the C-promotor (discussed in Reference 2). See text for further details.

SV40 origin.[42] These regions are thought to be important in the control of SV40 late expression and, by analogy, are presumed to be important in the regulation of S-mRNA transcription. As mentioned in Section II.A.2, the HBV S-mRNA has a major 5′ end at position 1255 and two minor starts further downstream. This feature of multiple starts may be of regulatory importance (see Section II.B.2). Recently it was shown that the activity of the isolated S-promotor region in vitro is regulated by the presence of a high-affinity Nuclear Factor I binding site approximately 190 bases upstream from the S-gene promotor (Figure 4). Deletion of this sequence leads to a drastic reduction in the activity of the S-promotor.[57] The minor, *Pre*-S1-mRNA, not detected by Northern blotting, but only by 5′ end mapping,[43] has its 5′ end at position 910. A TATA box sequence can be detected approximately 30 nucleotides upstream of this 5′ end and is presumably a component of the *Pre*-S1 promotor.[58,59] Further sequences regulating *Pre*-S1 mRNA expression are not known to date.

Considering now the regulation of transcription of the C-mRNA/pregenome, there is again no perfect TATA box homology; rather, only a so-called "ATA-box" is present at the appropriate distance from the mRNA start. It is conserved in all hepatitis B viruses and is presumably a component of the promotor (discussed in Cattaneo et al.[2]). A putative enhancer has been localized approximately 450 base-pairs upstream of the C-gene promotor.[60,61] As with other enhancers, the HBV enhancer acts in an orientation-independent manner and is tissue specific.[62] A further regulatory sequence, separate from the enhancer, is a glucocorticoid-responsive element which maps upstream of the enhancer[63] (see Figure 4).

2. *Translational Control*

As described above, mRNAs initiating within, or just upstream of, the *Pre*-S/S region have been identified, namely, the minor *Pre*-S1-mRNA with a 5′ end at position 910 and the S-mRNA with a major 5′ end at position 1255. Thus, transcriptional control can account for the relative amounts of *Pre*-S1/S and HBsAg proteins found in vivo where *Pre*-S1/S is a minor component found mainly in Dane particles and less in subviral S-particles and filaments.[20] The unique *Pre*-S1 region of the *Pre*-S1/S protein most likely carries the host cell receptor binding site.[25] *Pre*-S2/S proteins are also minor conponents in comparison to HBsAg[20] and, in this case, the control of gene expression lies, at any rate, partially, at the level of translation. The major S-mRNA initiates approximately 15 nucleotides upstream of the *Pre*-S2 AUG. This first AUG is, however, used inefficiently, presumably because it is unusually near the 5′ end of the S-mRNA and because it lies in a sequence context which is unfavorable for ribosome recognition (discussed in Cattaneo et al.[2] and Kozak[64]). Presumably HBsAg arises by translation initiating at the second AUG which lies in a favorable sequence context.[2] However, it is not clear what role the minor mRNAs, which lack the *Pre*-S2 AUG, play in the synthesis of HBsAg.

The mechanisms regulating the gene expression of the *Pre*-C/C region are not completely understood. There is, at any rate, a control at the transcriptional level (see above), since the major C-mRNA initiates within the *Pre*-C-region, and only in WHV and GSHV have transcripts, initiating upstream of *Pre*-C, been detected. Whether, superimposed on this transcriptional control, a translational control is operating remains to be established. The C-mRNA also appears to function as the messenger for the translation of the gene products from the P-ORF. However, since the P gene products have not yet been identified, it is only possible to speculate on the mechanisms controlling their synthesis (Section II.C.4).

C. Model for HBV Replication

From the data described above, the following model for HBV replication can be formulated.

1. *Production of Pregenome*

After entry into the cell and uncoating, the viral genome is transported to the nucleus where the incomplete plus strand is completed and the genome-bound protein, the RNA plus strand primer and short redundancy on the minus strand removed by cellular repair mechanisms. The nicks in the DNA strands are ligated to yield a supercoiled, circular molecule. This cccDNA is transcribed into unspliced, capped, polyadenylated mRNA species which, on the one hand, are used for the production of viral proteins. In addition, the C-mRNA which initiates within DR1 in the *Pre*-C region and is terminally redundant, also has the function of RNA pregenome.

2. *Minus Strand Synthesis*

Evidence points to the reverse transcription of the DNA minus strand from the RNA pregenome occurring in intracellular core particles.[40] The signals which trigger capsid assembly are unknown, but must lead to cores containing the RNA pregenome, the minus strand protein primer, nascent minus strand and polymerase, ribonuclease H activity, etc. The 5′ end of the minus strand lies within the DR1 sequence which is present twice on the

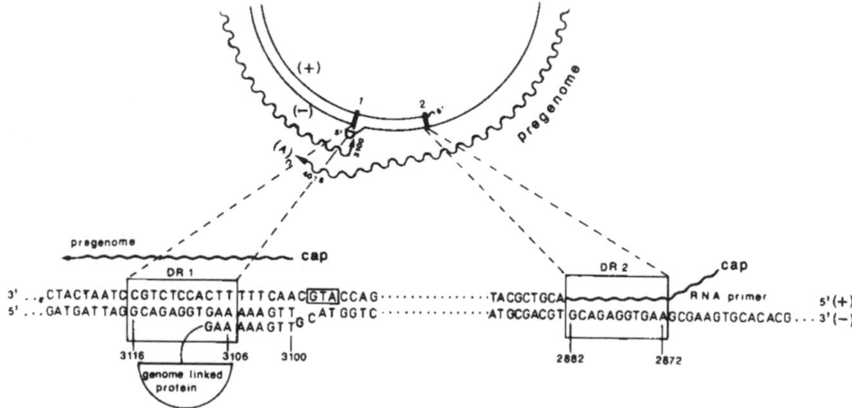

FIGURE 5. The cohesive overlap region of HBV. The double semi circle on the top shows the viral minus strand with the 5′ bound protein (small half circle) and the incomplete viral plus strand with the RNA primer at the 5′ end. The black bars 1 and 2 show the positions of DR1 and DR2, respectively. The RNA pregenome is shown as a wavy line. On the lower part, the DNA sequences in the vicinity of DR1 and DR2, which are important for replication, are shown. The relative position of the 5′ end of the RNA pregenome and its use as an RNA primer for plus-strand DNA synthesis are also shown. The 3′ and 5′ end of the minus strand and the 5′ end of the plus strand, found in virions, are given. See text for further details.

pregenomic RNA. However, initiation at the DR1 near the 5′ end would yield a very short (HBV, nine nucleotides) "strong-stop" DNA and require a very early template switch. No template switch is necessary if DNA minus strand initiation occurs within the DR1 sequence at the 3′ end of the pregenome and proceeds to the 5′ end of this molecule. The result is a DNA minus strand with a 3′ end corresponding exactly to the 5′ end on the RNA pregenome and a terminal redundancy of nine nucleotides (Figure 5). The primer for minus-strand synthesis is presumed to be the protein which remains covalently attached to the 5′ end.[26,27] It remains unknown if this protein is of viral or cellular origin.

3. Plus Strand Synthesis

Concomitant to reverse transcription of the pregenomic RNA, a ribonuclease H activity is required which removes the RNA from the RNA-DNA hybrid. This removal of RNA does not occur completely randomly, but rather an important function of the ribonuclease H is to generate a capped oligoribonucleotide from the 5′ end of the pregenomic RNA with a 3′ end mapping to the 3′ end of DR1 (see Figure 5). This oligoribonucleotide is removed from DR1 and transferred to the homologous DR2 near the 3′ end of the minus strand. The mechanism by which this transfer, which reduces base-pairing from 17 (DR1 plus six adjacent bases) to 11 (DR2), is not yet known, but probably involves protein factors. Plus strand synthesis proceeds from this primer and after reaching the 3′ end of the minus strand DNA, a template switch is required and made possible by the short terminal redundancy on the minus strand (Figure 5). At some point just prior to, or during, plus strand synthesis, cores become associated with membrane, containing *Pre*-S/S proteins, and are released from the infected cell. Presumably as a result of deoxynucleoside triphosphates no longer being available (the membrane is impermeable to these), plus strand synthesis stops before completion. Thus, released virions contain the partially single-stranded circular DNA genome peculiar to hepatitis B viruses.

4. Proteins Involved in the Replication Cycle

The replication cycle described in the last section requires several protein activities, the most central of which is the viral DNA polymerase. As mentioned previously, HBV virions contain a polymerase activity which can complete the unfinished viral DNA plus strand. It is presumed that this activity is a product of the long P-ORF which can code for a protein of approximately 90,000 daltons, the approximate size of many polymerase molecules. The HBV polymerase functions as a reverse transcriptase, and there is a region in the translation product of the P-ORF which shares amino acid sequence homology with a region of the reverse transcriptase of retroviruses,[65] a fact which supports the notion that the P-ORF codes for the HBV polymerase. As mentioned in Section II.A.2, no RNA transcript initiating directly upstream of the P-ORF has been detected, so that it seems likely that P-gene products are translated from the overlength C-mRNA. It has not yet been possible to detect such products in acutely infected liver, but in the livers of a number of patients with hepatocellular carcinomas, two gene products containing sequences of the P-ORF could be detected.[66] Using a number of sequence-specific antisera, these products could be identified as being C-pol fusion proteins of molecular weight 35,000 and 69,000 daltons. These products contain C-specific sequences, as well as P-specific sequences, which are derived from the N terminal region of the P-ORF. It has been proposed that these products arise by ribosomal frameshifting as occurs in the biosynthesis of retroviral reverse transcriptase[67] and in the synthesis of a related protein encoded by the yeast retrotransposon Ty1.[68,69] Whether these HBV C-pol fusion products, found in significant amounts only in a few hepatocellular carcinomas, and the biosynthetic mechanisms implicated by them, can be applied to the situation in an acute HBV infection, remains to be established.

A further activity required by the proposed replication scheme (Figures 2 and 5) is a ribonuclease H responsible for removing the RNA frm the RNA-DNA hybrid and generating the primer oligoribonucleotide required for plus strand synthesis. As is the case with retroviruses, it is likely that this activity is also a product of the P-ORF, either as part of the polymerase protein or as a separate protein entity. Integration of the hepatitis B genome, as found in hepatocellular carcinomas, is an extremely rare event in the acute infection so that it is unclear if a viral endonuclease activity, as required by retroviruses, plays any role at any stage in HBV replication. On the other hand, if the proteins of the P-ORF are synthesized as C-pol fusion proteins which are subsequently processed, a viral protease activity may be required.

III. COMPARISON OF THE LIFE CYCLE OF HEPATITIS B VIRUS WITH OTHER RETROVIRUSES AND RETROTRANSPOSONS

In the past few years it has become clear that hepatitis B viruses are members of a large family of related viruses and mobile genetic elements which include retroviruses,[70] cauliflower mosaic virus,[71,72] and retrotransposons.[73] The central feature common to these viruses is the reverse transcription of a terminally redundant RNA species of plus polarity. This process was first discovered and has been analyzed in great detail in retroviruses,[70] and, thus, we shall first compare the life cycle of hepatitis B viruses with that of retroviruses (Table 1). For minus DNA strand synthesis, retroviruses employ a cellular t-RNA as primer at a single primer site, which results in a template switch having to occur very early in replication. The primer for HBV minus strand synthesis is probably the protein which remains attached to the 5′ end of this strand. Although there are two potential primer sites, it appears likely that only one site is used (near the 3′ end of the RNA pregenome) so that, in contrast to retroviral replication, a template switch is not necessary for minus strand synthesis. In both virus systems, the primer for plus strand synthesis is an oligoribonucleotide which has been generated by the action of ribonuclease H. In the case of retroviruses, the primer is a polypurine tract which remains associated with the minus DNA strand. However, with hepatitis B viruses, the oligoribonucleotide primer is displaced from its complementary minus strand and reannealed at another position which is homologous to only part of the primer.

Table 1
SIMILARITIES AND DIFFERENCES BETWEEN HBV-LIKE VIRUSES AND RETROVIRUSES

	HBV-like viruses	Retroviruses
Gene Organization	CORE X / POL / SURFACE	GAG ENV / POL
Overlap of genes	Extensive	Limited
Redundancies RNA	Yes	Yes
Redundancies DNA	Short	Long
Transcription	Unidirectional	Unidirectional
DNA template	Extrachromosomal	Integrated
Promotors	Two or three	One
Splicing in major RNAs	No	Yes
Ribosomal frameshifting	Likely	Yes
Reverse transcription	Yes	Yes
Primer for (−)-strand	Protein, two potential primer sites	tRNA, one primer site
Primer for (+)-strand	"Displaced" 5′ end of genomic RNA	Polypurine tract
Template switch in (−)-strand synthesis	Unlikely	Essential
Site of virus budding	Endoplasmatic reticulum	Plasma membrane
Nucleic acid in extracellular virus	Partially double-stranded DNA	Single-stranded RNA (diploid)
Integration	Very seldom	Obligatory

Note: For references on the retroviral life cycle see Varmus and Swanstrom.[70] The gene organization of Rous-associated virus (RAV) is indicated as a retroviral prototype.

In both cases, a double-stranded (retroviruses) or partially double-stranded (HBV) DNA molecule is generated from the RNA (pre)genome, but at completely different stages in the replication cycles. With HBV, the replication cycle is completed at this stage and the partially double-stranded DNA is released in infectious virions. However, with retroviruses, the double-stranded DNA molecule represents a relatively early intermediate. There is an obligatory integration of this DNA provirus in the host genome and subsequent transcription, etc., finally resulting in the release of infectious virions containing an RNA genome.

Another point of similarity between HBV and retroviruses is the gene organization, i.e., the order of the genes on the genome (Table 1). The HBV genes C (core), P (polymerase), and S (surface) are present in a condensed form on the HBV genome, i.e., the genes are extensively overlapping. In retroviruses, the three genes involved in replication gag (core), pol (polymerase), and env (surface) have the same order on the genome, but overlap only to a limited extent. There is no evidence that the HBV X-gene plays any role similar to retroviral oncogenes. Transcription of these similarly organized genomes is, however, totally different. Whereas retroviruses have only one promotor and subsequent splicing gives rise to subgenomic mRNAs, hepatitis B viruses have several promotors and the RNA transcripts are not spliced. A point of similarity, however, may exist in the biosynthesis of the P-gene products. In retroviruses a gag–pol precursor protein is produced by ribosomal frameshifting,[67] and this may also occur with hepatitis B viruses.

In summary, there are some basic mechanisms which are shared between retroviruses and hepatitis B viruses which, however, occur at different stages in the replication cycles. That is, the replication cycles are similar in many points, but are temporally displaced with respect to each other.

In recent years, the main steps in the transposition mechanism of a number of eukaryotic retrotransposons have been elucidated[73] (discussed in detail in the following chapter). Briefly, retrotransposons consist of two long terminal repeats (LTRs) bordering usually two (but sometimes three) genes which are similar to the retroviral gag and pol (and env) genes. That is, the genome organization is similar to those of retroviruses and hepatitis B viruses (hepatitis B viruses do not, however, have LTRs). Transposition involves transcription to an RNA intermediate which is reverse transcribed intracellularly to yield a DNA species which is subsequently integrated into the cell genome. As in retroviruses, synthesis of the retrotransposon reverse transcriptase involves ribosomal frameshifting between two adjacent ORFs[68,69] to yield a fusion protein which is subsequently proteolytically processed. Thus, the replication strategy of retrotransposons closely resembles the replication cycle of retroviruses, but without an extracellular phase. Thus, the similarities and differences between retroviruses and hepatitis B viruses, discussed above, also apply to hepatitis B viruses and retrotransposons. Recently, Tuttleman et al.[39] could show that in duck hepatitis B virus infected hepatocytes, the pool of cccDNA in the nucleus can be expanded directly by reverse transcription of RNA pregenomes and reimport from the cytoplasm without an extracellular phase. As mentioned above, this mechanism is also used by retrotransposons, but not by retroviruses. In the case of hepatitis B viruses, where integration of functional genomes into the cellular genome does not occur, this mechanism of production of the cccDNA is probably important to maintain persistant infection.[39]

The replication mechanism of cauliflower mosaic virus, discussed in detail in the previous chapter,[71,72] can be regarded as being intermediate between hepatitis B and retroviral replication. Like HBV, CaMV is a DNA virus which replicates via an RNA pregenome. However, the primer for minus-strand synthesis is a tRNA as with retroviruses and not a protein as with HBV. The organization of the three genes for core (IV), polymerase (V), and inclusion body protein (VI) is similar to gag (C), pol (P), and env (S). The RNA transcripts, however, appear to be unspliced and represent a terminally redundant overlength transcript (presumably the pregenome) and a subgenomic transcript which is translated to the inclusion body protein. This transcription strategy is reminiscent of hepatitis B viruses.

The mechanistic and genome organizational similarities between hepatitis B viruses, retroviruses, retrotransposons, and CaMV described until now, point to an evolutionary relationship between these elements. As mentioned previously, a segment of the gene product of the HBV P-ORF shows amino acid sequence homology to a region of the retroviral reverse transcriptases.[65] This homology could also be found in a region of the gene V product of CaMV and in the translation product encoded by the appropriate ORF of retrotransposons (for a review, see Reference 74). Sequence homology is, however, not restricted only to the pol-gene products. As described recently by Miller and Robinson,[75] there are further sequences shared between hepatitis B viruses and the other reverse-transcribing elements, so that it may be that these elements have all evolved from a common ancestor.

REFERENCES

1. **Tiollais, P., Pourcel, C., and Dejean, A.,** The hepatitis B virus, *Nature (London)*, 317, 489, 1985.
2. **Cattaneo, R., Sprengel, R., Will, H., and Schaller, H.,** Molecular biology of hepatitis B virus, in *The Molecular Genetics of Mammalian Cells, A Primer in Developmental Biology*, Shepard, M., Simonsen, C., and Malacinski, G., Eds., MacMillan, New York, 1986.
3. **Howard, C. R.,** The biology of hepadnaviruses, *J. Gen. Virol.*, 67, 1215, 1986.

3a. **Ganem, D. and Varmus, H. E.**, The molecular biology of the hepatitis B viruses, Annu. Rev. Biochem., 56, 651, 1987.

4. **Sureau, C., Roup-Lemonne, J.-L., Mullins, J. I., and Essex, M.**, Production of hepatitis B virus by a differentiated human hepatoma cell line after transfection with cloned circular HBV DNA, *Cell*, 47, 37, 1986.

5. **Chang, C., Jeng, K.-S., Hu, C.-P., Lo, S. J., Su, T.-S., Ting, L.-P., Chou, C.-K., Han, S., Pfaff, E., Salfeld, J., and Schaller, H.**, Production of hepatitis B virus in vitro by transient expression of cloned HBV DNA in a hepatoma cell line, *EMBO J.*, 6, 675, 1987.

6. **Galibert, F., Mandart, E., Fitoussi, F., Tiollais, P., and Charnay, P.**, Nucleotide sequence of the hepatitis B virus genome (subtype ayw) cloned in E. coli, *Nature (London)*, 281, 646, 1979.

7. **Pasek, M., Goto, T., Gilbert, W., Zink, B., Schaller, H., MacKay, P., Leadbetter, G., and Murray, K.**, Hepatitis B virus genes and their expression in E. coli, *Nature (London)*, 282, 575, 1979.

8. **Valenzuela, P., Gray, P., Quiroga, M., Zaldivar, J., Goodman, H. M., and Rutter, W. J.**, Nucleotide sequence of the gene coding for the major protein of hepatitis B surface antigen, *Nature (London)*, 280, 815, 1979.

9. **Sninsky, J. J., Siddiqui, A., Robinson, W. S., and Cohen, S. N.**, Cloning and endonuclease mapping of the hepatitis B viral genome, *Nature (London)*, 279, 236, 1979.

10. **Ono, Y., Onda, H., Sasada, R., Igarashi, K., Sugino, Y., and Wishioka, K.**, The complete nucleotide sequences of cloned hepatitis B virus DNA subtype *adr* and *adw*, *Nucleic Acids Res.*, 11, 1747, 1983.

11. **Fujiyama, A., Miyanohara, A., Nozaki, C., Yoneyama, T. R., Ohtomo, N., and Matsubara, K.**, Cloning and structural analyses of hepatitis B virus DNAs, subtype *adr*, *Nucleic Acids Res.*, 11, 4601, 1983.

12. **Will, H., Kuhn, C., Cattaneo, R., and Schaller, H.**, Structure and function of the hepatitis B virus genome, in *Primary and Tertiary Structure of Nucleic Acids and Cancer Research*, Miwa, M., Nishimura, S., Rice, A., Söll, D. G., and Sugimara, T., Eds., Japan Science Society Press, Tokyo, 1982, 237.

13. **Galibert, F., Chen, T. N., and Mandart, E.**, Nucleotide sequence of a cloned woodchuck hepatitis B virus genome: comparison with the hepatitis B virus sequence, *J. Virol.*, 41, 51, 1982.

14. **Kodama, K., Ogasawara, N., Yoshikowa, H., and Murakami, S.**, Nucleotide sequence of a cloned woodchuck hepatitis virus genome: evolutional relationship between hepadnaviruses, *J. Virol.*, 56, 978, 1985.

15. **Seeger, C., Ganem, D., and Varmus, H. E.**, Nucleotide sequence of an infectious molecularly cloned genome of ground squirrel hepatitis virus, *J. Virol.*, 51, 367, 1984.

16. **Mandart, E., Kay, A., and Galibert, F.**, Nucleotide sequence of a cloned duck hepatitis B virus genome: comparison with woodchuck and human hepatitis B virus sequences, *J. Virol.*, 19, 782, 1984.

17. **Sprengel, R., Kuhn, C., Will, H., and Schaller, H.**, Comparative sequence analysis of duck and human hepatitis B virus genomes, *J. Med. Virol.*, 15, 323, 1985.

18. **Eble, B. E., Lingappa, V. R., and Ganem, D.**, Hepatitis B surface antigen: an unusual secreted protein initially synthesized as a transmembrane polypeptide, *Mol. Cell. Biol.*, 6, 1454, 1986.

19. **Patzer, E. J., Nahamura, G. R., Simonsen, C. C., Levinson, A. D., and Brands, R.**, Intracellular assembly and packaging of hepatitis B surface antigen particles occur in the endoplasmic reticulum, *J. Virol.*, 58, 884, 1986.

20. **Heermann, K. H., Goldmann, V., Schwartz, W., Seyffarth, T., Baumgarten, H., and Gerlich, W. H.**, Large surface proteins of hepatitis B virus containing the Pre-S sequence, *J. Virol.*, 52, 396, 1984.

21. **Pfaff, E., Klinkert, M.-Q., Theilmann, L., and Schaller, H.**, Characterization of large surface proteins of hepatitis B virus by antibodies to Pre S-S encoded amino acids, *Virology*, 148, 15, 1986.

22. **Neurath, A. R., Kent, S. B. H., Strick, N., Taylor, P., and Stevens, C. E.**, Hepatitis B virus contains pre-S gene-encoded domains, *Nature (London)*, 315, 154, 1985.

23. **Wong, D. T., Nath, N., and Sninsky, J. J.**, Identification of hepatitis B virus polypeptides encoded by the entire preS open reading frame, *J. Virol.*, 55, 223, 1985.

24. **Neurath, A. R., Kent, S. B. H., Strick, N., and Parker, K.**, Identification and chemical synthesis of a host cell receptor binding site on hepatitis B virus, *Cell*, 46, 429, 1986.

25. **Ondora, S., Ohori, H., Yamaki, M., and Ishida, N.**, Electron microscopy of human hepatitis B virus cores by negative staining-carbon film technique, *J. Med. Virol.*, 10, 147, 1982.

26. **Gerlich, W. and Robinson, W. S.**, Hepatitis B virus contains protein covalently attached to the 5′-terminus of its complete DNA strand, *Cell*, 21, 801, 1980.

27. **Molnar-Kimber, K. L., Summers, J., Taylor, J. M., and Mason, W. S.**, Protein covalently bound to minus-strand DNA intermediates of duck hepatitis B virus, *J. Virol.*, 45, 165, 1983.

28. **Kaplan, P. M., Greenman, R. L., Gerin, J. L., Purcell, R. H., and Robinson, W. S.**, DNA polymerase associated with human hepatitis B antigen, *J. Virol.*, 12, 995, 1973.

29. **Takahashi, K., Machida, A., Funatsu, G., Nomura, M., Usuda, S., Aoyagi, S., Tachibana, K., Miyamoto, H., Imai, M., Nakamura, T., Miyakawa, Y., and Mayumi, M.**, Immunochemical structure of hepatitis B e antigen in the serum, *J. Immunol.*, 130, 2903, 1983.

30. **Ou, J.-H., Laub, O., and Rutter, W. J.,** Hepatitis B virus gene function: the precore region targets the core antigen to cellular membranes and causes the secretion of the e-antigen, *Proc. Natl. Acad. Sci. U.S.A.*, 83, 1578, 1986.
31. **Roossinck, M. J., Jameel, S., Loukin, S. H., and Siddiqui, A.,** Expression of hepatitis B viral core region in mammalian cells, *Mol. Cell. Biol.*, 6, 1393, 1986.
31a. **Schlicht, H. J., Galle, P., and Schaller, H.,** The duck hepatitis B virus pre-C region encodes a signal which is essential for the synthesis and secretion of processed core proteins but not for virus formation, *J. Virol.*, 61, 00, 1987.
32. **Moriarty, A. M., Alexander, H., Lerner, R. A., and Thornton, G. B.,** Antibodies to peptides detect new hepatitis B antigen: serological correlation with hepatocellular carcinoma, *Science*, 227, 429, 1985.
33. **Meyers, M. L., Trepo, L. V., Nath, H., and Sninski, J. J.,** Hepatitis B virus polypeptide X: expression in Escherichia coli and identification of specific antibodies in sera from hepatitis B virus-infected humans, *J. Virol.*, 57, 101, 1986.
34. **Elfassi, E., Haseltine, W. A., and Dienstag, J. L.,** Detection of hepatitis B virus X product using an open reading frame Escherichia coli expression vector, *Proc. Natl. Acad. Sci. U.S.A.*, 83, 2219, 1986.
35. **Pfaff, E., Salfeld, J., Gmelin, K., Schaller, H., and Theilmann, L.,** Synthesis of the X-protein of hepatitis B virus in vitro and detection of anti-X antibodies in human sera, *Virology*, 158, 456, 1987.
35a. **Twu, J.-S. and Schloemer, R. H.,** Transcriptional transactivating function of hepatitus B virus, *J. Virol.*, 61, 3448, 1987.
36. **Ruiz-Opazo, N., Chakraborty, P. R., and Shafritz, D. A.,** Characterization of viral genomes in the liver and serum of chimpanzee long-term hepatitis B carriers: a possible role for supercoiled HBV-DNA in persistant HBV infection, *J. Cell. Biochem.*, 19, 281, 1982.
37. **Weiser, B., Ganem, D., Seeger, C., and Varmus, H. E.,** Closed circular viral DNA and asymmetrical heterogeneous forms in livers from animals infected with ground squirrel hepatitis virus, *J. Virol.*, 48, 1, 1983.
38. **Mason, W. S., Halpern, M. S., England, J. M., Seal, G., Egan, J., Coates, L., Aldrich, C., and Summers, J.,** Experimental transmission of duck hepatitis B virus, *Virology*, 131, 375, 1983.
39. **Tuttleman, J. S., Pourcel, C., and Summers, J.,** Formation of the pool of covalently closed circular viral DNA in hepadnavirus-infected cells, *Cell*, 47, 451, 1986.
40. **Summers, J. and Mason, W. S.,** Replication of the genome of a hepatitis B-like virus by reverse transcription of an RNA intermediate, *Cell*, 29, 403, 1982.
41. **Cattaneo, R., Will, H., and Schaller, H.,** Hepatitis B virus transcription in the infected liver, *EMBO J.*, 3, 2191, 1984.
42. **Cattaneo, R., Will, H., Herdandez, N., and Schaller, H.,** Signals regulating hepatitis B surface antigen transcription, *Nature (London)*, 305, 336, 1983.
43. **Will, H., Reiser, W., Weimer, T., Pfaff, E., Buescher, M., Sprengel, R., Cattaneo, R., and Schaller, H.,** Replication strategy of human hepatitis B virus, *J. Virol.*, 61, 904, 1987.
44. **Buescher, M., Reiser, W., Will, H., and Schaller, H.,** Characterization of transcripts and the potential RNA pregenome of duck hepatitis B virus: implications for replication by reverse transcription, *Cell*, 40, 717, 1985.
45. **Moroy, T., Etiemble, J., Trepo, C., Tiollais, P., and Buendia, M. A.,** Transcription of woodchuck hepatitis B virus in chronically infected liver, *EMBO J.*, 4, 1507, 1985.
46. **Enders, G. H., Ganem, D., and Varmus, H. E.,** Mapping the major transcripts of ground squirrel hepatitis virus: the presumptive template for reverse transcription is terminally redundant, *Cell*, 42, 297, 1985.
47. **Molnar-Kimber, K. L., Summers, J. W., and Mason, W. S.,** Mapping of the cohesive overlap of duck hepatitis B virus DNA and of the site of initiation of reverse transcription, *J. Virol.*, 51, 181, 1984.
48. **Lien, J.-M., Aldrich, C. E., and Mason, W. S.,** Evidence that a capped oligoribonucleotide is the primer for duck hepatitis B virus plus strand DNA synthesis, *J. Virol.*, 57, 229, 1985.
49. **Seeger, C., Ganem, D., and Varmus, H. E.,** Biochemical and genetic evidence for the hepatitis B virus replication strategy, *Science*, 232, 477, 1986.
50. **Asselsberg, F. A. M., Will, H., Wingfield, P., and Hirschi, M.,** A recombinant chinese hamster ovary cell line containing a 300-fold amplified tetramer of the hepatitis-B genome together with a double selection marker expresses high levels of viral protein, *J. Mol. Biol.*, 189, 401, 1986.
51. **Pourcel, C., Louise, A., Gervais, M., Chenciner, N., Dubois, M.-F., and Tiollais, P.,** Transcription of the hepatitis B surface antigen gene in mouse cells transformed with cloned viral DNA, *J. Virol.*, 42, 100, 1982.
52. **Christman, J. K., Gerber, M., Price, P. M., Flordellis, C., Edelman, J., and Acs, G.,** Amplification of expression of hepatitis B surface antigen in 3T3 cells cotransfected with a dominant-acting gene and cloned viral DNA, *Proc. Natl. Acad. Sci. U.S.A.*, 79, 1815, 1982.
53. **Wang, Y., Schäfer-Ridder, M., Stratowa, C., Wong, T. K., and Hofschneider, P. H.,** Expression of hepatitis B surface antigen in unselected cell culture transfected with recircularized HBV DNA, *EMBO J.*, 1, 1213, 1982.

54. **Cattaneo, R., Will, H., Darai, G., Pfaff, E., and Schaller, H.,** Detection of an element of the SV40 late promotor in vectors used for expression studies in COS cells, *EMBO J.*, 2, 511, 1983.
55. **Gough, N. M. and Murray, K.,** Expression of the hepatitis B virus surface, core and e antigen genes by stable rat and mouse cell lines, *J. Mol. Biol.*, 162, 43, 1982.
56. **Will, H., Cattaneo, R., Pfaff, E., Kuhn, C., Roggendorf, H., and Schaller, H.,** Expression of hepatatis B antigens with a simian virus 40 vector, *J. Virol.*, 50, 335, 1984.
57. **Shaul, Y., Ben-Levy, R., and De-Medina, T.,** High affinity binding site for nuclear factor I next to hepatitis B virus S gene promotor, *EMBO J.*, 8, 1967, 1986.
58. **Malpiece, Y., Michel, M. L., Carloni, G., Revel, M., Tiollais, P., and Weissenbach, J.,** The gene S promotor of hepatitis B virus confers constitutive gene expression, *Nucleic Acids Res.*, 11, 4645, 1983.
59. **Rall, L. B., Standring, D. N., Laub, O., and Rutter, W. J.,** Transcription of hepatitis B virus by RNA polymerase II, *Mol. Cell. Biol.*, 3, 1766, 1983.
60. **Shaul, Y., Rutter, W. J., and Laub, O.,** A human hepatitis B viral enhancer element, *EMBO J.*, 4, 427, 1985.
61. **Tognoni, A., Cattaneo, R., Serfling, E., and Schaffner, W.,** A novel expression selection approach allows precise mapping of the hepatitis B virus enhancer, *Nucleic Acids Res.*, 13, 7457, 1985.
62. **Jameel, S. and Siddiqui, A.,** The human hepatitis B virus enhancer activity requires transacting cellular factor(s) for activity, *Mol. Cell. Biol.*, 6, 710, 1986.
63. **Tur-Kaspa, R., Burk, R. D., Shaul, Y., and Shafritz, D. A.,** Hepatitis B virus DNA contains a glucocorticoid-responsive element, *Proc. Natl. Acad. Sci. U.S.A.*, 83, 1627, 1986.
64. **Kozak, M.,** Comparison of initiation of protein synthesis in procaryotes, eucaryotes and organelles, *Microbiol. Rev.*, 47, 1, 1983.
65. **Toh, H., Hayashida, H., and Miyata, T.,** Sequence homology between retroviral reverse transcriptase and putative polymerase of hepatitis B virus and cauliflower mosaic virus, *Nature (London)*, 305, 827, 1983.
66. **Will, H., Salfeld, J., Pfaff, E., Manso, C., Theilmann, L., and Schaller, H.,** Putative reverse transcriptase intermediates of human hepatitis B virus in primary liver carcinomas, *Science*, 231, 594, 1986.
67. **Jacks, T. and Varmus, H.,** Expression of the Rous Sarcoma virus pol gene by ribosomal frame shifting, *Science*, 230, 1237, 1985.
68. **Mellor, J., Fulton, S. M., Dobson, M. J., Wilson, W., Kingsman, S. M., and Kingston, A. J.,** A retrovirus-like strategy for expression of a fusion protein encoded by yeast transposon Ty1, *Nature (London)*, 313, 243, 1985.
69. **Wilson, W., Malim, M. H., Mellor, G., Kingsman, A. J., and Kingsman, S. M.,** Expression strategies of the yeast retrotransposon Ty: a short sequence directs ribosomal frameshifting, *Nucleic Acids Res.*, 14, 7001, 1986.
70. **Varmus, H. E. and Swanstrom, R.,** Replication of retroviruses, in *RNA Tumor Viruses*, Weiss, R., Teich, N., Varmus, H., and Coffin, J., Eds., Cold Spring Harbor Laboratory, Cold Spring Harbor, N.Y., 1984.
71. **Hohn, T., Hohn, B., and Pfeiffer, P.,** Reverse transcription in CaMV, *TIBS*, 10, 205, 1985.
72. **Bonneville, J. M., Hohn, T., and Pfeiffer, P.,** Reverse transcription in the plant virus, cauliflower mosaic virus, in *RNA Genetics*, Vol. 2, Domingo, E., Holland, J., and Ahlquist, P., Eds., CRC Press, Boca Raton, Fla., 1987, in press.
73. **Baltimore, D.,** Retroviruses and retrotransposons: the role of reverse transcription in shaping the eukaryotic genome, *Cell*, 40, 481, 1985.
74. **Hull, R. and Covey, S. N.,** Genome organization and expression of reverse transcribing elements: variations and a theme, *J. Gen. Virol.*, 67, 1751, 1986.
75. **Miller, R. H. and Robinson, W. S.,** Common evolutionary origin of hepatitis B virus and retroviruses, *Proc. Natl. Acad. Sci. U.S.A.*, 83, 2531, 1986.

Chapter 4

RETROTRANSPOSONS

Jef D. Boeke

TABLE OF CONTENTS

I. Introduction ... 60
- A. Definition of Terms ... 61
- B. General Properties ... 61
- C. General Structure ... 64
  - 1. DNA and RNA ... 64
  - 2. Open Reading Frames ... 64
  - 3. Heterogeneity ... 65
- D. A Model for Retrotransposition ... 65
  - 1. Transcription ... 65
  - 2. Encapsidation ... 65
  - 3. Reverse Transcription ... 70
  - 4. Integration ... 70

II. Distribution ... 70
- A. Yeasts ... 72
- B. Molds and Slime Molds ... 72
- C. *Drosophila* ... 75
- D. Lepidoptera ... 75
- E. Plants ... 75
- F. Rodents and Other Mammals ... 75

III. Structural Comparison ... 76
- A. Open Reading Frames and Their Expression ... 76
  - 1. Yeast and *Drosophila* Elements ... 76
    - a. Expression of pol and env ... 76
    - b. Protein Products ... 78
  - 2. Mouse Elements ... 79
  - 3. DIRS1 ... 79
- B. *Cis*-Acting Sequences ... 79
  - 1. LTR Sequences ... 79
    - a. Transcriptional Signals ... 79
    - b. Integration Signals ... 81
    - c. Alignment Signals ... 81
  - 2. Primer Binding Sites ... 81
  - 3. Host Target Site Duplications ... 82
  - 4. Packaging Signals ... 82
  - 5. Enhancers and Silencers ... 82

IV. Mechanism of Retrotransposition ... 83
- A. Copia ... 83
- B. Other *Drosophila* Elements ... 84
- C. Ty Element of Yeast ... 84
- D. DIRS1 (Dictyostelium) ... 85

V. Regulation of Retrotransposition 85
  A. Transcriptional Control 85
    1. Ty and DIRS Elements 88
    2. *Drosophila* Elements 88
    3. IAP Elements 88
  B. Translational Control 88
  C. Posttranslational Regulation 88
  D. Defective and Functional Elements 89

VI. Effects on Neighboring Genes 89
  A. Class I — Insertions Within the Coding Sequence 89
  B. Class II — Inactivating Insertions Within the 5′ Noncoding Region 91
  C. Class III — Insertions Within Introns 91
  D. Class IV — Insertions Which Activate the Target Gene 92
  E. Specificity of Integration 92
    1. Ty Elements 92
    2. *Drosophila* Elements 92
    3. Other Elements 92

VII. Recombination Between Retrotransposons 92
  A. LTR-LTR Recombination 93
  B. Gene Conversion 93
  C. Reciprocal Recombination Events 93
  D. Interactions Between Hybrid Dysgenesis and *Drosophila* Retrotransposons 93

VIII. Suppressors of Retrotransposon Insertion 94
  A. *SPT3* 94
  B. su(Hw) 95

Acknowledgments 96

References 96

I. INTRODUCTION

Retrotransposons constitute an extensively distributed family of eukaryotic transposable elements. The retrotransposons share a number of structural features which reflect the mechanism by which new copies of the transposon are generated; this process is similar to that of retroviral reverse transcription and integration. In this review, several aspects of retrotransposon biology in light of recent investigations will be considered. Specifically, this review will summarize recent findings on the mechanism and regulation of retrotransposition, the structure of the elements, roles of the encoded gene product(s) in transposition, interactions between retrotransposons and adjacent host genes, and interactions between retrotransposons and host gene products (i.e., the insertion suppressors). Several other recent reviews of these topics are also available.[1-12] This review will concern itself primarily with yeast and insect elements, although where possible, results from vertebrate systems are also incorporated. Not discussed is another major class of eukaryotic transposons, typified by the P and Ac elements, which are characterized by short inverted terminal repeats.

A. Definition of Terms

In recent years the number of new systems in which reverse transcription is known or inferred has increased dramatically. Reverse transcriptase, which was once though to be associated only with the life cycle of retroviruses, has been demonstrated or implicated in the replication of hepadna- and caulimo-viruses as well as numerous transposable elements and in the generation of processed pseudogenes. Only retrotransposons, defined here as those *noninfectious* transposable elements which *have* long terminal repeat (LTR) sequences, and transpose or are thought to transpose by a retrovirus-like reverse transcription mechanism, will be considered in this review. It should be pointed out that members of a second class of widely distributed transposons which probably transpose by reverse transcription, typified by the mammalian Alu sequences, the *Drosophila* F elements, murine L1 sequences, and processed pseudogenes (reviewed recently by Vanin[13] and by Weiner et al.[12]), *lack* LTRs. This class of elements, called "retroposons" by Rogers[14] and "retrotranscripts" by Temin,[8] is beyond the scope of this review. Strictly speaking, all retroviruses could be considered retrotransposons because they are formally transposable elements;[5] (Figure 1) however, retroviruses which were originally isolated as infectious agents will not be considered here, except for comparative purposes. It may well be that some "retrotransposons" will in fact later turn out to be infectious agents (retroviruses), even though they were originally discovered by virtue of the fact that their insertion led to a mutant phenotype or because their proviruses were found as a component of the middle repetitive DNA fraction. The fact that many *Drosophila* retrotransposons have a reading frame corresponding to a retroviral env gene[15-17] lends support to this notion.

Transposition is a term which is sometimes used loosely to describe the movement of a sequence to a new location in the genome. Here transposition will refer only to the insertion of a copy of a transposable element into a *nonhomologous* site. Such insertion is almost always accompanied by the duplication of a few nucleotides of target DNA. Transposable elements can also move to faraway sites in the genome by homology-dependent mechanisms (see Section VII), but these are not transpositions in the strict sense of the word.

B. General Properties

Retrotransposons generally exist as families containing a few[18] to over a thousand[19] members per host genome. This repetitiveness makes these gene families somewhat difficult to study because (1) there is considerable heterogeneity among members of the families (some elements may be functional while others are not), and (2) traditional genetic analysis is always much more difficult to apply to a multigene family. The repetitiveness of these gene families is further complicated by the fact that in most retrotransposon families, there exists an even larger family of solo LTRs, derived from intact retrotransposons, in addition to the family of intact elements. Solo LTRs arise from retrotransposons through homologous recombination between the two directly repeated LTR sequences which flank the core region, with resultant loss of the core sequence (Figure 2). The repetitiveness and ubiquity of retrotransposons is testimony to a long-standing relationship with the "host" organisms. The relationship between retrotransposon(s) and host has resulted in the evolution of various intricate mechanisms which interconnect the regulation of the host with that of its retrotransposon(s). Many of these regulatory interconnections probably arise as a consequence of the fact that retrotransposons, like retroviruses, are potent mutagens. Retrotransposons are able to inactivate host genes by various types of insertional inactivation, but they can also activate the expression of previously silent genes in a very specific manner (i.e., they

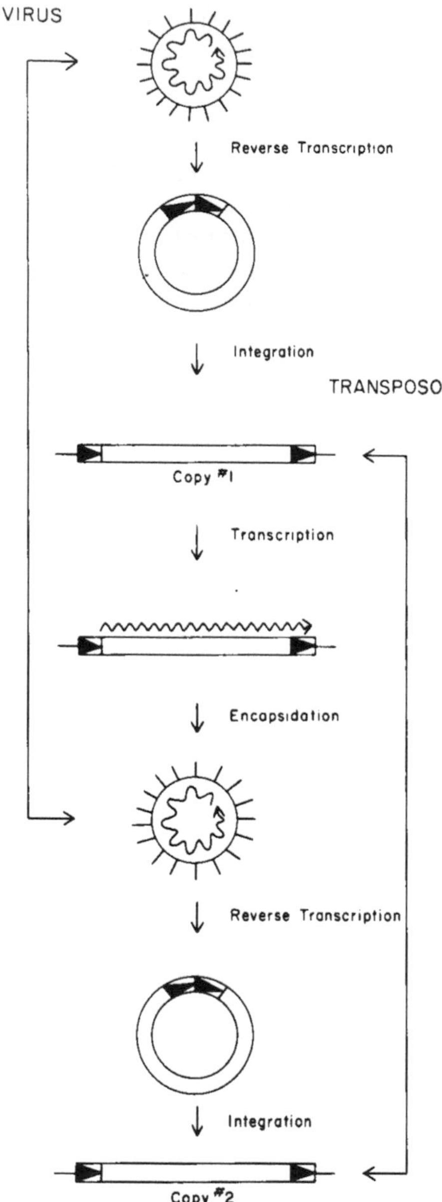

FIGURE 1. Retrovirus vs. retrotransposon life cycles. Circles with spikes represent viral (or virus-like) particles; the wavy lines represent RNA. Boxed triangles represent LTR sequences, the open box between them the internal portion of the element, and simple lines represent the host DNA. The circular form represents the possible integration or transposition intermediate.[74] The life cycles of retroviruses and retrotransposons are probably a simple permutation of the same process.[6] It should be emphasized, however, that it is not proven that virus-like particles are an intermediate in retrotransposition. The retroviral life cycle is usually thought of as starting and ending with viral particles; two-LTR circular DNA and the proviral DNA are intermediates. However, the provirus is essentially a transposon as well; it begins as copy #1 in one cell, and after one cycle of infection, another cell bearing copy #2 is produced. If the original cell is reinfected, it will bear copy #1 and copy #2. Retrotransposons probably leave the cell as a particle rarely if at all; after transposition, the cell contains an additional copy of the transposon.

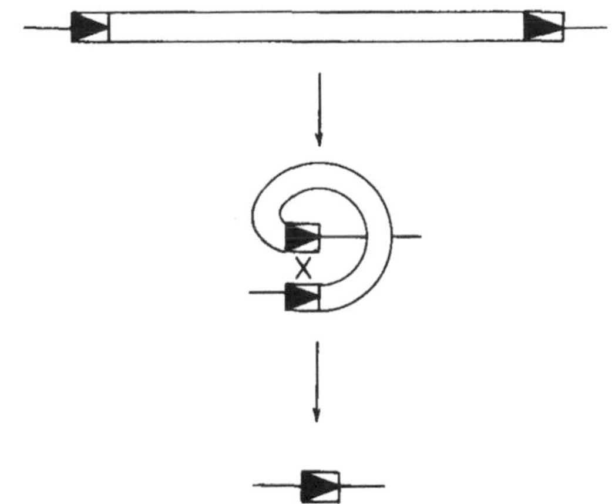

FIGURE 2. LTR-LTR recombination. A frequent event apparently catalyzed by host recombination systems is the "looping out" of a copy of a retrotransposon. This requires a single crossover event (X) occurring between the two copies of the LTR (boxed triangle). The result is a single copy or solo LTR. The same reaction can be the result of a gene conversion event involving a donor solo LTR and a recipient intact element.

can subject the gene to a new type of regulation). This suggests that retrotransposons are maintained over evolutionary time because they can cause insertion mutations which are beneficial to the host as well as detrimental insertions. For example, the evolution of a regulated promoter sequence by the step-wise accumulation of point mutations in a constitutive promoter would be a slow process indeed. A retrotransposon insertion, on the other hand, can introduce a regulated promoter (or at least a regulated enhancer) into the vicinity of a target gene in a single step. Thus, in an evolutionary sense, retrotransposons may provide their host cells or organisms with an important selective advantage in that in retrotransposons, they possess a highly specific internal mutagen which allows them to quickly adapt at the genetic level to a new environment or stress.[11,20-26]

The presence of any multigene family which is dispersed throughout the genome offers the possibility of the creation of deletions, inversions, and translocations of large pieces of chromosomes. These chromosome aberrations are produced as the result of homologous recombination events which can occur between widely separated copies of the element. Theoretically, such aberrations could result in the alteration of expression patterns of relatively large numbers of genes in a single step. If such events occurred at a very high frequency, this would clearly lead to destabilization or scrambling of the genome and consequent loss of fitness. On the other hand, the rare occurrence of such events might provide a selective advantage. Indeed, it can be argued that the occasional occurrence of such chromosome rearrangements provides the molecular basis for the punctuated equilibrium theory of evolution, in which there is a change in the expression of a large set of genes in a single step, resulting in a "hopeful monster" which differs significantly from its predecessor and, in some cases, may be able to exploit a previously unoccupied ecological niche. Clearly, the organism must have evolved mechanisms to limit the frequencies of both transposition and genomic rearrangements to a level which is evolutionarily beneficial, yet not detrimental to normal cellular function.

In yeast,[27-30] *Drosophila,*[31] and the mouse,[32] suppressor mutations which specifically reverse the phenotype of retrotransposon insertion mutations have been isolated. Generally, these are mutations in single-copy genes whose products act specifically upon retrotransposons, their associated solo LTRs, and the genes adjacent to such elements. It thus seems

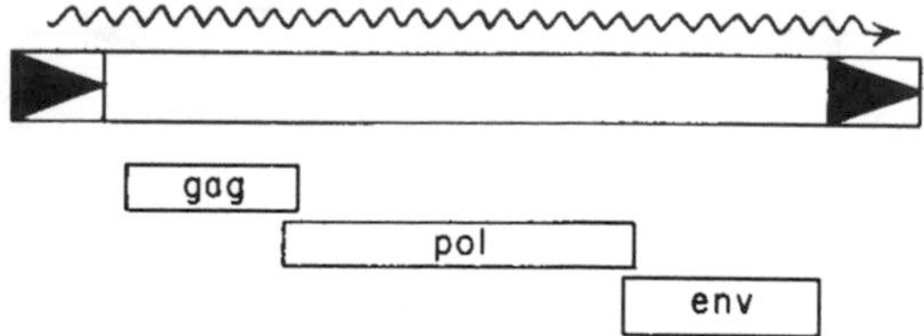

FIGURE 3. General structure of a retrotransposon. The boxed triangles represent LTR sequences, and the box between them the core sequence of the retrotransposon. The boxes beneath the retrotransposon represent open-reading frames which correspond to the retroviral genes gag, pol, and env. The wavy line symbolizes the end-to-end transcript.

likely that networks of genes whose products interact with retrotransposons have evolved within the host to help poise the level of potentially deleterious mutagenesis and recombination events. The ideal level of transposon activity would be one in which the gross deleterious effects are minimized, but the possibility of introduction of beneficial rearrangements (in the evolutionary sense) is retained.[24] It also seems reasonable that the transposons themselves (perhaps in conjunction with cellular gene products) may have the capacity to detect stress in the environment and raise their transposition frequency to higher levels in response to such stress.

C. General Structure

1. DNA and RNA

The structure of a generic retrotransposon is indicated in Figure 3. The DNA consists of a pair of directly repeated LTR sequences, a few hundred base-pairs long, flanking a unique core sequence of several thousand base-pairs. The core sequence which lies between the LTRs usually contains one to three open reading frames. An abundant end-to-end transcript initiating within the 5′ LTR sequence and terminating in the 3′ LTR sequence can often be found in the host cells;[32-35] this transcript is synthesized by RNA polymerase II.[37] The structure of this transcript in all cases is very similar; it initiates about halfway into the 5′ LTR and continues through the element, passing the point in the 3′ LTR homologous to the initiation point so that the RNA is terminally redundant, by some dozens of base-pairs. Thus, this transcript is very similar structurally to the genomic RNA of retroviruses, and available evidence suggests that it is the genetic material for transposition.[38] In many cases, additional retrotransposon transcripts can be found, but their analysis is difficult because of the repetitive nature of retrotransposons. It is often difficult to be certain whether a given transcript represents the product of a single, aberrant element, or is a transcript which is produced by all members of the retrotransposon family.

2. Open Reading Frames

The open reading frames (ORFs) in most retrotransposons sequenced to date can be correlated with the well-known genes of avian and murine retroviruses; gag, pol, and env. The gag proteins (multiple protein products are derived from retroviral primary translation products by proteolytic processing) function primarily as core virion structural proteins in all retroviruses. The mouse retrotransposon, IAP (intracisternal A-particle), produces particles in which the major protein is encoded by the gag reading frame.[39] There is considerable evidence that this is also the case in copia family elements[40] and in yeast Ty elements.[203-205] The pol ORF contains several domains which encode different functions, with the predominant one being the reverse transcriptase enzymatic function (RT domain). Also encoded in the pol ORF are an integrase or endonuclease function, thought to catalyze the integration of the reverse transcript into the target site[41-50] (by analogy to phage lambda, this is called the int domain) a protease which processes the gag, gag/pol,[51,203-205] and env primary translation products (pro domain), and an RNAse H function (encoded within the RT domain). The env gene encodes glycoproteins found in the lipid envelope of true retroviruses; it is responsible for the attachment of the virus particles to host cells and hence is a determinant of host range. Its role (in any) in retrotransposons is thus far unknown.

3. Heterogeneity

Nearly all families of repeated genes demonstrate structural variability, and retrotransposons are no exception. Numerous examples of restriction site polymorphisms, deletion, and insertion variants have been described for virtually every retrotransposon examined in detail.[52-55] It seems likely that one source of such heterogeneity may be the relatively error-prone nature of reverse transcriptase.[56] The high rate of recombination during the reverse transcription process[57] would then tend to generate an even larger array of variability by recombining different mutant forms. Nevertheless, when different elements within a family have been sequenced, conservation of open reading frames is often observed,[58-65] suggesting that there is selective pressure to retain some degree of function in these elements.

D. A Model for Retrotransposition

A model describing the retrotransposition process can be proposed by drawing heavily on the proposed models and experimental evidence for the mechanism of retroviral reverse transcription and integration. The available information for retrotransposons can be incorporated into the retroviral model (see Figure 4 for a general overview of the model for retrotransposition). It should be emphasized that this model makes many assumptions and that although parts of it are likely to be true, the model as a whole is by no means proven. Moreover, it incorporates results obtained from quite different systems and so may well be wrong in detail for any give retrotransposon. The special case of DIRS1[63,66,67] is undoubtedly different (see below). As detailed below, the retrotransposition process can be divided into several steps: transcription, particle formation, reverse transcription, and integration.

1. Transcription

The end-to-end retrotransposon transcript (Figure 3) is transcribed by cellular RNA polymerase II. This transcript can have at least two different fates: (1) it may be used as the template for translation into element-encoded proteins (the end-to-end transcript is defined as (+) strand = coding strand), or (2) it may serve as the genetic material for transposition.

By aligning the sequence of the end-to-end transcript of retrotransposons with their DNA copies (Figure 5), it is possible to define functional domains of the LTR sequence in the same manner as in retroviruses. That is, a repeated sequence, R, is found at both ends of the RNA. This sequence is typically a few dozen nucleotides in length. The transcribed portion of the 5′ LTR sequence is found only in the *5′* end of the RNA; hence its name U5. Likewise, the transcribed portion of the 3′ LTR is found only in the *3′* end of the RNA, called the U3 region.

2. Encapsidation

The RNA which is to act as genetic material is encapsidated by element encoded protein(s). The resulting particles probably contain several such proteins, including reverse transcriptase. At least one cellular tRNA which serves as the primer for reverse transcription in (−) strand synthesis is also encapsidated; there may be other "host" components in the virions as well. Whether the reverse transcription process takes place in the cytoplasm or the nucleus is unclear; in the case of Ty elements, there seem to be predominantly cytoplasmic particles[68,69] (Figure 6), whereas in the case of *Drosophila*, both cytoplasmic[70] and nuclear[40] virus-like

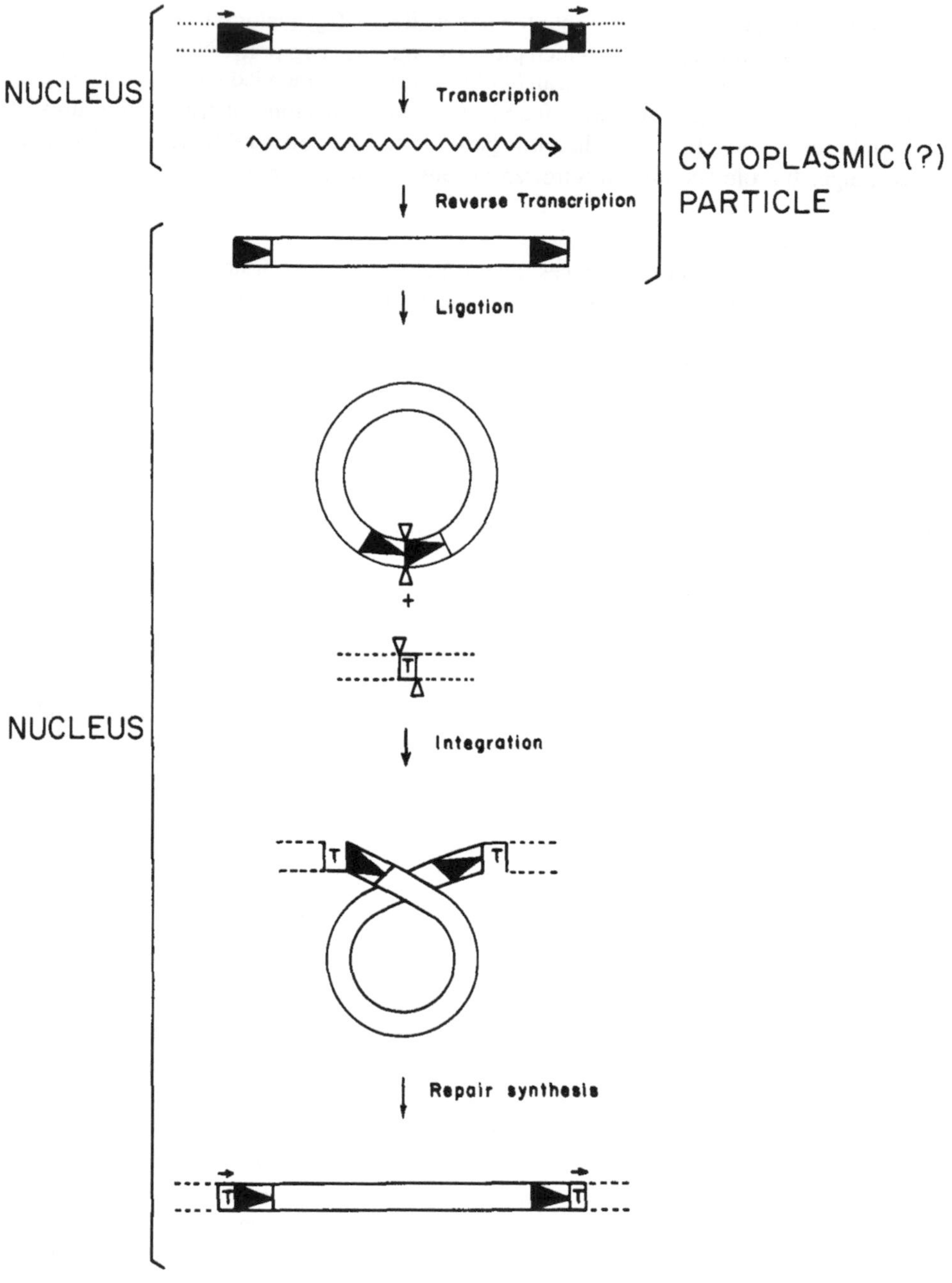

FIGURE 4. A model for retrotransposition. The boxed triangles represent LTR sequences, and the box between them the core sequence of the retrotransposon. The flanking host cell target repeats are symbolized by boxes containing a "T" (the arrows signify that these are directly repeated). The wavy line represents the end-to-end transcript. The circle represents a putative transposition intermediate;[74] the site of action of "integrase", or "att" site, is indicated by open triangles. The part of the cell in which the diagrammed events are thought to occur are included in brackets.

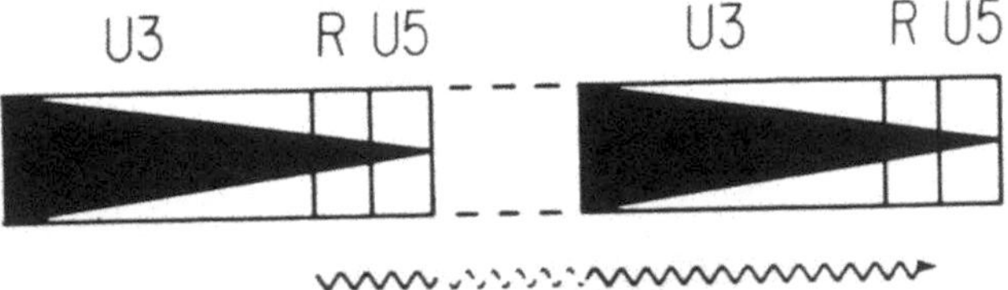

FIGURE 5. Domains of the LTR. The domains of retrotransposon LTRs are defined in the same way as are retroviral LTR domains. The boundaries between U3, R, and U5 sequences are defined by position of the ends of the end-to-end transcript. (From Boeke, J. D., Garfinkel, D. J., Styles, C. A., and Fink, G. R., *Cell*, 40, 491, 1985. With permission.)

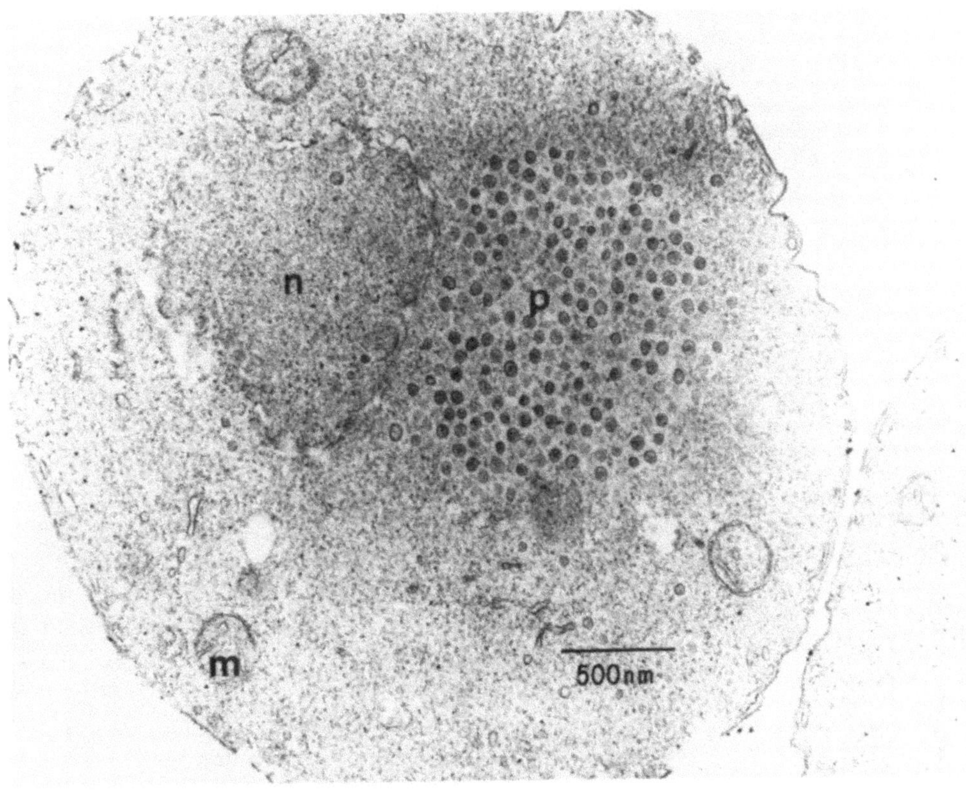

A

FIGURE 6. Virus-like particles (VLPs) produced by retrotransposons. (A) Ty VLP particles produced in yeast cells in response to overproduction of Ty RNA from a recombinant plasmid. Note the cytoplasmic location of the particles. (B) Cytoplasmic virus-like particles from *Drosophila melanogaster*; (*a*) thin section of *Drosophila* tissue culture cell, (*b*) VLPs seen at higher magnification, (*c*) partially purified preparation of VLPs which contain reverse transcriptase activity. (From Heine, C. W., Kelly, D. C., and Avery, R. J., *J. Gen. Virol.*, 49, 385, 1980. With permission.) (C) Nuclear VLPs from adult gut cells; similar particles are seen in nuclei of *Drosophila* brain tumor cells. (From Akai, H., Gateff, E., Davis, L. E., and Schneiderman, H. A., *Science*, 157, 810, 1967. With permission.)

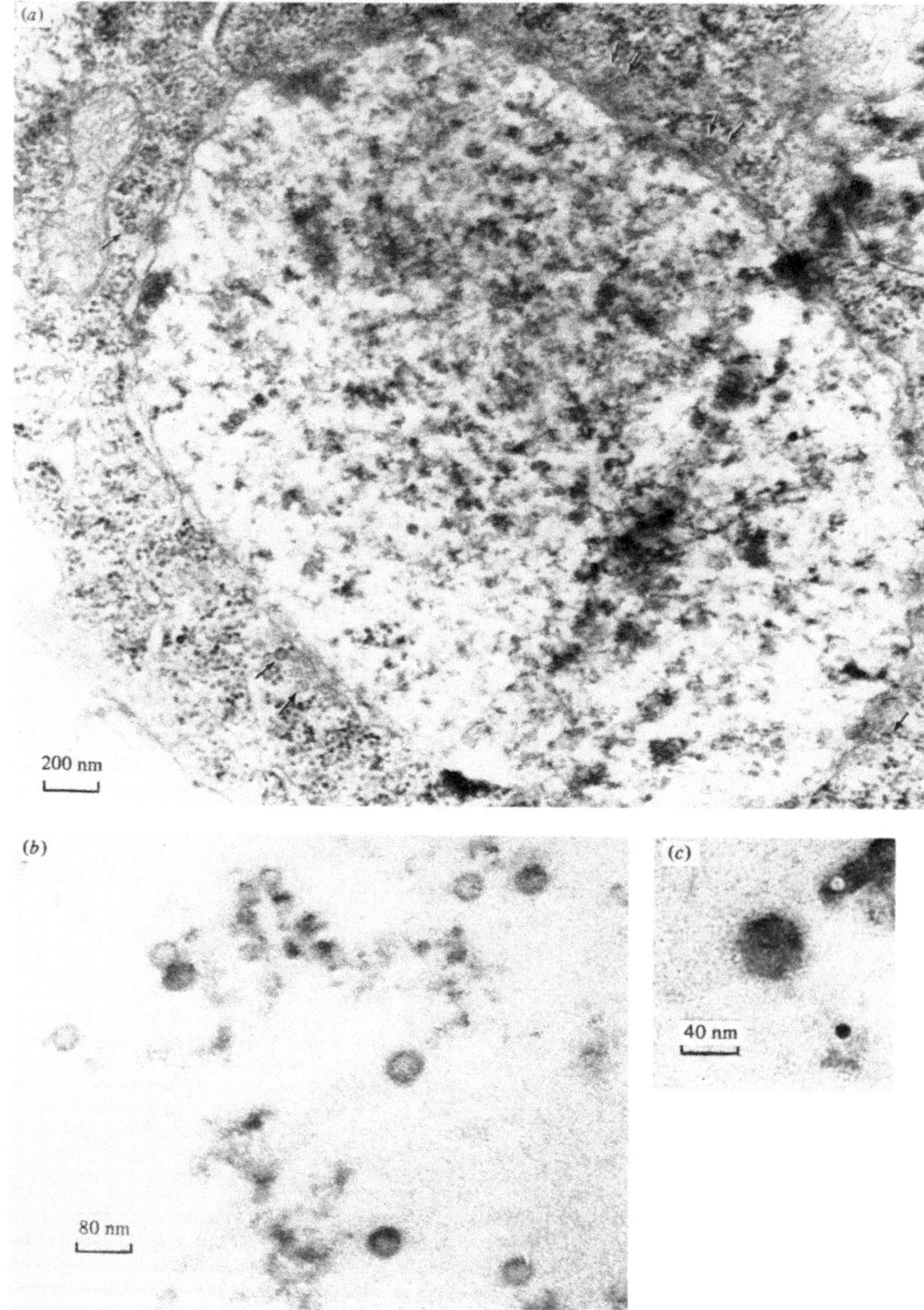

FIGURE 6B

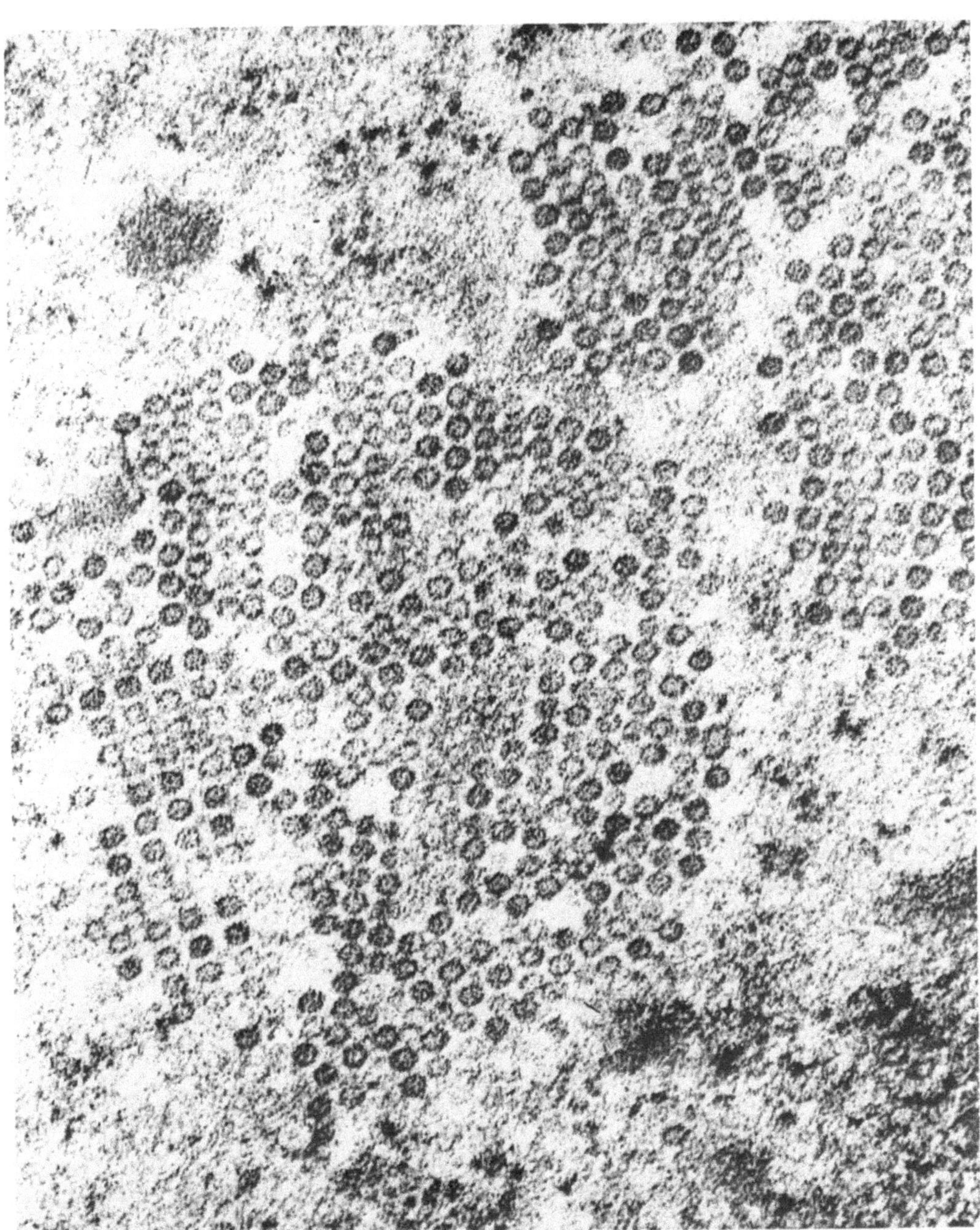

FIGURE 6C

particles (VLPs) containing reverse transcriptase activity have been observed, although the copia particles are said to be nuclear.[40] Murine IAP particles are localized in the cisternae of the endoplasmic reticulum.[71]

It should be pointed out that there is as yet no direct evidence that encapsidation is essential to the process of retrotransposition. However, reverse transcription of element-specific RNA does take place in vitro in isolated retrotransposon particles,[40,68-69] which suggests that the formation of VLPs is an important intermediate step in the transposition process. Particle formation could be important for at least two reasons. First of all, it could protect the cell from potentially deleterious levels of free reverse transcriptase, which might result in an intolerably high level of reverse transcription of cellular mRNAs and, potentially, the accumulation of large numbers of processed pseudogenes. Furthermore, maturation of the reverse transcriptase within the particle (by retrotransposon-encoded protease action) may activate an enzymatically inactive gag/pol precursor to an enzymatically active reverse transcriptase only after incorporation into the particle. Second, it is likely that the packaging of all of the "reagents" for transposition in a VLP makes the transposition process more efficient; it is possible that many reagents (tRNA, retrotransposon RNA, reverse transcriptase, etc.) need to be brought together in a high local concentration in order for the process to occur at the necessary frequency. The cleavage of precursor proteins within the particle might also ensure the proper stoichiometry of element-encoded proteins.

3. Reverse Transcription

Reverse transcription requires at least two specific priming events: (−) strand priming by a cellular tRNA, which results in an RNA/DNA hybrid known as (−) strong-stop DNA and (+) strand priming by an unknown entity (probably an oligoribonucleotide which is the product of RNAse H digestion) which yields (+) strong-stop DNA (see Figure 7). Strong-stop DNA molecules have recently been isolated from several *Drosophila* retrotransposons[36] and from Ty elements.[72] The model for the detailed mechanism of reverse transcription of retrotransposons is essentially identical to that proposed by Gilboa et al.[73] (Figure 7) for retroviruses.

4. Integration

The linear reverse transcript is circularized by an as yet unidentified activity(s), resulting in one of at least two types of product: a one- (probably formed by a recombinational process) or two-LTR circle (formed by ligation) (Figure 8). It is presumed that the two-LTR circles are the active precursors to integration and that the "circle junction" between the LTRs in the two-LTR circles form an "att" site at which the integrase or endonuclease can act.[41,74] The two-LTR circles may enter the nucleus complexed to integrase, or perhaps as a form of VLP. Recent evidence suggests that linear forms of DNA may be intermediates in the transposition process as well.[206] Once in the nucleus, the transposition intermediates can find a target site in the DNA and are integrated, perhaps in concert with cellular protein(s). A consequence of this integration is the duplication of a few nucleotides of host DNA, which suggests that the integrase makes a staggered cut of just that size in the target DNA, which is later filled in by repair enzymes or perhaps reverse transcriptase.

II. DISTRIBUTION

While true (i.e., infectious) retroviruses are widely distributed among vertebrates including fish, reptiles, birds, and mammals,[75] virus-like particles resembling retroviruses have been identified by electron microscopy in a variety of invertebrates including tapeworms,[76] various insects, (particularly *Drosophila,*[40,70,77-79]) and yeast.[68,69] Only in two cases[40,68,69] have these particles been directly correlated with the activity of the previously identified retrotransposons copia and Ty. As genetic or molecular biological entities, retrotransposons (or repetitive elements with retrotransposon-like structures) have now been isolated from many invertebrate eukaryotes including yeast (Ty1 and Ty2), slime molds (DIRS1, *Hpa*II repeat), maize (cin1 and bs1), *Drosophila* (copia, B104, gypsy, etc.), and a lepidopteran species (TED). Most bacterial transposons have inverted repeat termini, unlike most (but not all — the DIRS1 element has inverted repeats) retrotransposons. The notable prokaryotic exception to the inverted repeat generality, Tn9, is unlikely to be a retrotransposon because it is a composite transposon, the "LTRs" of which consist of IS1 elements, which are themselves transpos-

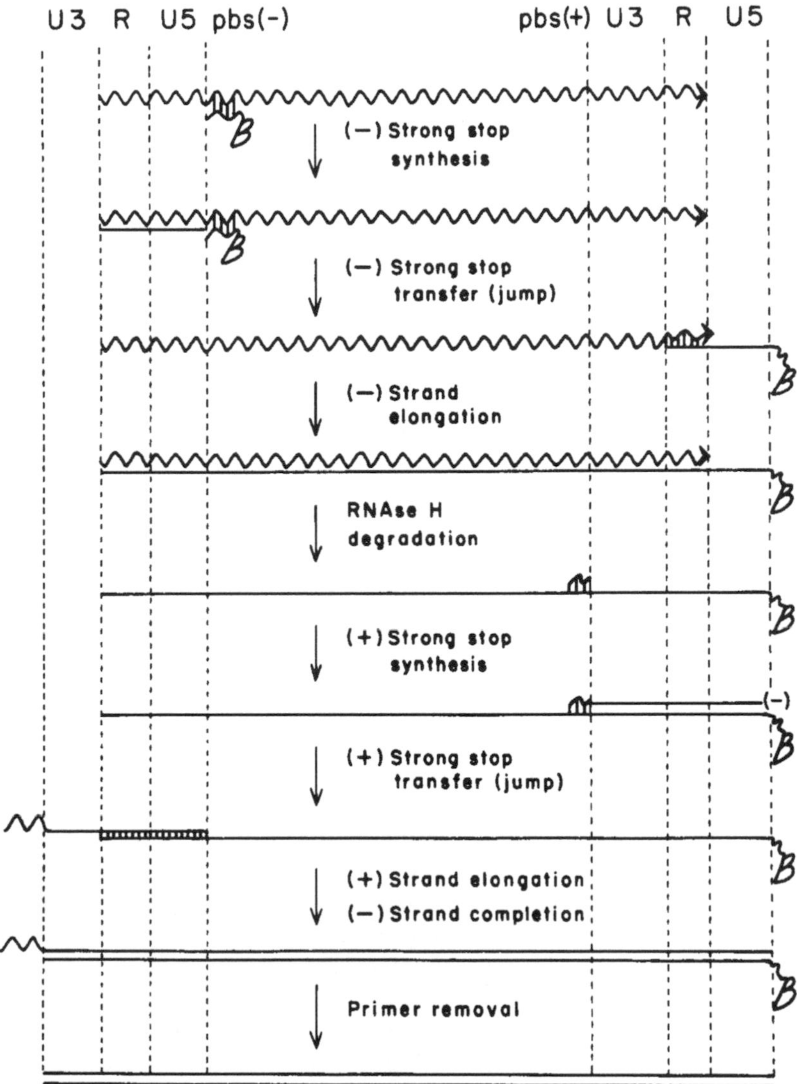

FIGURE 7. The steps of reverse transcription. A simplified model for the steps required to generate a full-length DNA copy from the end-to-end transcript by reverse transcription is presented. Straight lines represent DNA, and wavy lines RNA. The domains of the LTRs, U3, R, and U5, as well as the (−) and (+) strand primer binding sites [pbs(−)] and [pbs(+)] are separated by dotted lines. The long wavy line symbolizes the end-to-end transcript. Base pairing to the tRNA (β), to the (+) strand primer, and to the newly transferred strong-stop DNAs is indicated by hyphens. Tor the sake of simplicity, the model assumes that one template RNA is used; however, it is likely that two templates are used and that RNAseH degradation begins immediately after (−) strong-stop synthesis. See References 6, 73 and 202 for a review of more detailed models of reverse transcription.

able, unlike retrotransposon LTR sequences. There is considerable evidence that bacterial transposition proceeds directly via a DNA intermediate.

Among vertebrates, there are many endogenous retroviruses which qualify as retrotransposons by the above definition. These include the IAP sequences, which have been shown to cause insertion mutations[80-82] and hence must be considered transposable.[7] Moreover, IAP particles apparently never leave the cell, and efforts to demonstrate their infectivity have

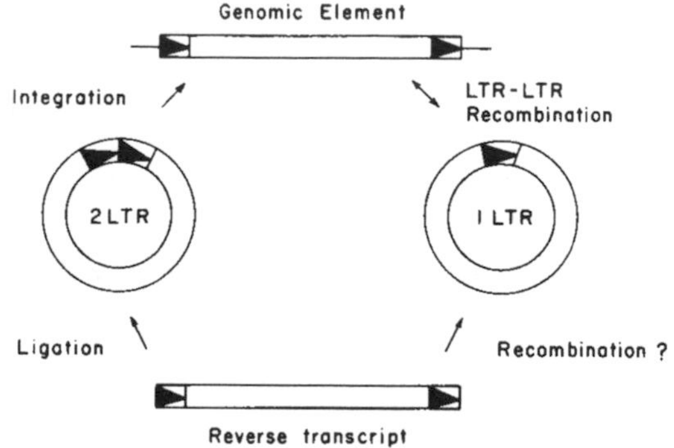

FIGURE 8. Circular DNA forms of retrotransposons. The possible interrelationships between various forms of retrotransposon DNA which have been isolated are indicated.

failed.[71] The murine VL30 sequence family also has many of the characteristics of retrotransposons.[83-85]

The distribution of known or suspected retrotransposons is summarized in Table 1.

A. Yeasts

Saccharomyces cerevisiae is the only yeast in which retrotransposons have been described thus far. In another yeast which is well studied at the molecular level, *Schizosaccharomyces pombe,* no candidates for retrotransposons have yet been unearthed, although many gene cloning and molecular mutation studies have been carried out in this organism. Perhaps *S. pombe* lacks retrotransposons.

In *S. cerevisiae,* two subfamilies of retrotransposons have been found, called Ty1 and Ty2.[52-53] These two classes of element are very closely related structurally, differing primarily within the coding regions. The gag region and a portion of the pol reading frame which lies between the int and RT domains are the regions of highest divergence between Ty1 and Ty2.[60] The LTR sequences for these two elements are very similar; the extent of heterology between a Ty1 LTR and a Ty2 LTR is similar to the extent of heterology seen between two LTRs from the same family. This, together with the fact that the int homology region seems relatively well conserved between Ty1 and Ty2, suggests that the two elements use similar integration components. Oddly, hybrids between the two elements have not been found, in spite of the propensity of Ty elements to recombine (see Section VII).

B. Molds and Slime Molds

Although both molds and slime mold species are genetically well studied, thus far, elements resembling retrotransposons have been described only in the slime molds. The DIRS1 element of *Dictyostelium* is possibly a retrotransposon, although it has a most unusual structure. DIRS1 elements (1) have *inverted* rather than direct LTR sequences, (2) contain a small internal complementary region (ICR), complementary to the termini of the inverted LTRs, and (3) are not flanked by short direct repeats of host sequence, but contain a 3′ AT-rich sequence of variable length.[63,66]

The *Hpa*II repeat of *Physarum* was recently shown to have many of the hallmarks of more standard retrotransposon structure, including LTR sequences and flanking regions containing terminal sequences very similar to those found in Ty and copia.[86] There is no evidence for duplication of host target sites upon insertion of *Hpa*II repeats, and the element appears to prefer insertion into inverted repeat target sequences. However, it should be emphasized that the sequences studied did not necessarily represent primary transposition

Table 1
KNOWN OR SUSPECTED RETROTRANSPOSONS

Element name	Other name(s)	Host organism	Total length	LTR length	Extent of characterization (structural)	Target dupl'n.	Ref.
Yeast							
Ty1	Ty912, Ty1-15, TyH3	*Saccharomyces cerevisiae*	5.9	334	Nucleotide sequence	5	58, 59, 65, 180
Ty2	Ty917, Ty1-17	*S. cerevisiae*	5.9	334	Nucleotide sequence	5	60, 64, 180
Slime molds							
DIRS1		*Dictyostelium discoideum*	4.8	—	Nucleotide sequence	—	63, 66, 67
*Hpa*II repeat		*Physarum polycephalum*	8.6	277	LTR and flanks	0	86
Drosophila							
copia		*Drosophila*	5.1	276	Nucleotide sequence	5	61, 62, 181, 182
297		*Drosophila*	6.5	415	Nucleotide sequence	4	16, 160, 183
412		*Drosophila* spp.	7.0	481 or 571	Nucleotide sequence	4	170, 184, 207
17.6		*Drosophila*	7.4	512	Nucleotide sequence	4	15, 161, 185, 186
B104	roo	*Drosophila*	8.7	429	LTR and flanks	5	187, 188
gypsy	mdg4	*Drosophila*	7.5	482	Nucleotide sequence	4	17, 162
HMS Beagle		*Drosophila*	7.3	266	LTR and flanks	4	151
mdg1		*Drosophila*	7.3	442	LTR and flanks	4	189, 190
mdg3		*Drosophila*	5.4	267	LTR and flanks	4	131, 191
1731		*Drosophila*	4.3	~350	Map		F. Peronnet in Ref. 10
springer		*Drosophila*	8.8	405	LTR	6	163
3518		*Drosophila*	6.6	~500	Map		192
calypso		*Drosophila*	7.2	?	Map		W. Bender in Ref. 10
flea		*Drosophila*	5.2	?	Map, partial LTR sequence	6	154

Table 1 (continued)
KNOWN OR SUSPECTED RETROTRANSPOSONS

Element name	Other name(s)	Host organism	Total length	LTR length	Extent of characterization (structural)	Target dupl'n.	Ref.
Lepidoptera							
TED		*Trichoplusia ni*	7.3	270	LTR and flanks	4	90, 193
Plants							
bs1		*Zea mays, Teosinte guerrero*	3.3	304	LTR and flanks	6	18, 148
cin1		*Z. mays and T. guerrero*	?	691	LTR	5	91
Rodents							
IAP		*Mus* spp.			LTR and flanks	6	54, 82, 93, 107, 194
		Hamster, rat	8.0	376	Nucleotide sequence		106
VL30		*Mus* spp.		556	LTR and flanks	4	84, 85
MuRRS		*Mus* spp.	5.7	509	Nucleotide sequence	4	94—96, 165, 196
Mys		*Peromyscus*	2.8	343	Nucleotide sequence	6	97
Human							
THE-1		*Homo sapiens*	2.3	350	Nucleotide sequence	5	98
hsRTVL-H		*Homo sapiens*	6.2	415	LTR and flanks	5	99

events. The sequences studied were excised from a complex region of DNA which was composed of multiple copies of the transposon, in which it had appeared to have transposed into itself. A similar type of transposition (of a putative retrotransposon into another copy of itself) was observed with DIRS1.[66] It is difficult to be certain whether these events represent true transpositions or whether homologous recombination processes were involved in generating the DNA sequences studied.

C. Drosophila

By far, the greatest variety of retrotransposons (and retrotransposon names!) which have been isolated derive from the fruit fly *Drosophila melanogaster.*[10] This is undoubtedly due in part to the rich genetic legacy of this organism and to the propensity of retrotransposons to cause insertion mutations, many of which happen to have developmentally interesting phenotypes. Many of these insertions were isolated long ago as spontaneous mutants showing some instability. A happy marriage of classical genetics and molecular biology enabled a correlation between such mutant loci and cloned DNA sequences from the "middle repetitive" fraction. The powerful technique of *in situ* hybridization on polytene chromosomes allowed the direct visualization that many middle repetitive sequences which later turned out to be retrotransposons were members of multicopy dispersed gene families.

The *Drosophila* retrotransposable elements fall into three groups. The first group, represented by copia and 412, resembles the yeast Ty element in its terminal TG . . . CA sequences and in the lack of an env gene. The second group, containing mdg1 and B104, probably contain a third open reading frame corresponding to env (based on their larger overall size) and have TG . . . CA termini. Members of a third group, containing 17.6, 297, HMS Beagle, and gypsy, contain env genes and have unusual termini, usually the sequence AGT . . . A(C/T)T (see Table 1 for references on individual *Drosophila* elements).

A considerable amount of work has been done on determining the distribution of these elements in other *Drosophila* species, as well as among different laboratory and wild strains. The copia element (originally isolated from *D. melanogaster*) is found in *D. simulans* and *D. mauritiana,* but not at all in *D. yakuba* and *D. erecta.*[87] *D. melanogaster* appears to have as much as seven times more dispersed middle repetitive DNA sequences (largely retrotransposons) than the other species studied.[88] Perhaps this is related to a long history of culture in the laboratory; wild yeast strains generally also tend to have a lower Ty copy number than their laboratory counterparts.[89]

D. Lepidoptera

A single retrotransposon-like element has been isolated from cells of the moth *Trichoplusia ni* by Miller and Miller.[90] The element was isolated as a mutation which arose upon growth of Nuclear polyhedrosis virus (NPV baculovirus) upon *T. ni.* The element transposed into the viral DNA; solo LTR derivatives of TED are thrown off from the mutant virus at high frequency. The extent of distribution of this element in other insect species is unknown.

E. Plants

Two retrotransposon-like elements have been described from maize, the cin1 (which is apparently a solo LTR in search of a core sequence)[91] and bs1[18] elements. The termini of the cin1 and bs1 LTR sequences are quite similar to copia and Ty. However, the size of the cin1 LTR is unusually large (691 base-pairs). The bs1 element was shown to be responsible for an *Adh1* mutation in maize.[92] Both elements have been found both in *Zea mays* and *Teosinte guerrero.*

F. Rodents and Other Mammals

The IAP elements are fairly well conserved between mice and hamsters.[93] IAP sequences have also been cloned from rat DNA; these are somewhat more distantly related than the former two.[93] Three other families of apparent retrotransposons, VL30,[83] MuRRS,[94-96] and mys,[97] have been isolated from various murine species.

At least two endogenous retrovirus-like elements (THE-1[98] and hsRTVL-H[99]) have been isolated from human DNA. It is not known whether these represent viruses or transposons; they have not been associated with insertion mutations.

III. STRUCTURAL COMPARISON

A. Open-Reading Frames and Their Expression

The general structures of some sequenced retrotransposons is summarized in Figure 9. With the exception of the special case of DIRS1 (Figure 11), all of these have the same general layout as the generic element in Figure 3. The three genes found in all retroviurses, gag, pol, and env, seem to have counterparts (in many cases) in retrotransposons.

1. Yeast and Drosophila Elements

In terms of genetic organization, there appear to be two classes of retrotransposons. The first class, containing the yeast Ty elements (Ty1 and Ty2 are very similar), and the *Drosophila* element copia lacks anything resembling an analogue to the retroviral env gene.[58-62] An additional distinguishing feature of this family of elements is that the order of functional domains within the pol reading frame (as judged by sequence homology to retrovirus pol genes) is different.[61] In Ty and copia, the order is pro–int–RT, whereas retroviruses and another sequenced elements have the order pro–RT–int.

The latter, more abundantly represented class, contains other sequenced *Drosophila* elements. Many of these also contain a third ORF, corresponding to env. While the significance of the "env" gene in retrotransposons is not clear at the moment, the fact that it is conserved among several *Drosophila* retrotransposon families (17.6, 297, gypsy, and possibly some of the other large retrotransposons) suggests that the env-containing elements may have an extracellular phase.[15-17]

a. Expression of pol and env

In all of the retrotransposons which express an end-to-end transcript, there is the question of how the pol and env reading frames are expressed. Typically, the first AUG is the one utilized by the eukaryotic translation machinery. There is now considerable evidence that (1) the pol reading frame in retroviruses and retrotransposons is expressed at a much lower level than the gag reading frame and (2) this attenuation of expression occurs in most cases at the translational level. There is no good evidence that the two frames are joined by a splicing event. The cases which are understood best include MuLV (Moloney murine leukemia virus), RSV (Rous sarcoma virus), and the Ty elements of yeast. In MuLV, a single UAG codon separates gag and pol. Readthrough from gag to pol occurs about 10% of the time via insertion of a glutamine residue at this amber codon.[51] This readthrough is presumably mediated by a cellular tRNA which suppresses termination. In RSV and most retrotransposons where the gag and pol frames overlap (i.e., are out of frame with respect to each other but overlap), a different translational mechanism is thought to operate. Again, about 90% of the gag proteins terminate normally, but 10% of the time, a cellular tRNA catalyzes either +1 or −1 frameshifting at an as yet unidentified point within the region of overlap between the gag and pol open reading frames (Table 2), thereby allowing the synthesis of a gag-pol readthrough product. Similar tRNAs have been implicated in prokaryotic systems where frameshifting is thought to operate. Whereas it is clear that pol is made as a readthrough protein in the case of Ty elements,[58,100] only with RSV has translational frameshifting been directly demonstrated as the mechanism of achieving readthrough. Jacks

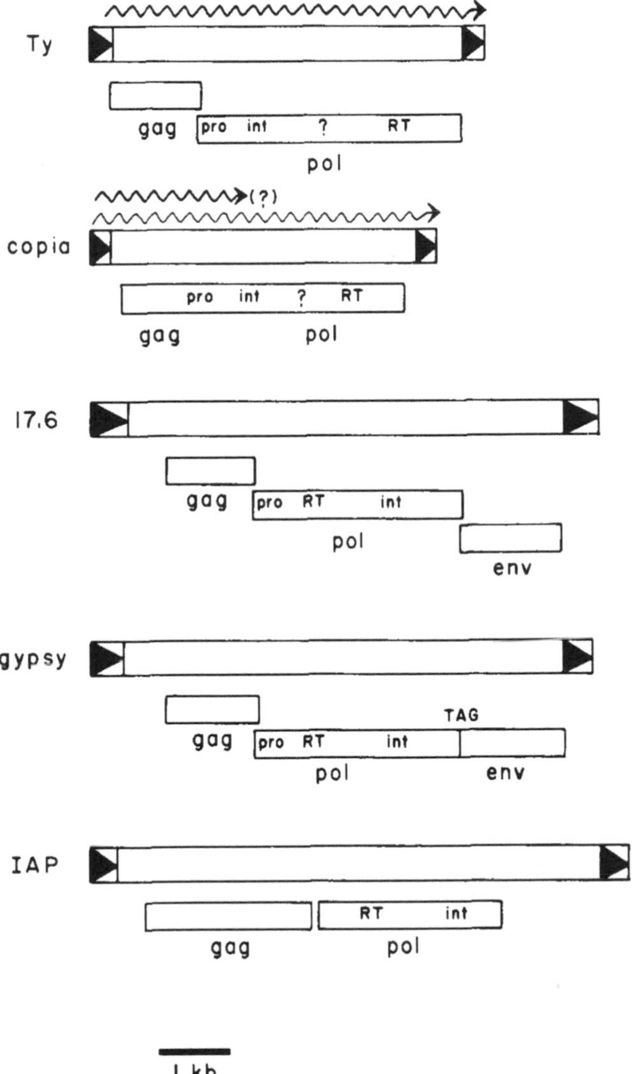

FIGURE 9. Summary of the structure of some sequenced retrotransposons. The symbols are the same as those used in Figure 3. RT, reverse transcriptase domain; int, integrase domain; pro, protease domain; ?, domain of unknown function. The domains have in most cases been identified by sequence similarity to retroviruses. The single amber codon separating pol and env in the gypsy element is indicated by TAG. In the case of IAP, both the gag and pol "genes" contain several interruptions in the reading frame.[106] In the other cases, the boxes represent true ORFs. Transcripts of Ty and copia elements are indicated by wavy lines (the ? next to the abundant transcript indicates uncertainty about the detailed structure of the 3′ end of the RNA).

and Varmus[101] showed that in vitro synthesized SP6/RSV RNA could direct the synthesis of small but significant amounts of gag/pol readthrough products in a eukaryotic (reticulocyte) translation system, but that the same synthetic RNA specified only gag product in an *Escherichia coli* system. The exact sequence of the gag/pol junction in the fusion proteins has not yet been determined. Such sequence data would implicate the specific tRNAs involved as well as defining the precise site of frameshifting. The site may be the termination codon

Table 2
DIRECTION OF FRAMESHIFTING IN VARIOUS RETROTRANSPOSONS AND RETROVIRUSES

Element	Junction	Frameshift type
RSV	gag/pol	−1
17.6	gag/pol	−1
Gypsy	gag/pol	−1
297	gag/pol	−1
DIRS1	ORF1/ORF3	−1
Ty1 and Ty2	gag/pol	+1

of the gag gene; the balance between action of termination factors and the tRNA might determine the extent of frameshifting. At the present time, however, there is nothing to preclude frameshifting by a frameshift-suppressing tRNA at a site within the gag/pol overlap region.

In copia, which has but one reading frame, the appropriate ratio of "gag" product to "pol" product is probably obtained by a transcriptional mechanism. In addition to the end-to-end transcript shared by all retrotransposons, copia produces a 5′ coterminal 2-kilobase transcript (Figure 9) which is much more abundant than the full-length RNA. At present, the detailed structure of the 3′ end of this transcript remains unknown. It has been argued that this transcript may be spliced from a position about 2 kilobases into the element to a microexon at the 3′ end of the element, because appropriate splicing signals exist to direct such an event.[61] Alternatively, it may be that the transcript is simply polyadenylated at a high frequency at a site about 2 kilobases into the element and that the poly(A) is added just after a T residue in the first position of a specified codon, generating a UAA translational termination signal in the mRNA. Such a mechanism apparently operates in the expression of many human mitochondrial genes.

b. Protein Products

Protein products have been identified for the copia and Ty elements, but the picture is undoubtedly far from complete. Although there are only one or two reading frames, respectively, the products from both frames are probably proteolytically processed. Furthermore, judging from the situation with retroviruses, many of the final products of proteolytic processing may be quite small. Thus, in addition to the many potential final products of processing, there may by many intermediates in the pathway. Indeed, candidates for several such intermediates have been reported.

In the case of Ty elements, the products of the gag gene have been partially characterized.[102-103,203-205] Interestingly, some of these proteins have DNA binding activity even when they are bound to nitrocellulose. Proteins of 50 to 52-kdalton Mr have been identified in yeast strains overproducing the Ty gag gene which are present in far lower levels in yeast cells lacking such a construct.[103-104] The 50-kdalton protein is phosphorylated. The gel mobilities of the gag gene products are apparently not very reproducible; this may be due to their high proline content and/or their posttranslational modifications.

Shiba and Saigo[40] showed that the copia elements of *Drosophila* direct the synthesis of VLPs composed of element-specific RNA and proteins. The most abundant protein in the particles has a molecular weight of about 30 kdaltons. It and several less abundant proteins ranging between about 25 and 90 kdaltons are identical in size and are immunologically related to a family of proteins synthesized in an in vitro translation system primed with copia RNA. Most of the in vitro-synthesized proteins could be immunoprecipitated with antibodies directed against the virus-like particles. A 51-kdalton protein and a family of smaller proteins were synthesized in vitro using copia 2-kilobase RNA as template by another group.[105] The reason for the discrepancy in molecular weights is unclear, but could be due to proteolytic processing by the "protease" domain.

2. Mouse Elements

The DNA sequence for one IAP element from Syrian hamster was recently reported.[106] While the authors were able to find regions of open reading frame with significant homology to retroviral gag and pol proteins, these regions spanned a total of at least five open reading frames. Two possible explanations for this finding are that (1) the ORFs are joined by splicing events or (2) the element sequenced represents a defective copy. The former seems unlikely in that no other retrovirus or retrotransposon has gag or pol genes derived from multiple ORFs. Moreover, mouse IAP was recently found to have intact ORFs.[208] The second possibility seems quite reasonable in light of the fact that there are numerous examples of endogenous defective viruses in murine systems. Deleted variants of IAP appear to be common as well.[107] Between the IAP pol region and the 3′ LTR lies a long (about 1700 base-pairs) region without long ORFs. Ono et al.[106] ascribe this to a vestigial env region which has accumulated many point mutations (even more than gag and pol), perhaps through disuse. There is no corresponding long, noncoding region in Ty or copia (which also lack env). It will be most interesting to compare this sequence to that of other members of the IAP family. That IAP is functional as a transposon seems certain as IAP elements have been shown to cause mutations in the immunoglobulin kappa chain genes.[80-82] Although the lack of an env gene may make IAP resemble Ty and copia, the order of functional domains in the pol gene is dissimilar to that of Ty and copia and similar to that of retroviruses.

Mys1, a recently discovered family of murine retrotransposons, has a curious ORF with homology to retroviral reverse transcriptases. Unlike all known pol genes, this ORF is encoded on what corresponds to the (−) strand of all other retroviruses and retrotransposons.[97]

3. DIRS1

The DIRS1 element of *Dictyostelium* has an unusual coding frame arrangement in addition to its different overall structure.[63] Since essentially nothing is known about the gene products, except that ORF3 has homology to the RT domain of retroviral pol genes, it seems premature to call the ORFs anything but by number. ORF1 overlaps both ORF2 and ORF3 in a somewhat gag/pol-like manner; ORF3 (which is much shorter than a "normal" pol ORF) is wholly embedded within ORF2. This is highly reminiscent of the coding arrangement of hepadnaviruses and suggests a possible evolutionary relationship between these widely separated elements. Neither DIRS1 nor hepadnaviruses contain any region homologous to the retroviral int domain. Hepadnaviruses do not normally integrate into the genome of the host, but DIRS1 elements clearly do; perhaps they use a fundamentally different integration mechanism from retroviruses. Consistent with this possibility, DIRS1 elements have a very different end structure than retroviruses and more "conventional" retrotransposons.

B. *Cis*-Acting Sequences

Numerous DNA (or RNA) sequences have been proposed to act as important sites for retrotransposon transcription, priming events in reverse transcription, LTR alignment in the RNA transcript (to allow for efficient "jumping" of strong-stop DNAs), and integration of the completed reverse transcript. Some of these sites are described below and are compared among retrotransposons and retroviruses in Table 3.

1. LTR Sequences

a. Transcriptional Signals

By aligning the sequence of the end-to-end transcript of retrotransposons with their DNA

Table 3
LTR AND FLANKING SEQUENCES IN RETROTRANSPOSONS[a]

Retrotransposon	Purine-rich sequence	U3	R	U5	(—)Primer binding site
Fungi					
Ty	GGGTGGTA	TGTTGGAATAGAAA		CACCCATTTCTCA	TGGTAGCGCC
*Hpa*II repeat	AGAGGGA	TGTTGGAATCTGT		AACTATTTCTAACA	GGTTATGGGCCC
Insects					
copia	GAGGGGGCG	TGTTGGAATATACTA		TTATAAATTACAACA	GGTTATGGGCCCAGT
412	AAAAGGAGGGAGA	TGTAGTATGTGCCTA		GAGTTCATCATTACA	TGGCGACCGTGACAG
B104		TGTTCACACATGAA		TTTGGGATTTTACA	TTTGGTCAATCGA
mdg1	AAAAGGAGGGAGA	TGTAGTTAATTTGAATTC		TTCGTATATACTACA	TGGCGACCGTGACAA
mdg3		TGTAGTAGGCTGCT		TGTAAACGGCGGCTAAA	
297	AAGGGAAGGGG	AGTGACGTATTTGGG		AACAACAATTTTACT	GGCGCAGTCGGTAGGAT
17.6	AAGGGAAGGGA	AGTGACATATTCACATAC		TTATTTGCAATT	GGCGCAGTCGATGTGAT
gypsy	GAGGGGGGAGT	AGTTAACAACTAACAA		ATTGGAACTTATATAATT	GGCGCCCAACCAACAATCT
HMS Beagle		AGTTATTGCCCTGCAATT		ACTGTAATTATTAACT	
flea		GTAAGATTGTTT		GCCTTACTCCATGTT	
springer		AATTAATTAAATGTAT		GGTTAACTTAGTTAACT	
TED		TGTTAGGTATAGCC		GTATACGTAATT	GGCGCAGTCGGTAGGAT
Plants					
bs1	GATACCAAT	TGTTAGCAACCCAAT		CCCATCTATATATCTGAC	TGTAACGGGGGGTGGAATTT
cin1		TGTTGGGGACCTTTCTCT		CGAGAATAAGGACCAACA	
Mammals					
IAP	AGGGGGAGA	TGTGAGGAGCCGCCCTCG		CGGTAGCGCGGGACA	TCTGGTGCCGAAACCCGGGA
VL30	GAAAGA	TGAAAATTACTGGCCTGTT		GGGTCTTTCA	TTTGGGGGCTCGTCCGGGAT
					TGGAGGTTCCACTGAGAT
					TGGTGCATTGGCCGGGAA
MuRRS	AGAAGAAGTGGGGAA	TGAAAGATCCTGGCATAAT		CCTCTCGGAGTCTTTCA	TTTGGGGGCTCGTCCCGGG
Mys	AAAAAAAGAAGAAAGGGGAAG	TGATGTGGATATC		AGAAGCCCTCCAGCAACA	TGGCGCCCAACGGCTCGA
THE-1	AGGGTGGAA	TGATATGGTTTGGCTGTGT		AAAACGGACTAATACA	GTAAATTAGTACCAGTAGA
hsRTVL	AAGAAGACAGGAA	TGTCAGGCCTCTGAGCCC		TGGACGCGCATGAAA	TTTGGTGCCGTGACTCAGAT

[a] References to the sequences may be found in Table 1.

copies (Figure 5), it is possible to define functional domains of the LTR sequence in the same manner as in retroviruses. These are named U3, R, and U5, as described above (Figure 5 and Section I.D.1). The R region is essential to the transfer of the strong-stop DNAs from one end of the RNA template to the other in the formation of a reverse transcript (Figure 7). The U3 region of the 5′ LTR must contain the promoter sequences for the retrotransposon transcript and indeed this segment of retrotransposon LTRs contains the usual TATA and CAAT homologies required for the initiation of transcription in the appropriate locations (see review by Varmus[5]). In the 3′ LTR, the U3 region, probably in conjunction with R and U5 sequences, also functions in specifying the polyadenylation and termination of the retrotransposon transcript. This raises the conundrum of how two identical LTR sequences (in most copies of retrotransposons, the 5′ and 3′ LTRs are identical in sequence) can behave in a transcriptionally "opposite" manner, with the 5′ LTR serving as promoter and the 3′ LTR as terminator. The available evidence suggests that when transcripts read into the LTR from upstream, the LTR functions strongly as a terminator and only weakly, if at all, as a promoter. If, however, there are few or no transcripts reading into an LTR from the 5′ direction, that LTR functions as a strong promoter.[108] Presumably, if the promoter sites within U3 are frequently traversed by a transcribing RNA polymerase arriving from upstream, the efficiency of utilization of the LTR promoter will decrease.

b. Integration Signals

At the 5′ and 3′ tips of the LTRs lie short sequences which show a remarkable degree of conservation between retroviruses and most retrotransposons. All retrovirus and many retrotransposon LTRs share the terminal DNA sequences TG . . . CA. There is in addition a family of retrotransposons whose LTRs share the sequence AGT . . . A(C/T)T. In both cases, these terminal sequences are the ones which would ultimately form the "att" site (or part of the "att" site) at the circle junction in two-LTR circle molecules, produced by circularization of the reverse transcript.

In many cases, there is considerable sequence homology adjacent to these terminal nucleotides between retrotransposons from widely separated hosts (e.g., compare Ty and copia 5′ ends [Table 3]).

c. Alignment Signals

In the yeast Ty element, there is a well-conserved, 14-base-pair inverted repeat, of which one copy lies in the gag region and the other copy lies within the 3′ LTR. It has been proposed that these sequences allow for alignment of the ends of the Ty RNA for efficient transfer of strong-stop DNA ("jumping") to the opposite end of the template during reverse transcription.[60]

2. Primer Binding Sites

In order for (−) and (+) strong-stop DNAs to be made, specific priming events must occur. These priming events require sequences located just inside the LTR sequences (Table 3). The primer for the (−) strong-stop DNA is usually a cellular tRNA; approximately 8 to 18 nucleotides of homology to the 3′ end of a given cellular tRNA may be found just 3′ to the 5′ LTR of most retrotransposons. As in retroviruses, the first modified base reached in the tRNA corresponds to the end of the homology between retrotransposon and tRNA. The (−) strong-stop primer binding site sequence can often be recognized by the sequence complementary to the CCA trinucleotide which forms the 3′ end of all tRNAs. Of the sequenced retrotransposons, only mdg3 lacks this CCA homology. Yet, Arkhipova et al.[36] have shown that even in this case, the primer for (−) strong-stop is a tRNA-sized, alkali-labile molecule. It may be that mdg3 utilizes a tRNA precursor which does not yet have the CCA attached to the 3′ terminus. Three other retrotransposons have a CCA homology situated at some distance 3′ to the 5′ LTR (copia, *Hpa*II repeat, and bs1 [see Table 3]). Perhaps these elements do not utilize tRNA primers, or utilize them in a different way than retroviruses do. Recently, copia was shown to use an initiator methionine tRNA half-molecule as primer.[209]

The mechanism of priming of (+) strong-stop DNA is not well understood in retrotransposons. There is fairly good conservation of a purine-rich stretch just 5′ to the 3′ LTR, however, the extent of purine-rich sequences in Ty and bs1 is not very impressive (6 purines in 8 base-pairs, and 5 purines in 9 base-pairs, respectively) in comparison to those seen in most retroviruses (e.g., 18 purines in 18 base-pairs in MMTV; 11 purines in 11 base-pairs in RSV). It has been hypothesized that an RNAse H-resistant oligonucleotide serves as the primer for (+) strong-stop DNA within this polypurine stretch; a short oligoribonucleotide primer is indeed found at the 5′ end of RSV (+) strong-stop DNA.[109] Although (+) strong-stop DNA molecules were also identified by Arkhipova et al.[36] in *Drosophila* tissue cultures, these molecules showed no evidence of an RNA primer component.

3. *Host Target Site Duplications*

All retrotransposons and retroviral proviruses are flanked by short duplications of a characteristic length (usually 4 to 6 base-pairs, see Table 1). These sequences presumably arise as a necessary consequence of the integration event. Target site duplications also seem to be a consequence of the insertion of virtually all types of transposons as well as the integration events accompanying processed gene formation.[12-13]

4. *Packaging Signals*

It has been shown that the transcripts of IAP, copia, and Ty elements are found packaged inside VLPs. The packaging of retrotransposon RNA is quite element specific. This strongly implies that retrotransposon transcripts contain a packaging signal similar to those found in retroviruses. In the retroviral case, sequences primarily responsible for packaging lie primarily just 3′ to the 5′ LTR in a noncoding region (called the leader region);[110-112] an additional sequence near to the 3′ LTR is also thought to be essential for RSV packaging.[113] In the case of Ty elements, there is no long noncoding region between the LTRs, so the packaging signal must lie either within the LTRs or within a coding region.

5. *Enhancers and Silencers*

Retrotransposons often are associated with the ability to activate the transcription of adjacent genes. Moreover, this activation is often under the same type of regulation as is the retrotransposon itself. A molecularly well-studied case is the yeast Ty element, in which the sequences responsible for this enhancement have been dissected out. Although the details of the picture are somewhat controversial, it is clear that a block of sequences within 1 kilobase of the 5′ LTR are essential to the activation of the adjacent gene.[114-117] Within this block of sequence there are several specific sequences which have been identified by different workers as having fairly significant homologies to the SV40 and core enhancer sequences. Overlapping these sequences and also within the LTR sequence itself, are sequences with homology to the proposed consensus a1/α2 control sequences.[118-119] Indeed, Ty transcription itself and, in many cases, the transcription of genes adjacent to Ty elements, are under mating type (a1/α2) regulation.[120,121] Perhaps the differences in results obtained by the different groups working on Ty enhancers are due to the fact that (1) they work on different copies of Ty elements, (2) they work with different host strains of yeast, and (3) the activated gene under study is different.

Kapakos et al.[116] have shown that the enhancer-containing segment within the Ty sequence is responsible for transcription of the Ty itself by deleting the sequence from the element. Although the stability of the Ty RNA was not affected by this maneuver, the transcription was markedly decreased (sixfold) compared to wild type.

Kapakos et al.[116] have also proposed the presence of a "silencer" region within certain members of the Ty2 family. The element studied, Ty917, apparently contains two domains which affect adjacent gene function, a site which increases expression, and 3′ to that, a silencer region which decreases it. When both regions are present, an intermediate level of expression is seen.

IV. MECHANISM OF RETROTRANSPOSITION

In this section, the reader will find reviews of some of the evidence supporting the model presented in Section I.D from the variety of systems currently under study. The first hints that retrotransposons might transpose in a retrovirus-like manner came from the examination of their DNA and RNA structure (reviewed above). The fact that in most cases retrotransposon LTRs have identical sequences suggested that they were homogenized in some way during the transposition process. Such homogenization is predicted by the model for retroviral reverse transcription (when only one RNA template [or two identical templates] is used for reverse transcription).

The demonstration by Panganiban and Temin[74] that a circle junction "att" fragment derived from a circular two LTR intermediate of spleen necrosis retrovirus can permute the integration of the virus strongly suggests that the two LTR molecules are an intermediate in its integration. Therefore, the finding of two LTR circles derived from retrotransposons[122] suggests that they may utilize such an intermediate. Nevertheless, direct integration of linear reverse transcripts has also been suggested as a possible mechanism[123,206] and remains a formal possibility.

A. Copia

A prediction of the retroviral model was that circular DNA forms and DNA/RNA hybrid forms of retrotransposons might be present in host cells; the structure of these was compared to the retroviral analogues and many similarities were found. The most convincing work in this area has been done with *Drosophila* elements, particularly copia. Flavell and Ish-Horowicz[122] isolated and cloned such molecules from *Drosophila* tissue cultures, and found that both one- and two-LTR circles were present. The two-LTR circles, however, were never perfect joins of the linear element found in the *Drosophila* genome; they often either contained one or more nucleotides of indeterminate origin, or bore deletions at the "circle junction".[124] Circular forms of Ty DNA have also been reported,[125,126] but the detailed structure of these has not been presented.

More recently, Flavell[127] studied the biosynthesis and end structure of the copia linear molecules isolated from *Drosophila* tissue cultures. He showed that the extrachromosomal copia linear DNA molecules were not synthesized semiconservatively within one cycle of cellular replication (by measuring the kinetics of heavy-heavy copia DNA accumulation in bromodeoxyuridine-containing medium). This is consistent with the generation of the extrachromosomal linears by reverse transcription.

The structure of the ends of these linear copia molecules was examined by cloning the molecules directly and sequencing the resultant constructs.[128] The end structure of these elements is exactly identical to the end structure in the integrated copia elements found in the chromosomes (less the host repeats, of course). This result implies that the two-LTR circle which would be formed by ligation would have a precise joint (without extra nucleotides); this is just the class of molecules Flavell and Ish-Horowicz[124] failed to see in the same line of tissue culture cells. Hence, the active two-LTR intermediate which is believed to be the precursor to the integration event may be short lived, becoming integrated into the genome very soon after it is formed. The finding that the copia linears are essentially equivalent to free proviruses[128] is also somewhat unexpected in light of the fact that the

CCA homology in the (−) primer binding site lies several base-pairs from the 3′ end of the U5 region. Recently, it has been found that the 5′ half-molecule of initiator methionine tRNA serves as the (-) strand primer; this contains an internal CCA.[209]

B. Other *Drosophila* Elements

Arkhipova et al.[36] have reported the isolation of DNA/RNA hybrid molecules derived from mdg1, mdg3, and mdg4 (gypsy) mobile elements as well as the corresponding (+) and (−) strong-stop DNAs. One- and two-LTR circular forms of several *Drosophila* retrotransposons have been found in both *Drosophila* tissue cultures and in embryos.[131-133] Curiously, the amount of free circles did not always correspond well to the amount of retrotransposon RNA. The number of copies of DNA circle per cell varied between <0.01 to as many as three to four copies per cell in some cell lines, with typical values of between 0.01 and 0.1 per cell.

C. Ty Element of Yeast

Mechanistically, the Ty1 element of yeast is well studied. Fusion of a particular Ty1 element to an active, inducible (*GAL1*) promoter and cloning of the fusion construct on a 2-μm (high-copy-number) plasmid in yeast led to a highly manipulable system in which very high levels of Ty element transposition could be induced or shut off at will.[38] Moreover, the transposon sequences could be marked within a specific restriction site within the Ty sequence (but outside of the gag and pol reading frames) with heterologous sequences such as *lacO* from *E.coli.* The marked transposons were themselves proficient for transposition, generating progeny transposons which carried the marker. Detailed studies of the marked progeny transposons (which could be "captured" on a second plasmid and shuttled into *E. coli* for easy structural analysis) revealed many features of the Ty element transposition process.

It was shown directly that Ty RNA was an intermediate in the transposition process by inserting an intron-containing segment as marker; the intron segment was excised precisely, leaving behind only the flanking exon sequences in the transposed Ty.

The fusion of the *GAL1* promoter sequences to the Ty element required the removal of the U3 region of the 5′ LTR. Nevertheless, deletion of these sequences did not prevent the efficient transposition of this Ty element. This is in stark contrast to bacterial transposons such as Tn10, which transpose directly through DNA and have an absolute requirement for both sets of terminal nucleotides. As predicted by the accepted model for retroviral reverse transcription (Figure 7), the U3 region of the 5′ LTR was regenerated during the transposition event.

Sequence polymorphisms within the U5 region and at the border between the U3 and R regions were inherited in the progeny transposons in a manner suggesting that the mRNA emanating from the recombinant plasmid was used as template for both LTR sequences. The results were again completely consistent with the reverse transcription mechanism outlined in Figure 7. A summary of the results obtained with *GAL1*/Ty fusion plasmids is shown in Figure 10.

The marked progeny transposons occasionally (about 20% of the time) showed evidence of heterogeneity with respect to the parental plasmid-borne element. That is, they carried the *lacO* marker, but differed at one or more polymorphic restriction sites within the element. Recently, using an *spt3* mutant, it was shown that such heterogeneity is abolished when production of the full-length transcript from chromosomal Tys is shut off (see Section VII.A), suggesting that the source of the heterogeneity was the result of some form of recombination between reverse transcripts or by template switching of reverse transcriptase between RNA templates.[132] Similar high frequencies of recombination have been reported for retroviruses.

Finally, it was shown that in yeast cells containing such plasmids, large numbers of VLPs

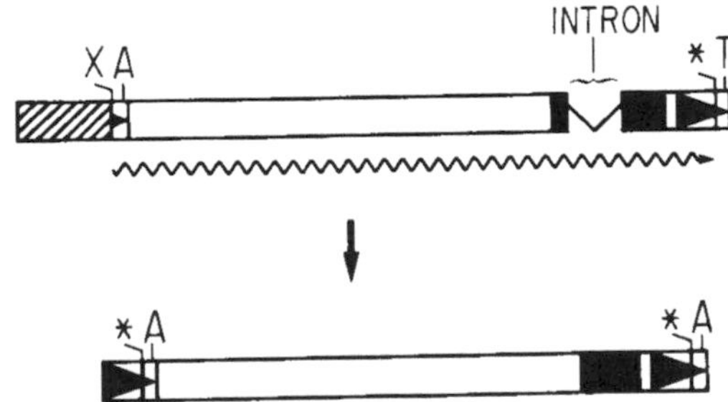

FIGURE 10. Ty summary. The salient features of a system developed to study the retrotransposition of the yeast Ty elements is summarized. The hatched box represents the *GAL1*-inducible promoter. The solid box represents a marker segment introduced into the Ty element; this marker fragment contains an intron (symbolized by a "V"). Otherwise the symbols are the same as in Figure 3. The "A", "T", "X", and "*" refer to sequence polymorphisms between the 5′ and 3′ LTRs. Note that their inheritance is in accord with the predictions of Figure 7. (From Boeke, J. D., Garfinkel, D. J., Styles, C. A., and Fink, G. R., *Cell*, 40, 491, 1985. With permission.)

accumulated under conditions where Ty transposition occurred at high frequency (i.e., when the plasmid was induced).[68] A constitutive overproducing construct made by other workers also directed the synthesis of such particles.[69] The VLPs contain a reverse transcriptase activity and protein molecules apparently derived from both the gag and pol reading frames.

D. DIRS1 *(Dictyostelium)*

A highly speculative but stimulating model for the retrotransposition of the DIRS1 element of *Dictyostelium discoideum* has been proposed by Cappello et al.[63] The model (described in detail in Figure 11) is strictly based on the structure of the element and the fact that one of its open reading frames shows homology to the RT domain of known reverse transcriptases. There is as yet little experimental support for the model.

V. REGULATION OF RETROTRANSPOSITION

Retrotransposition can clearly be highly detrimental to the host cell if allowed to occur at high rates. On the other hand, certain conditions of environmental stress may require higher transposition levels if adequate degrees of variability are to be introduced into the population as a means of coping with the new, stressful environment. Hence, it seems reasonable to assume that the frequency of retrotransposition is tightly regulated. Regulation of retrotransposition could occur at several levels: transcriptional, translational, or post-translational.

A. Transcriptional Control

There is abundant evidence for control or retrotransposon expression at the RNA level. Regulation by such environmental and cellular factors such as UV light, heat shock, carbon source, carcinogens, the tumorous state, and other types of stress have been reported. In addition, developmental stage-specific regulation of retrotransposon expression has been reported for many *Drosophila* retrotransposons and for IAPs. It should be emphasized that in many of these cases it is not completely clear that the rate of transcription is affected rather than the stability of the RNA.

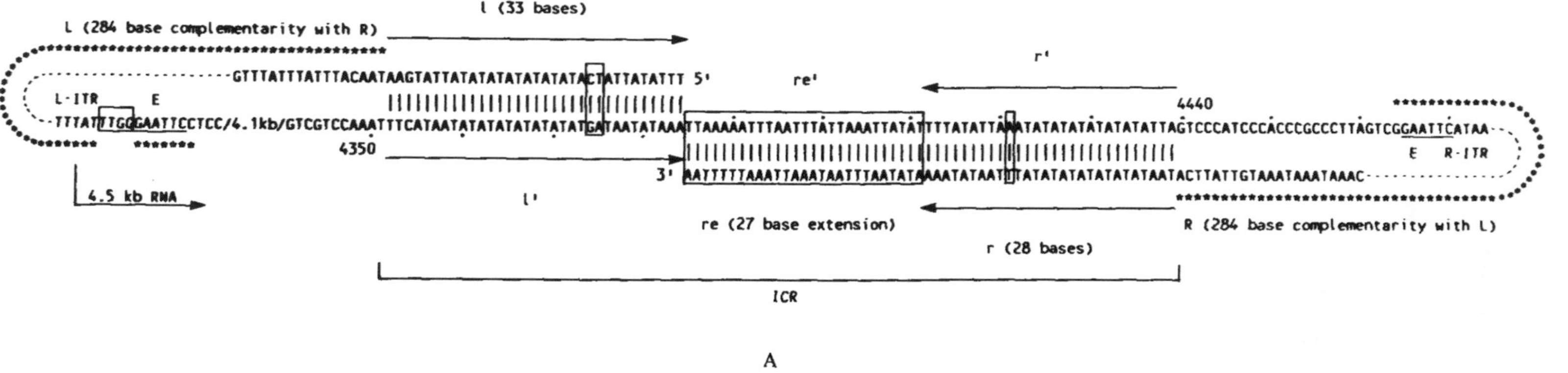

FIGURE 11. DIRS1: model for its retrotransposition. (A) The complementarity of the distal terminal segments of both ITRs (inverted terminal repeats) with the internal DIRS1 segment, ICR (internal complementary region), is shown. All sequences are from the mRNA strand (coding strand). Complementarity between the ITR sequences and the ICR is indicated by a vertical line between complementary bases. The star above and below the left (L) and right (R) ITRs depicts complementarity throughout L and R. The *Eco*RI sites (E) define the internal boundaries of the ITRs. The distal terminal segment of the left ITR, l, and its complement in the ICR, ' are indicated by arrows. The distal terminal segments of the right ITR, r and re, and their complements r' and re' (re is the 27-base pair extension found only in the right ITR). Boxed bases are different between left and right ITRs. The transcription initiation site of the major transcript (4.5 kilobases) is indicated. (B) The diagram proposes a model by which the major transcript (line 2) can direct the synthesis of a putative transposition intermediate (line 9) that contains all of the sequences found in transposed DIRS1 elements (line 1). The hatched areas of L and R indicate the regions of the ITRs containing sequence polymorphisms. Filled boxes indicate segments carrying the specific sequence differences noted above (Figure 11A). RT, reverse transcriptase; wavy lines, RNA; straight lines, DNA. (From Cappello, J., Handelsman, K., and Lodish, H. F., *Cell*, 43, 105, 1985. With permission.)

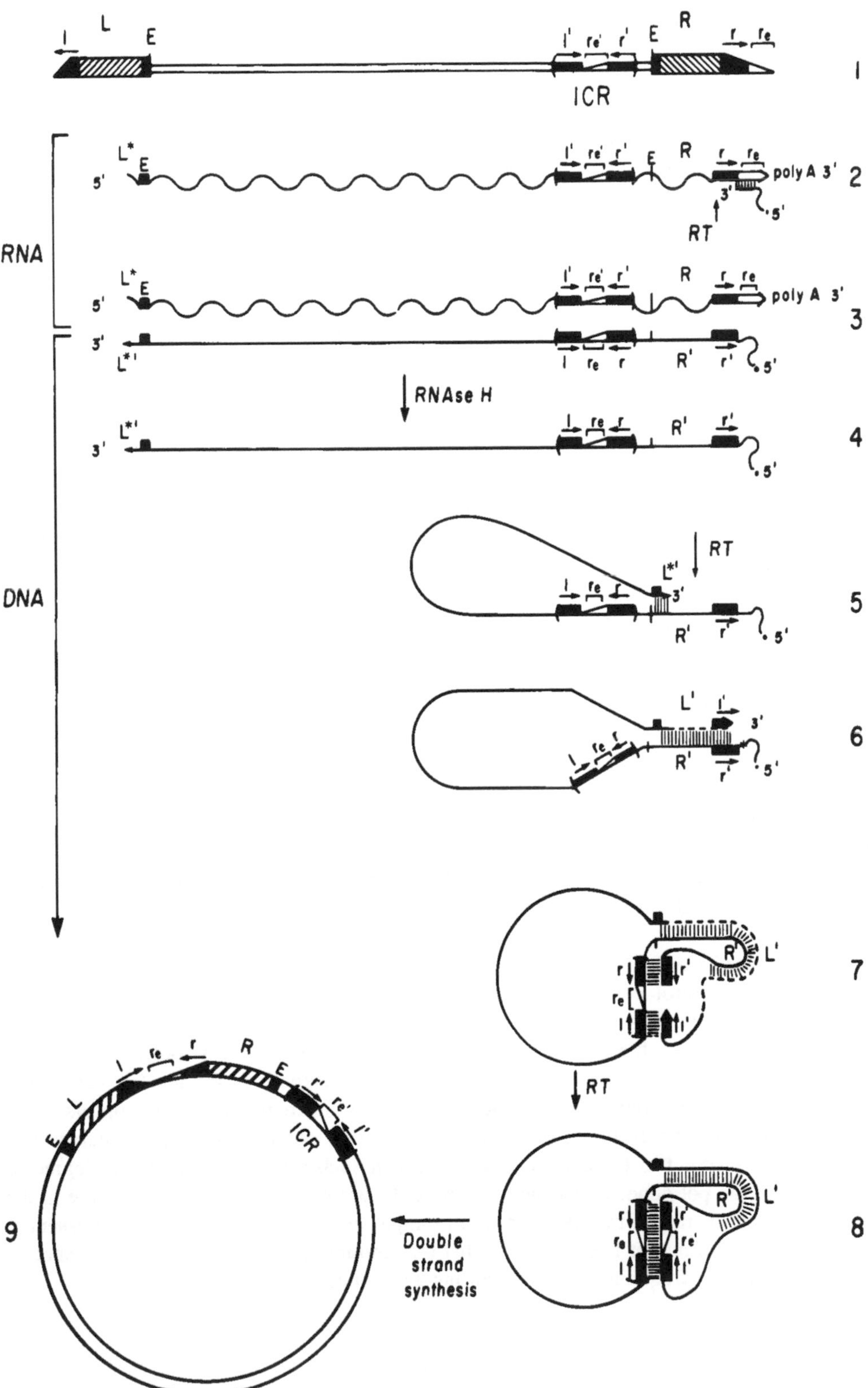

L
E
R
ICR
RNA
DNA
poly A 3'
RT
RNAse H
Double strand synthesis
1
2
3
4
5
6
7
8
9

FIGURE 11B

1. Ty and DIRS Elements

The transcription of Ty elements as a whole is under the regulation of the MAT locus.[35] In mating-proficient (a or α) cells, Ty transcription is abundant, whereas in a/α diploid (nonmater) cells, the transcription is reduced by five- to tenfold. Paquin and Williamson[133] demonstrated that the amount of transposition in mater vs. nonmater cells corresponds to the levels of RNA seen in those same strains. Additionally, Ty transcription (in certain strains at least) is controlled by carbon source. Taguchi et al.[121] reported a severalfold higher level of Ty RNA on glucose than on glycerol medium. These workers also reported that repression of Ty transcription in a/α cells was far more evident on the glycerol medium. Rolfe et al.[134] detected approximately tenfold induction of Ty RNA after exposure of yeast strains to UV light.

DIRS1 element transcription is induced strongly by heat shock in both its natural host, *Dictyostelium*, and in yeast.[67] There are several excellent matches to the heat shock consensus promoter sequence appropriately located within the DIRS1 element. The consequence of the heat shock response on DIRS1 transposition is unknown.

2. Drosophila Elements

Schwartz et al.[135] and Flavell et al.[105] demonstrated the developmental regulation of copia transcription. Embryos produced less copia RNA than other stages. Since then, many *Drosophila* retrotransposons have been examined as to developmental transcriptional regulation. The gypsy element, for example, is transcribed at a low level in embryos and adults, but abundantly in prepupal to mid-late pupal stages.[136]

Copia element transcript level was recently shown to increase 3- to 13-fold in response to heat shock, hydrogen peroxide, and sodium azide treatments (all inducers of heat shock proteins in *Drosophila*).[137] Heat shock consensus sequences were found in the copia LTR, but also in the nonheat-shock-inducible 297 LTR.

3. IAP Elements

The transcription of IAP elements is very much elevated in most tumor cell lines.[138] This corresponds with original observations on the presence of IAP particles in various cell lines. The presence of the particles shows a marked developmental regulation in normal mice; the particles are seen only in early embryos.[139]

B. Translational Control

Translational control of some type probably operates as well in retrotransposons, for retrotransposon transcripts seem to be very inefficient templates for translation both in vivo and in vitro. However, the reason for this inefficiency is unclear. One potential means of regulation which may be relatively unique to retrotransposons is at the level of expression of the second reading frame. As was described above, in many cases this appears to be catalyzed by a specific frameshifting event. The cellular components which catalyze this process could respond to the cellular environment in such a way as to limit the degree of expression of the pol gene. Presumably, limiting the amount of reverse transcriptase in the cell would limit the rate of transposition. For example, if a specific tRNA species is essential to proper frameshifting, it is conceivable that the amount of charged tRNA of that species might influence the frequency of frameshifting and hence transposition.

C. Posttranslational Regulation

At least one retrotransposon gene product (p1[Ty1-15] of Ty elements) was shown to be phosphorylated.[103] The kinase responsible for this phosphorylation has not been identified, but may well be a cellular gene product which regulates transposition. Another potential step at which retrotransposition may be regulated is at the level of processing of the primary translation products into smaller proteins. Finally, cellular components of the reverse transcription process, such as the priming tRNA or other tRNA, may also play a role in regulating transposition frequency.

Ty element transposition, like the transposition of many transposons, is a temperature-sensitive process. Garfinkel et al.[68] suggested that this is due to thermosensitivity of the reverse transcriptase encoded by Ty. Ty VLPs contain reverse transcriptase activity which is much more active at low temperatures. The plot of reverse transcriptase activity vs. temperature is quite similar to the plot of transposition vs. temperature. Ty transposition is virtually shut off at 36°C in yeast.[132,140]

D. Defective and Functional Elements

An additional layer of regulation may be provided by the fact that retrotransposons are all members of multigene families. Some or, indeed, most of the family members may be defective in some way.[141] For example, a certain set of elements may produce nonfunctional gene products. In the case of IAP, a good many copies of the element carry a substantial deletion;[107] moreover, even the nondeleted copy which was sequenced contains many nonsense and frameshift codons separating various portions of the gag and pol genes. A yeast Ty element, Ty173, which has a normal overall structure, carries a point mutation in its pol gene which renders it nonfunctional.[142] Finally, one of the three copies of the copia elements which have been sequenced bears a frameshift mutation which is presumably inactivating.[62] Thus, it may be that only a relatively small fraction of elements specifies gene products which are competent to promote transposition.

All elements of a family are probably affected transcriptionally by the sequences adjacent to them. Evidence for such regulation has been obtained in retroviral systems. As discussed above, the presence of a transcript reading into the 5′ LTR of an element can cause the LTR promoter to be expressed at a much lower level. Thus, those elements finding themselves 3′ to active cellular promoters will undoubtedly be transcribed at a low level, whereas others which are in a transcriptional "vacuum" will be expressed at a high level. Dzumagaliev et al.[143] have recently shown that transcribed and nontranscribed variants of the mdg3 element can be found in different cell lines. Similar effects have been seen in retroviral systems.[144] It may be that cells shuffle elements of varying degrees of functionality behind promoters of various strengths, using cellular homologous recombination systems in order to provide just the optimal level of transposition. Thus, the balance between functional and nonfunctional elements in the cell may determine the transposition frequency, and cells which attain this balance by recombinational "shuffling" may have a selective advantage.

VI. EFFECTS ON NEIGHBORING GENES

Insertion of a transposon into a piece of DNA can be situated in several ways with respect to the cellular sequences affected. There are several classes of transposon insertion events which have been observed. For the purpose of this review, these are divided into four classes based on where within the gene or its flanking regions the transposon is inserted. The classes are subdivided into two subclasses in which the transposon is inserted such that its transcriptional orientation is in (A) the same or (B) the opposite direction as that of the target gene. The classes of insertion events are summarized in Figure 12.

A. Class I — Insertions Within the Coding Sequence

When a retrotransposon lands within the coding sequence of a gene, the gene is inactivated. Examples of this type of insertion are known from yeast and *Drosophila*. In yeast, the *ura3-52* mutation[145] is an example of this class. This mutation does not revert at a detectable frequency, suggesting that precise excision is extremely rare in Ty elements. The *ura3-52*

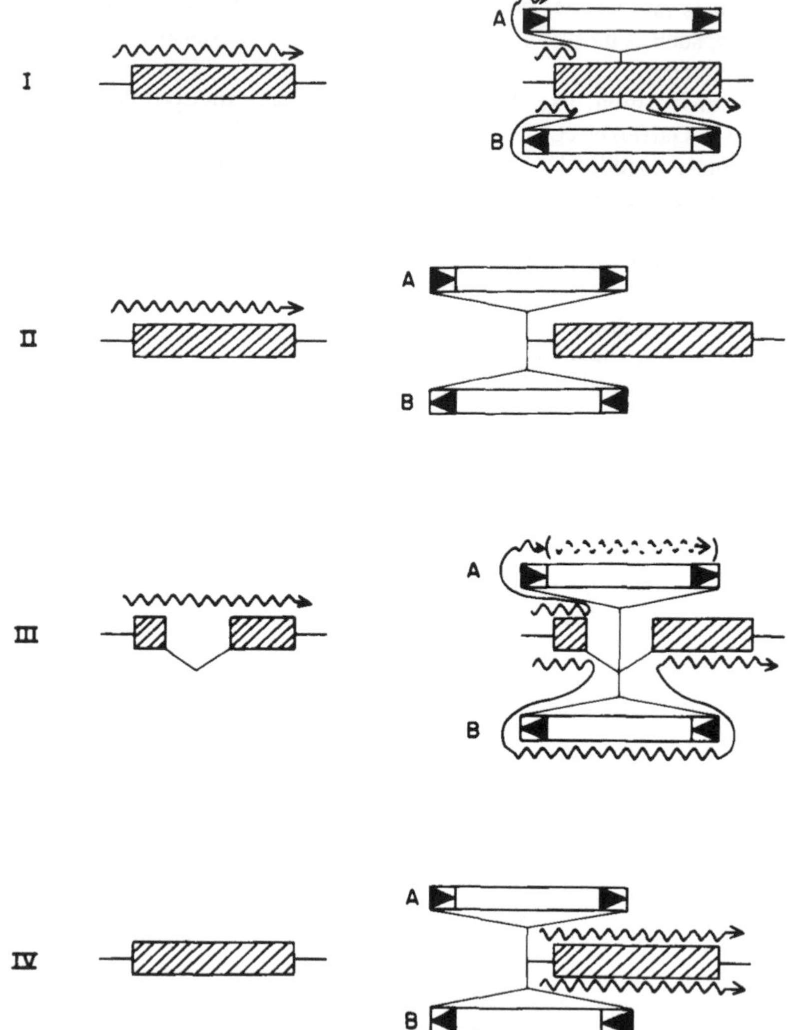

FIGURE 12. Classes of retrotransposon insertion. The hatched box represents the coding sequences of a target gene. Its flanking DNA is symbolized by a simple line. The "V" in case III represents an intron sequence. The broken wavy line in case III represents the w^a mutation of *Drosophila* (see text). The symbols are otherwise the same as in Figure 3.

insertion results in complete uracil auxotrophy. Additional Ty insertions located within the *URA3* coding sequence behave similarly.[146]

The copia insertion $w^{hd81b11}$ is an example from *Drosophila* of a class I insertion mutation. It lies within a small exon of the *white* gene, whose product determines the amount and distribution of eye pigments, and is dispensable to the fly. The $w^{hd81b11}$ mutation presumably totally abolishes gene function as the mutant flies have a bleached white eye phenotype which is also seen in the case of P element insertions into coding regions of the *white* locus.[147] The *bs1* mutation in the *Adh1* gene of maize is also of this type.[148]

Insertion of a B104 element within the *Glued* locus of *Drosophila* leads to the dominant *Gl* mutation. This insertion is near the 3′ end of the gene; the transcription of the gene and the retrotransposon are in the same (A) orientation. This results in a truncated *glued* transcript. The results can be easily explained by an insertion within the coding sequence, which leads to the production of an altered "poisonous" polypeptide.[149] The data is also consistent with the possibility that this insertion is actually within an intron (see Section C below).

B. Class II — Inactivating Insertions Within the 5′ Noncoding Region

In the second class, the transposon falls in the 5′ noncoding region of the gene and, again, the expression of the gene is prevented, presumably because upstream regulatory elements of the gene are separated by the length of the transposon from the body of the target gene. Examples of this class from yeast include the Ty912 and Ty917 mutations in the *HIS4* noncoding region.[150] In both cases, the upstream sequences which are the site of action of positive activators of *HIS4* transcription, are separated from the TATA box and the *HIS4* structural gene by the 5.9-kilobase Ty insert. Similarly, a *Drosophila* cuticle protein gene (CP3) is disrupted by an HMS Beagle insertion directly within the TATA box of the gene, destroying its expression, presumably at the level of transcription.[151] Other insertions of *Drosophila* elements into TATA boxes have been reported, but their effect on expression is uncertain.

C. Class III — Insertions Within Introns

In the third class, the transposon falls within an intron of the gene with four possible outcomes: (1) no phenotype, (2) partial gene inactivation, (3) complete inactivation, or (4) activation of the target gene. This class of insertion is quite common in *Drosophila* and has also been found in the case of murine IAP.[80-82] In subclass A, where the retrotransposon and the target gene are transcribed in the same orientation, a truncated target gene mRNA usually results due to premature termination within the 5′ or 3′ LTR. In the $w^{hd81b11}$ copia element insertion in the *white* gene (see below), termination reportedly occurs within the 5′ LTR. Oddly, in the case of the w^{a} copia insertion mutation, the 3′ LTR, rather than the 5′ LTR, serves to terminate the *white* transcript. A solo LTR at the same position also fails to terminate the *white* transcript.[152] In subclass B, in which the retrotransposon and its target gene are in transcriptionally opposite orientations, there is often no observable effect on the transcription of the target gene (the retrotransposon sequences are presumably "spliced out"). Often, phenotypes suggestive of a developmental defect result from such insertions. Such phenotypes may result from the inappropriate developmental- or tissue-specific expression of the target gene. These alterations of expression are presumably a consequence of the retrotransposon insertion, which might contain enhancers or promoters which respond to developmental cues and in some subtle way (e.g., only in specific cell types) alter the appropriate expression of the target gene. Extensively studied *Drosophila* genes with several examples of such mutations are the *white* and *notched* loci.

Four class III *white* mutations have been extensively characterized molecularly: w^{sp}, w^{zm}, w^{a}, and w^{bf}.[147,153] In all of these cases, the retrotransposon insertion results in only a partial (if any) loss of function of the gene product as some eye color develops in these flies. In the case of w^{sp} and w^{zm} "speckled" and "mottled" phenotypes are seen (the w^{zm} mutant phenotype is only expressed in the *zl* mutant background; in normal flies, the insertion mutation has no phenotype), whereas the other two retrotransposon insertions lead to the development of a lighter than normal eye pigmentation.

Kidd and Young[154] have recently studied eight independent retrotransposon insertions into the *notch* locus; all were class IIIB insertions and several were within a few base-pairs of each other. Five of the insertions were flea element insertions. All of these gave rise to a characteristic "glossy" eye phenotype, whereas two insertions of other retrotransposons into very similar locations (in one case only 2 base-pairs away from a "glossy" flea insertion) resulted in a strikingly different "facet" phenotype. It is possible that the flea insertions resulted in the inappropriate expression or repression of the *notch* gene at one particular stage or in one cell type, resulting in one characteristic flea-determined developmental phenotype, whereas the other retrotransposons caused a morphologically distinguishable developmental abnormality, possibly reflecting a subtle difference in the developmental regulation of the retrotransposons in question.

D. Class IV — Insertions Which Activate the Target Gene

In the fourth class, the retrotransposon lands in the 5′ noncoding region of a previously unexpressed gene; the result is expression of the previously silent gene. Examples of this class of insertion abound in yeast. In nearly every case, the gene adjacent to a Ty element is under the regulation of the MAT locus. The gene is expressed well in mater cells (a or α), but poorly in a/α diploids, just like Ty itself.[120,133,155]

It may be that insertions within an intron can activate the target gene in certain cases. A B104 insertion in the *Drosophila Antennipedia* locus *Antp*^{Ns} was interpreted as resulting in a gain of function (i.e., an increase in gene expression) by Scott et al.[156] The B104 element is in the "B" orientation in this mutant. It may be that the "glossy" and "facet" phenotype of the notch mutations reviewed above result from inappropriate hyper-expression of notch transcription. Insertion of gypsy and copia elements into the achaete-scute locus apparently results in a gain of function in spite of the fact that their insertion gives rise to a shortened transcript.[157]

E. Specificity of Integration

Other than the specific cases noted below, there is as yet little evidence for marked specificity of insertion of retrotransposons as far as target site selection is concerned. There is no evidence for a clear-cut "consensus target sequence", except in a few cases. There may be a slight preference for noncoding regions, but few, if any, conclusions can be drawn from the existing body of knowledge on retrotransposons because little systematic collection of insertions mutations has been done.

1. Ty Elements

Ten Ty insertions which inactivate or partially inactivate the yeast *LYS2* gene have been isolated and mapped. Nine of these fall within a 0.6-kilobase-pair fragment which contains the 5′ noncoding region of the gene; only one of the insertions is well within the 4-kilobase *LYS2* structural gene.[158,159] This result suggests that Ty elements prefer to insert in transcription initiation regions. Insertions of Ty elements in the 5′ noncoding region of several yeast genes has been described. There are relatively few cases of insertions within structural genes, but this probably reflects the bias of the investigators who were searching for certain types of mutations (e.g., unstable mutations or mutations which restored function to a target gene).

2. Drosophila Elements

Several elements from *Drosophila* seem to have a marked preference for the target sequences TATA, TACA, and TATATA. These include 297,[160] 17.6,[161] HMS Beagle,[151] gypsy,[17,162] springer,[154,163] and flea.[154] The reason for this preference is unclear, but many transposons seem to prefer AT-rich target sites in general.

3. Other Elements

The slime mold elements DIRS1 and *Hpa*II repeat both seem to have a propensity for transposing into themselves.[66,86] This property is shared with elements of the P and Ds/Ac (nonretrotransposon) transposon family.

VII. RECOMBINATION BETWEEN RETROTRANSPOSONS

The cellular homologous recombination systems operate on retrotransposon sequences, catalyzing the formation of various types of chromosomal aberration. Aside from the fundamental process of transpositional insertion, most or all of the chromosome aberrations in which retrotransposons participate appear to be unrelated to retrotransposition.

A. LTR-LTR Recombination

When homologous recombination occurs between the two LTR sequences of a retrotransposon, the intervening (core) sequence can be lost, resulting in a solo LTR residing at the original location. This event (Figure 2) has now been reported in many systems[90,152,164-168] and often has dramatic consequences on the expression of the gene adjacent to the retrotransposon. It must be kept in mind, however, that in many cases these events were selected for study precisely because they caused a change in the phenotype of the organism. This LTR-LTR recombination reaction is probably the major mechanism by which the cell removes excess copies of retrotransposons from its genome. Presumably, the rates of new transposition and LTR-LTR excision are roughly equal in populations with a stable transposon copy number. In yeast, the frequency of LTR-LTR excision at a single retrotransposon copy has been estimated at about 10^{-5}.[27]

LTR-LTR excision of both Ty elements[169] and *Drosophila* 412 elements[170] cloned on plasmids occurs at a very high frequency upon transformation into yeast. Oddly, only in the case of the Ty experiment is the LTR-LTR recombination process dependent on the function of the yeast *RAD52* gene. Only the incoming plasmid suffers the high-frequency recombination; resident chromosomal Ty elements do not.[169]

B. Gene Conversion

Gene conversion events involving widely separated Ty elements occur in yeast cells.[171] These events occur at a relatively high frequency, and there seems to be a preference for conversion events involving two elements on the same chromosome. It has also been demonstrated that a solo LTR can be converted to an intact element at relatively high frequency by gene conversion both in yeast Ty elements and in the *Drosophila* gypsy element.[168,172] In the case of the unstable ct^{MR2} mutation in *Drosophila*, a gypsy element was shown to be responsible for inactivating the *cut* gene. At a high frequency, this gypsy element is excised, often leaving behind a solo LTR and reverting the mutant phenotype. The solo LTR "revertants", however, now became mutant again at a frequency of 10^{-2} (a phenomenon referred to as "transposition memory" by these workers). Interestingly, two classes of mutants were obtained at high frequency; these corresponded to two structural variants of the gypsy sequence, an apparently normal gypsy and a copy of gypsy, themselves bearing an insertion of another transposable element called "jockey" (which is apparently not a retrotransposon). The two types of mutants have clearly distinguishable phenotypes.[168] Presumably, these interconversions were mediated by cellular homologous recombination systems.

C. Reciprocal Recombination Events

Reciprocal recombination events between two retrotransposon copies on the same or different chromosomes could lead to deletions, duplications, inversions, or translocations. Such aberrations have been observed during both mitosis and meiosis. In one well-studied case, a 20-kilobase segment of DNA from yeast chromosome III was deleted by a recombination event which occurred between two Ty elements.[171] Recently, Davis et al.[173] noted similar deletion and corresponding duplication events in *Drosophila*. These involved tandem copies of the B104 (roo) transposon which were separated by about 30 kilobases; they occurred at frequencies of about 10^{-4}.

D. Interactions between Hybrid Dysgenesis and *Drosophila* Retrotransposons

A high proportion of mutants is found among the surviving progeny of dysgenic crosses.

Table 4
SUPPRESSORS OF RETROTRANSPOSON INSERTIONS

Organism	Suppressor	Retrotransposon affected[a]	Ref.
Yeast	*spt1-spt15*	Ty and solo LTR	27—30,150
Drosophila	Su(Hw)	Gypsy	31
	Su(f)	Gypsy	179
	Su(s)	412	198, 199
	Su(w^a)	copia	200
Mouse	Su(dil)	Endogenous MuLV provirus	32, 201

[a] In some cases, the *Drosophila* suppressors appear to interact with more than one retrotransposon family.[177]

Most are caused by P-element (the causative agent of hybrid dysgenesis) insertion, but occasionally, retrotransposon-induced mutations have been found. Unfortunately, it is not firmly established whether the frequency of retrotransposon insertions is elevated following a dysgenic cross. Gerasimova et al.[174,175] proposed that retrotransposon transposition occurs at much higher frequencies during dysgenic crosses, although this feeling is not widely shared. Gerasimova et al.[175] report that the hybridization patterns of gypsy, copia, mdg1, and mdg3 elements all are markedly rearranged as though many or all of the retrotransposon copies moved in response to the dysgenic state. Even more curiously, in rare progeny of these flies (but not in the majority of the progeny), further massive rearrangements of retrotransposon distribution occur. This was frequently associated with appearance of mutations at a characteristic set of loci. This phenomenon has received the unfortunate name of "transposition explosion" or "transposition burst". However, at least some and possibly all of these events can be explained by the transient induction in certain premeiotic cells of a hyperrecombinogenic state in which homologous recombination events occur at a very high frequency. Such a state might be induced in response to dysgenesis. This would result in the high-frequency excision and reintegration by various retrotransposons at the corresponding solo LTR sequences scattered throughout the genome. There is as yet no direct evidence for an increase in the frequency of primary transposition events involving retrotransposons due to hybrid dysgenesis.

VIII. SUPPRESSORS OF RETROTRANSPOSON INSERTION

The variety of genes which interact with retrotransposons from yeast and *Drosophila* is somewhat bewildering and cannot be adequately reviewed here (see References 176 and 177 for reviews of suppression). Two well-studied examples, the *SPT3* gene of yeast[27,28] and su(Hw) of *Drosophila*,[31,136,178,179] will be reviewed here briefly. In both cases, the suppressors were isolated as mutations which reverse the effect of an inactivating Ty LTR (*spt3*) or gypsy (su[Hw]) insertion. In Table 4, other suppressor/retrotransposon interactions are summarized.

A. *SPT3*

The *SPT3* gene of yeast is a single-copy gene which has a profound effect on Ty transcription. In strains containing a null allele of *spt3*, the total amount of Ty transcript is reduced by some 20-fold. A small amount of a Ty transcript, which is 800 nucleotides shorter at the 5′ end, accumulates in the mutant strain. The *SPT3* gene appears to encode a factor required for proper initiation of transcription at Ty LTR promoters.[28] This explains the His^+ phenotype of the *spt3* mutant in combination with the *his4-912-LTR*(δ) allele (Figure 13). When a solo LTR is present at the *HIS4* locus of yeast, transcribed in the direction of

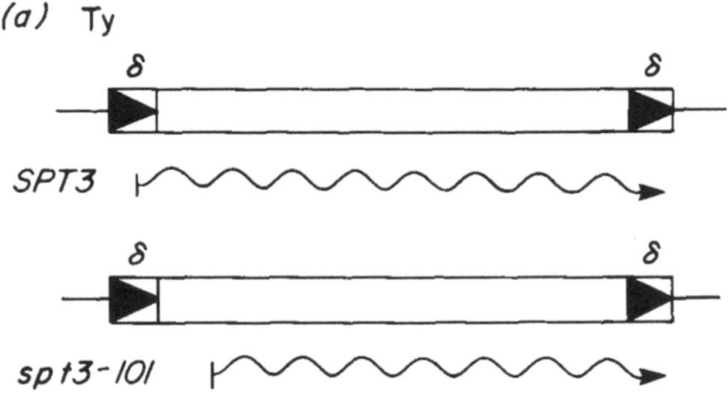

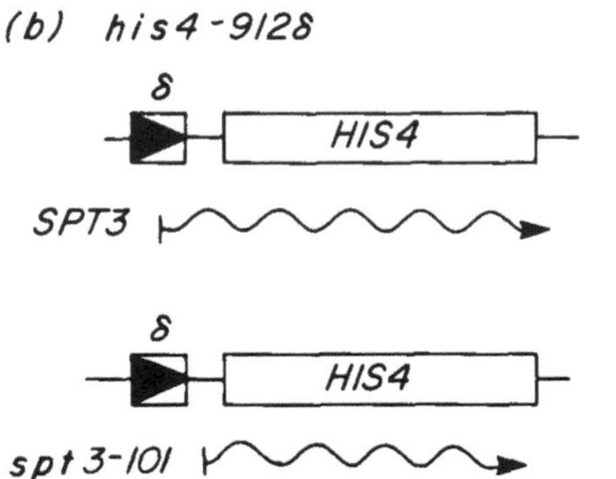

FIGURE 13. Suppression by *spt3*. In both the intact Ty(a) and the *his4-912δ* (b) cases, a long transcript initiating in the LTR δ is seen in the wild-type *SPI3* strain. In case (b) a nonfunctional *HIS4* transcript results (there are out-of-frame AUGs in the mRNA upstream of the *HIS4* AUG. In cells bearing the *spt3* mutation, transcription from the LTR promoters does not occur. Downstream promoters are now able to function. In case (a) this results in a nonfunctional Ty transcript (the AUG for its gag gene is missing in the shortened transcript). In case (b) a functional *HIS4* mRNA results. (From Winston, F., Durbin, K. J., and Fink, G. R., *Cell*, 39, 675, 1984. With permission. Copyright M. I. T.)

the *HIS4* gene, a longer than normal transcript of the *HIS4* gene is seen in *spt3* strains. This transcript, which initiates within the LTR, is apparently nonfunctional due to AUGs in the transcript upstream of the *HIS4* AUG. In the presence of the *spt3* mutation, however, the LTR-initiated transcript is not synthesized (the LTR promoter is silent). Instead, transcription initiates downstream; a normal-sized *HIS4* transcript is made which is translationally functional.

B. su(Hw)

This *Drosophila* suppressor behaves remarkably similar to *spt3*. It has been shown that this suppressor specifically reverses the effect of gypsy element insertions.[31] An insertion of gypsy into the *yellow* gene (y^2) results in aberrant pigmentation of the cuticle and in the loss of a specific *yellow* transcript only during the pupal stage. This is precisely the stage at which gypsy is transcribed most heavily.[136,178,179] In the absence of the su(Hw) gene product, the mutant phenotype is reversed. The transcription of gypsy elements in su(Hw) flies is reduced by about fivefold. A consequence of the decrease in gypsy transcription is the reactivation of the *yellow* gene transcript which has been shut off by the original gypsy insertion during pupation.

ACKNOWLEDGMENTS

The author gratefully acknowledges support by the Mellon Foundation and by a grant from the National Institutes of Health. JoAnn Olsen provided expert secretarial assistance. Steven Desiderio, Phil Hieter, and David Garfinkel kindly read the manuscript critically.

REFERENCES

1. **Roeder, G. S. and Fink, G. R.,** Transposable elements in yeast, in *Mobile Genetic Elements,* Shapiro, J. A., Ed., Academic Press, New York, 1983, 299.
2. **Williamson, V. M.,** Transposable elements in yeast, in *International Review of Cytology,* Vol. 83, Academic Press, New York, 1983, 1.
3. **Shapiro, J. A., Ed.,** *Mobile Genetic Elements,* Academic Press, New York, 1983, 1.
4. **Risser, R. and Horowitz, J. M.,** Endogenous mouse leukemia viruses, *Annu. Rev. Genet.,* 17, 185, 1983.
5. **Varmus, H. E.,** Reverse transcriptase rides again, *Nature (London),* 314, 583, 1985.
6. **Varmus, H. E.,** Retroviruses, in *Mobile Genetic Elements,* Shapiro, J. A. Ed., Academic Press, New York, 1983, chap. 10.
7. **Baltimore, D.,** Retroviruses and retrotransposons: the role of reverse transcription in shaping the eukaryotic genome, *Cell,* 40, 481, 1985.
8. **Temin, H. M.,** Reverse transcription in the eukaryotic genome: retroviruses, pararetroviruses, retrotransposons, and retrotranscripts, *Mol. Biol. Evol.,* 2, 455, 1985.
9. **Fink, G. R., Boeke, J. D., and Garfinkel, D. J.,** The mechanism and consequences of retrotransposition, *Trends Genet.,* 2, 118, 1986.
10. **Finnegan, D. J. and Fawcett, D. H.,** Transposable elements in *Drosophila melanogaster, Oxford Surv. Eukaryotic Gene,* 3, 1, 1986.
11. **McDonald, J. F., Strand, D. J., Lambert, M. E., and Weinstein, B.,** The responsive genome: evidence and evolutionary implications, in *Development as an Evolutionary Process,* Rauff, R. and Rauff, E., Eds., Alan R. Liss, New York, in press.
12. **Weiner, A. M., Deininger, P. L., and Efstratiadis, A.,** Nonviral retroposons: genes, pseudogenes, and transposable elements generated by the reverse flow of genetic information, *Annu. Rev. Biochem.,* 55, 631, 1986.
13. **Vanin, E. F.,** Processed pseudogenes: characteristics and evolution, *Annu. Rev. Genet,* 19, 253, 1985.
14. **Rogers, J.,** Retroposons defined, *Nature (London),* 301, 460, 1983.
15. **Saigo, K., Kugimiya, W., Matsuo, Y., Inouye, S., Yoshioka, K., and Yuki, S.,** Identification of the coding sequence for a reverse transcriptase-like enzyme in a transposable genetic element in *Drosophila melanogaster, Nature (London),* 312, 659, 1984.
16. **Inouye, S., Yuki, S., and Saigo, K.,** Complete nucleotide sequence and genome organization of a *Drosophila* transposable genetic element, 297, *Eur. J. Biochem.,* 154, 417, 1986.
17. **Marlor, R. L., Parkhurst, S. M., and Corces, V. G.,** The *Drosophila melanogaster* gypsy transposable element encodes putative gene products homologous to retroviral proteins, *Mol. Cell. Biol.,* 6, 1129, 1986.
18. **Johns, M. A., Mottinger, J., and Freeling, M.,** A low copy number, *copia*-like transposon in maize, *EMBO J.,* 4, 1093, 1985.
19. **Lueders, K. K. and Kuff, E. L.,** Sequences associated with intracisternal A particles are reiterated in the mouse genome, *Cell,* 12, 963, 1977.
20. **Adams, J. and Oeller, P. W.,** Structure of evolving populations of *S. cerevisiae:* adaptive changes are frequently associated with sequence alterations involving mobile elements belonging to the Ty family, *Proc. Natl. Acad. Sci. U.S.A.,* 83, 7124, 1986.
21. **Langley, C. H., Brookfield, J. F. Y., and Kaplan, N.,** Transposable elements in Mendelian populations. I. A theory, *Genetics,* 104, 457, 1983.
22. **Montgomery, E. A. and Langley, C. H.,** Transposable elements in Mendelian populations. II. Distribution of three *copia*-like elements in a natural population of *Drosophila melanogaster, Genetics,* 104, 473, 1983.
23. **Kaplan, N. L. and Brookfield, J. F. Y.,** Transposable elements in Mendelian populations. III. Statistical results, *Genetics,* 104, 485, 1983.
24. **Charlesworth, B. and Langley, C. H.,** The evolution of self-regulated transposition of transposable elements, *Genetics,* 112, 359, 1986.

25. **Montgomery, E., Charlesworth, B., and Langley, C. H.,** A test for the role of natural selection in the stabilization of transposable element copy number in a population of *Drosophila melanogaster, Genet. Res. Cambr.,* 49, 31, 1987.
26. **Golding, G. B., Aquadro, C. F., and Langley, C. H.,** Sequence evolution within populations under multiple types of mutation, *Proc. Natl. Acad. Sci. U.S.A.,* 83, 427, 1986.
27. **Winston, F., Chaleff, D. T., Valent, B., and Fink, G. R.,** Mutations affecting Ty-mediated expression of the *HIS4* gene of *Saccharomyces cerevisiae, Genetics,* 107, 179, 1984.
28. **Winston, F., Durbin, K. J., and Fink, G. R.,** The *SPT3* gene is required for normal transcription of Ty elements in *S. cerevisiae, Cell,* 39, 675, 1984.
29. **Winston, F., Dollard, C., Malone, E. A., Clare, J., Kapakos, J. G., Farabaugh, P., and Minehart, P. L.,** Three genes are required for *trans-* activation of Ty transcription in yeast, *Genetics,* 115, 649, 1987.
30. **Winston, F. and Minehart, P. L.,** Analysis of the yeast *SPT3* gene and identification of its product, a positive regulator of Ty transcription, *Nucleic Acids Res.,* 14, 6885, 1986.
31. **Modolell, J., Bender, W., and Meselson, M.,** *Drosophila melanogaster* mutations suppressible by the suppressor of hairy-wing are insertions of a 7.3-kilobase mobile element, *Proc. Natl. Acad. Sci. U.S.A.,* 80, 1678, 1983.
32. **Sweet, H. O.,** *Dilute suppressor,* a new suppressor gene in the house mouse, *J. Hered.,* 74, 304, 1983.
33. **Young, M. W. and Schwartz, H. E.,** Nomadic gene families in *Drosophila, Cold Spring Harbor Symp. Quant. Biol.,* 45, 629, 1981.
34. **Flavell, A. J., Levis, R., Simon, M. A., and Rubin, G. M.,** The 5′ termini of RNAs encoded by the transposable element *copia, Nucleic Acids Res.,* 9, 6279, 1981.
35. **Elder, R. T., Loh, E. Y., and Davis, R. W.,** RNA from the yeast transposable element Ty1 has both ends in the direct repeats, a structure similar to retrovirus RNA, *Proc. Natl. Acad. Sci. U.S.A.,* 80, 2432, 1983.
36. **Arkhipova, I. R., Mazo, A. M., Cherkasova, V. A., Gorelova, T. V., Schuppe, N. G., and Ilyin, Y. V.,** The steps of reverse transcription of *Drosophila* mobile dispersed genetic elements and U3–R–U5 structure of their LTRs, *Cell,* 44, 555, 1986.
37. **Jerome, J. F. and Jaehning, J. A.,** mRNA transcription in nuclei isolated from *Saccharomyces cerevisiae, Mol. Cell. Biol.,* 6, 1633, 1986.
38. **Boeke, J. D., Garfinkel, D. J., Styles, C. A., and Fink, G. R.,** Ty elements transpose through an RNA intermediate, *Cell,* 40, 491, 1985.
39. **Paterson, B. M., Segal, S., Lueders, K. K., and Kuff, E. L.,** RNA associated with murine intracisternal type A particles codes for the main particle protein, *J. Virol.,* 27, 118, 1978.
40. **Shiba, T. and Saigo, K.,** Retrovirus-like particles containing RNA homologous to the transposable element *copia* in *Drosophila melanogaster, Nature (London),* 302, 119, 1983.
41. **Panganiban, A. T. and Temin, H. M.,** The retrovirus *pol* gene encodes a product required for DNA integration: identification of a retroviral *int* locus, *Proc. Natl. Acad. Sci. U.S.A.,* 81, 7885, 1984.
42. **Duyk, G., Leis, J., Longiaru, M., and Skalka, A. M.,** Selective cleavage in the avian retroviral long terminal repeat sequence by the endonuclease associated with the alpha-beta form of avian reverse transcriptase, *Proc. Natl. Acad. Sci. U.S.A.,* 80, 6745, 1983.
43. **Duyk, G., Longiaru, M., Cobrinik, D., Kowal, R., DeHaseth, P., Skalka, A. M., and Leis, J.,** Circles with two tandem long terminal repeats are specifically cleaved by *pol* gene-associated endonuclease from avian sarcoma and leukosis viruses: nucleotide sequences required for site-specific cleavage, *J. Virol.,* 56, 589, 1985.
44. **Golomb, M. and Grandgenett, D.P.,** Endonuclease activity of purified RNA-directed DNA polymerase from avian myeloblastosis virus, *J. Biol. Chem.,* 254, 1606, 1979.
45. **Golomb, M., Grandgenett, D. P., and Mason, W.,** Virus-coded DNA endonuclease from avian retrovirus, *J. Virol.,* 38, 548, 1981.
46. **Grandgenett, D. P. and Vora, A. C.,** Site-specific nicking at the avian retrovirus LTR circle junction by the viral pp32 DNA endonuclease, *Nucleic Acids Res.,* 13, 6205, 1985.
47. **Grandgenett, D. P., Vora, A. C., Swanstrom, R. and Olsen, J. C.,** Nuclease mechanism of the avian retrovirus pp32 endonuclease, *J. Virol.,* 58, 970, 1986.
48. **Grandgenett, D.P., Golomb, M., and Vora, A. C.,** Activation of an Mg^{2+}-dependent DNA endonuclease of avian myeloblastosis virus alpha/beta DNA polymerase by in vitro proteolytic cleavage, *J. Virol.,* 33, 264, 1980.
49. **Grandgenett, D., Quinn, T., Hippenmeyer, P. J., and Oroszlan, S.,** Structural characterization of the avian retrovirus reverse transcriptase and endonuclease domains, *J. Biol. Chem.,* 260, 8243, 1985.
50. **Leis, J., Duyk, G., Johnson, S., Longiaru, M., and Skalka, A.,** Mechanism of action of the endonuclease associated with alpha-beta and beta-beta forms of avian RNA tumor virus reverse transcriptase, *J. Virol.,* 45, 727, 1983.

51. **Yoshinaka, Y., Katoh, I., Copeland, T. D., and Oroszlan, S.,** Murine leukemia virus protease is encoded by the *gag-pol* gene and is synthesized through suppression of an amber termination codon,, *Proc. Natl. Acad. Sci. U.S.A.*, 82, 1618, 1985.
52. **Cameron, J. R., Loh, E. Y., and Davis, R. W.,** Evidence for transposition of dispersed repetitive DNA families in yeast, *Cell*, 16, 739, 1979.
53. **Kingsman, A. J., Gimlich, R. L., Clarke, L., Chinault, A. C., and Carbon, J.,** Sequence variation in dispersed repetitive sequences in *Saccharomyces cerevisiae, J. Mol. Biol.*, 145, 619, 1981.
54. **Christy, R. J., Brown, A. R., Gourlie, B. B., and Huang, R. C. C.,** Nucleotide sequences of murine intracisternal A-particle gene LTRs have extensive variability within the R region, *Nucleic Acids Res.*, 13, 289, 1985.
56. **Gopinathan, K. P., Weymouth, L. A., Kunkel, T. A., and Loeb, L. A.,** Mutagenesis *in vitro* by DNA polymerase from an RNA tumour virus, *Nature (London)*, 278, 857, 1979.
57. **Skalka, A. M., Boone, L., Junghans, R., and Luk, D.,** Genetic recombination in avian retroviruses, *J. Cell. Biochem.*, 19, 293, 1982.
58. **Clare, J. and Farabaugh, P.,** Nucleotide sequence of a yeast Ty element: evidence for an unusual mechanism of gene expression, *Proc. Natl. Acad. Sci. U.S.A.*, 82, 2829, 1985.
59. **Hauber, J., Nelbock-Hochstetter, P., and Feldman, H.,** Nucleotide sequence and characteristics of a Ty element from yeast, *Nucleic Acids Res.*, 13, 2745, 1985.
60. **Warmington, J. R., Waring, R. B., Newlon, C. S., Indge, K. J., and Oliver, S. G.,** Nucleotide sequence characterization of Ty 1-17, a class II transposon from yeast, *Nucleic Acids Res.*, 13, 6679, 1985.
61. **Mount, S. M. and Rubin, G. M.,** Complete nucleotide sequence of the *Drosophila* transposable element *copia:* homology between *copia* and retroviral proteins, *Mol. Cell. Biol.*, 5, 1630, 1985.
62. **Emori, Y., Shiba, T., Kanaya, S., Inouye, S., Yuki, S., and Saigo, K.,** The nucleotide sequences of *copia* and *copia*-related RNA in *Drosophila* virus-like particles, *Nature (London)*, 315, 773, 1985.
63. **Cappello, J., Handelsman, K., and Lodish, H. F.,** Sequence of dictyostelium DIRS-1: an apparent retrotransposon with inverted terminal repeats and an internal circle junction sequence, *Cell*, 43, 105, 1985.
64. **Farabaugh, P. J.,** personal communication, 1986.
65. **Boeke, J. D., Eichinger, D., Castrillon, D., and Fink, G. R.,** The yeast genome contains functional and nonfunctional copies of transposon Ty1, *Mol. Cell. Biol.*, 1987, submitted.
66. **Cappello, J., Cohen, S. M., and Lodish, H. F.,** *Dictyostellium* transposable element DIRS-1 preferentially inserts into DIRS-1 sequences, *Mol. Cell. Biol.*, 4, 2207, 1984.
67. **Cappello, J., Zuker, C., and Lodish, H. F.,** Repetitive *Dictyostelium* heat-shock promoter functions in *Saccharomyces cerevisiae, Mol. Cell. Biol.*, 4, 591, 1984.
68. **Garfinkel, D. J., Boeke, J. D., and Fink, G. R.,** Ty element transposition: reverse transcriptase and virus-like particles, *Cell*, 42, 507, 1985.
69. **Mellor, J., Malim, M. H., Gull, K., Tuite, M. F., McCready, S., Dibbayawan, T., Kingsman, S. M., and Kingsman, A. J.,** Reverse transcriptase activity and Ty RNA are associated with virus-like particles in yeast, *Nature (London)*, 318, 583, 1985.
70. **Heine, C. W., Kelly, D. C., and Avery, R. J.,** The detection of intracellular retrovirus-like entities in *Drosophila melanogaster* cell cultures, *J. Gen. Virol.*, 49, 385, 1980.
71. **Kuff, E. L., Wivel, N. A., and Lueders, K. K.,** The extraction of intracisternal A-particles from a mouse plasma-cell tumor, *Cancer Res.*, 28, 2137, 1968.
72. **Bystrom, A. and Fink, G. R.,** unpublished results, 1985.
73. **Gilboa, E., Mitra, S. W., Goff, S., and Baltimore, D.,** A detailed model of reverse transcription and a test of crucial aspects, *Cell*, 18, 93, 1979.
74. **Panganiban, A. T. and Temin, H. M.,** Circles with two tandem LTRs are precursors to integrated retrovirus DNA, *Cell*, 36, 673, 1984.
75. **Teich, N.,** Taxonomy of retroviruses, in *RNA Tumor Viruses*, 2nd ed., Weiss, R., Teich, N., Varmus, H., and Coffin, J., Eds., Cold Spring Harbor Laboratory, Cold Spring Harbor, N.Y., 1984, chap. 2.
76. **Dougherty, R. M., DiStefano, H., Feller, U., and Mueller, J. F.,** On the nature of particles lining the excretory ducts of pseudophyllidean cestodes, *J. Parasitol.*, 61, 1006, 1975.
77. **Akai, H., Gateff, E., Davis, L. E., and Schneiderman, H. A.,** Virus-like particles in normal and tumorous tissues of Drosophila, *Science*, 157, 810, 1967.
78. **Gateff, E.,** Malignant neoplasm of genetic origin in *Drosophila melanogaster, Science*, 200, 1448, 1978.
79. **Saigo, K., Shiba, T., and Miyake, T.,** Virus-like particles of *Drosophila melanogaster* containing t-RNA and 5S ribosomal RNA. I. Isolation and purification from cultured cells and detection of low molecular weight RNAs in the particles, in *Invertebrate Systems In Vitro*, Kurstak, E., Maramorosch, K., and Dubendorfer, A., Eds., Elsevier/North-Holland, Amsterdam, 1980, 411.
80. **Hawley, R. G., Shulman, M. J., Murialdo, H., Gibson, D. M., and Hozumi, N.,** Mutant immunoglobulin genes have repetitive DNA elements inserted into their intervening sequences, *Proc. Natl. Acad. Sci. U.S.A.*, 79, 7425, 1982.

81. **Hawley, R. G., Shulman, M. J., and Hozumi, N.**, Transposition of two different intracisternal A particle elements into an immunoglobulin kappa-chain gene, *Mol. Cell. Biol.*, 4, 2565, 1984.
82. **Kuff, E. L., Feenstra, A., Lueders, K., Smith, L., Hawley, R., Hozumi, N., and Shulman, M.**, Intracisternal A-particle genes as movable elements in the mouse genome, *Proc. Natl. Acad. Sci. U.S.A.*, 80, 1992, 1983.
83. **Keshet, E., Shaul, Y., Kaminchik, J., and Aviv, H.**, Heterogeneity of "virus-like" genes encoding retrovirus-associated 30S RNA and their organization within the mouse genome, *Cell*, 20, 431, 1980.
84. **Itin, A. and Keshet, E.**, Nucleotide sequence analysis of the long terminal repeat of murine virus-like DNA (VL30) and its adjacent sequences: resemblance to retrovirus proviruses, *J. Virol.*, 47, 656, 1983.
85. **Itin, A. and Keshet, E.**, Primer binding sites corresponding to several tRNA species are present in DNAs of different members of the same retrovirus-like gene family (VL30), *J. Virol.*, 54, 236, 1985.
86. **Pearston, D. H., Gordon, M., and Hardman, N.**, Transposon-like properties of the major, long repetitive sequence family in the genome of *Physarum polycephalum*, *EMBO J.*, 4, 3557, 1985.
87. **Dowsett, A. P.**, Closely related species of *Drosophila* can contain different libraries of middle repetitive DNA sequences, *Chromosoma*, 88, 104, 1983.
88. **Dowsett, A. P. and Young, M. W.**, Differing levels of dispersed repetitive DNA among closely related species of *Drosophila*, *Proc. Natl. Acad. Sci. U.S.A.*, 79, 4570, 1982.
89. **Phillipsen, P.**, personal communication, 1984.
90. **Miller, D. W. and Miller L. K.**, A virus mutant with an insertion of a *copia*-like transposable element, *Nature (London)*, 299, 5883, 1982.
91. **Shepherd, N. S., Schwarz-Sommer, Z., vel Spalve, J. B., Gupta, M., Wienand, U., and Saedler, H.**, Similarity of the *Cin1* repetitive family of *Zea mays* to eukaryotic transposable elements, *Nature (London)*, 307, 185, 1984.
92. **Mottinger, J. P., Johns, M. A., and Freeling, M.**, Mutations of the *Adhl* gene in maize following infection with barley stripe mosaic virus, *Mol. Gen. Genet.*, 195, 367, 1984.
93. **Lueders, K. K. and Kuff, E. L.**, Comparison of the sequence organization of related retrovirus-like multigene families in three evolutionarily distant rodent genomes, *Nucleic Acids Res.*, 11, 4391, 1983.
94. **Schmidt, M., Gloggler, K., Wirth, T., and Horak, I.**, Evidence that a major class of mouse endogenous long terminal repeats (LTRs) resulted from recombination between exogenous retroviral LTRs and similar LTR-like elements (LTR-IS), *Proc. Natl. Acad. Sci. U.S.A.*, 81, 6696, 1984.
95. **Schmidt, M., Wirth, T., Kroger, B., and Horak, I.**, Structure and genomic organization of a new family of murine retrovirus-related DNA sequences (MuRRS), *Nucleic Acids Res.*, 13, 3461, 1985.
96. **Kohrer, K., Grummt, I., and Horak, I.**, Functional RNA polymerase II promoters in solitary retroviral long terminal repeats (LTR-IS elements), *Nucleic Acids Res.*, 13, 2631, 1985.
97. **Wichman, H. A., Potter, S. S., and Pine, D. S.**, Mys, a family of mammalian transposable elements isolated by phylogenetic screening, *Nature (London)*, 317, 77, 1985.
98. **Paulson, K. E., Deka, N., Schmid, C. W., Misra, R., Schindler, C. W., Rush, M. G., Kadyk, L., and Leinwand, L.**, A transposon-like element in human DNA, *Nature (London)*, 316, 359, 1985.
99. **Mager, D. L. and Henthorn, P. S.**, Identification of a retrovirus-like repetitive element in human DNA, *Proc. Natl. Acad. Sci. U.S.A.*, 81, 7510, 1984.
100. **Mellor, J., Fulton, S. M., Dobson, J. J., Wilson, W., and Kingsman, A. J.**, A retrovirus-like strategy for the expression of a fusion protein encoded by yeast transposon Ty1, *Nature (London)*, 313, 243, 1985.
101. **Jacks, T. and Varmus, H. E.**, Expression of the Rous sarcoma virus *pol* gene by ribosomal frameshifting, *Science*, 230, 1237, 1985.
102. **Dobson, M. J., Mellor, J., Fulton, A. M., Roberts, N. A., Bowen, B. A., Kingsman, S. M., and Kingsman, A. J.**, The identification and high level expression of a protein encoded by the yeast Ty element, *EMBO J.*, 3, 1115, 1984.
103. **Mellor, J., Fulton, A. M., Dobson, M., Roberts, N. A., Wilson, W., Kingsman, A., and Kingsman, S. M.**, The Ty transposon of *Saccharomyces cerevisiae* determines the synthesis of at least three proteins, *Nucleic Acids Res.*, 13, 6249, 1985.
104. **Garfinkel, D. J.**, unpublished data, 1985.
105. **Flavell, A. J., Ruby, S. W., Toole, J. J., Roberts, B. E., and Rubin, G. M.**, Translation and developmental regulation of RNA encoded by the eukayotic transposable element *copia*, *Proc. Natl. Acad. Sci. U.S.A.*, 77, 7107, 1980.
106. **Ono, M., Toh, H., Miyata, T., and Awaya, T.**, Nucleotide sequence of the Syrian hamster intracisternal A-particle gene: close evolutionary relationship of type A particle gene to types B an D oncovirus genes, *Virology*, 55, 387, 1985.
107. **Lueders, K. K. and Mietz, J. A.**, Structural analysis of type II variants within the mouse intracisternal A-particle sequence family, *Nucleic Acids Res.*, 14, 1495, 1986.
108. **Cullen, B. R., Lomedico, P. T., and Ju, G.**, Transcriptional interference in avian retroviruses — implications for the promoter insertion model of leukaemogenesis, *Nature (London)*, 307, 241, 1984.

109. **Resnick, R., Omer, C. A., and Faras, A. J.,** Involvement of retrovirus reverse transcriptase-associated RNase H in the initiation of strong-stop (+) DNA synthesis and the generation of the long terminal repeat, *J. Virol.*, 51, 813, 1984.
110. **Taylor, J. M. and Cywinski, A.,** A defective retrovirus particle (SE21Q1b) packages and reverse transcribes cellular RNA, utilizing tRNA-like primers, *J. Virol.*, 51, 267, 1984.
111. **Mann, R., Mulligan, R. C., and Baltimore, D.,** Construction of a retrovirus packaging mutant and its use to produce helper-free defective retrovirus, *Cell*, 33, 153, 1983.
112. **Katz, R. A., Terry, R. W., and Skalka, A. M.,** A conserved *cis*-acting sequence in the 5′ leader of Avian sarcoma virus RNA is required for packaging, *J. Virol.*, 59, 163, 1986.
113. **Sorge, J., Ricci, W., and Hughes, S. H.,** *Cis*-acting RNA packaging locus in the 115-nucleotide direct repeat of Rous sarcoma virus, *J. Virol.*, 48, 667, 1983.
114. **Roeder, G. S., Rose, A. B., and Pearlman, R. E.,** Transposable element sequences involved in the enhancement of yeast gene expression, *Proc. Natl. Acad. Sci. U.S.A.*, 82, 5428, 1985.
115. **Errede, B., Company, M., Ferchak, J. D., Hutchison, C. A., III, and Yarnell, W. S.,** Activation regions in a yeast transposon have homology of mating type control sequences and to mammalian enhancers, *Proc. Natl. Acad. Sci. U.S.A.*, 82, 5423, 1985.
116. **Kapakos, J. G., Clare, J. J., and Farabaugh, P. J.,** A Ty transposable element of yeast contains both transcriptional enhancer and silencer sites, *Mol. Cell. Biol.*, submitted.
117. **Company, M. and Errede, B.,** Transcriptional analysis of Ty1 deletion and inversion derivatives at *CYC7*, *Mol. Cell. Biol.*, 6, 3299, 1986.
118. **Miller, A. M., McKay, V. L., and Nasmyth, K. A.,** Identification and comparison of two sequence elements that confer cell-type specific transcription in yeast, *Nature (London)*, 314, 598, 1985.
119. **Siliciano, P. G. and Tatchell, K.,** Identification of the DNA sequences controlling the expression of the MATα locus of yeast, *Proc. Natl. Acad. Sci. U.S.A.*, 83, 2320, 1986.
120. **Errede, B., Cardillo, T. S., Sherman, F., Dubois, E., Deschamps, J., and Wiame, J. M.,** Mating signals control expression of mutations resulting from insertion of transposable repetitive element adjacent to diverse yeast genes, *Cell*, 25, 427, 1980.
121. **Taguchi, A. K. W., Ciriacy, M., and Young, E. T.,** Carbon source dependence of transposable element-associated gene activation in *Saccharomyces cerevisiae*, *Mol. Cell. Biol.*, 4, 61, 1984.
122. **Flavell, A. J. and Ish-Horowicz, D.,** Extrachromosomal circular copies of the eukaryotic transposable element *copia* in cultured *Drosophila* cells, *Nature (London)*, 292, 591, 1981.
123. **Errede, B., Company, M., and Swanstrom, R.,** An anomalous Ty1 structure attributed to an error in reverse transcription, *Mol. Cell Biol.*, 6, 1334, 1986.
124. **Flavell, A. J. and Ish-Horowicz, D.,** The origin of extrachromosomal circular *copia* elements, *Cell*, 34, 415, 1983.
125. **Ballario, P., Filetici, P., Junakovic, N., and Pedone, F.,** Ty1 extrachromosomal circular copies in *Saccharomyces cerevisiae*, *FEBS Lett.*, 155, 225, 1983.
126. **Newlon, C. S., Devenish, R. J., and Lipchitz, L. R.,** Mapping autonomously replicating segments on a circular derivative of chromosome, III, in Proc. Berkeley Workshop on Recent Advances in Yeast Molecular Biology: Recombinant DNA, Berkeley, Calif., 1982.
127. **Flavell, A. J.,** Role of reverse transcription in the generation of extrachromosomal *copia* mobile genetic elements, *Nature (London)*, 310, 514, 1984.
128. **Flavell, A. J. and Brierley, C.,** The termini of extrachromosomal linear *copia* elements, *Nucleic Acids Res.*, in press, 1986.
129. **Arkhipova, I. R., Gorelova, T. V., Ilyn, Y. V., and Schuppe, N. G.,** Reverse transcription of *Drosophila* mobile dispersed genetic element RNAs: detection of intermediate forms, *Nucleic Acids Res.*, 12, 7533, 1984.
130. **Ilyin, Y. V., Schuppe, N. G., Lyubomirskaya, N. V., Gorelova, T. V., and Arkhipova, I. R.,** Circular copies of mobile dispersed genetic elements in cultured *Drosophila melanogaster* cells, *Nucleic Acids Res.*, 12, 7517, 1984.
131. **Mossie, K. G., Young, M. W., and Varmus, H. E.,** Extrachromosomal DNA forms of copia-like transposable elements, F elements and middle repetitive DNA sequences in *Drosophila melanogaster*, *J. Mol. Biol.*, 182, 31, 1985.
132. **Boeke, J. D., Styles, C. A., and Fink, G. R.,** The yeast *SPT3* gene is required for transposition and transpositional recombination of chromosomal Ty elements, *Mol. Cell. Biol.*, 6, 3575, 1986.
133. **Paquin, C. E. and Williamson, V. M.,** Ty insertions at two loci account for most of the spontaneous antimycin A resistance mutations during growth at 15 degrees C of *Saccharomyces cerevisiae* strains lacking *ADH1*, *Mol. Cell. Biol.*, 6, 70, 1986.
134. **Rolfe, M., Spanos, A., and Banks, G.,** Induction of yeast Ty element transcription by ultraviolet light, *Nature (London)*, 319, 339, 1986.
135. **Schwartz, H. E., Lockett, T. J., and Young, M. W.,** Analysis of transcripts from two families of nomadic DNA, *J. Mol. Biol.*, 157, 49, 1982.

136. **Parkhurst, S. M. and Corces, V. G.,** Forked, gypsys, and suppressors in *Drosophila, Cell,* 41, 429, 1985.
137. **Strand, D. J. and McDonald, J. F.,** *Copia* is transcriptionally responsive to environmental stress, *Nucleic Acids Res.,* 13, 4401, 1985.
138. **Grigoryan, M. S., Kramerov, D. A., Tulchinsky, E. M., Revasova, E. S., and Lukanidin, E. M.,** Activation of putative transposition intermediate formation in tumor cells, *EMBO J.,* 4, 2209, 1985.
139. **Biczysko, W., Pienkowski, M., Solter, D., and Koprowski, H.,** Virus particles in early mouse embryos, *J. Natl. Cancer Inst.,* 51, 1041, 1973.
140. **Paquin, C. E. and Williamson, V. M.,** Temperature effects on the rate of Ty transposition, *Science,* 226, 53, 1984.
141. **Cullen, B. R., Skalka, A. M., and Ju, G.,** Endogenous avial retroviruses contain deficient promoter and leader sequences, *Proc. Natl. Acad. Sci.,* 80, 2946, 1983.
142. **Boeke, J. D., Eichinger, D., Castrillon, D., and Fink, G. R.,** The yeast genome contains functional and nonfunctional copies of transposon Ty1, *Mol. Cell. Biol.,* 1987, submitted.
143. **Dzumagliev, E. B., Mazo, A. M., Bayev, A. A., Jr., Gorelova, T. V., Arkhipova, I. R., Schuppe, N. G., and Ilyin, Y. V.,** The structure of long terminal repeats of transcriptionally active and inactive copies of *Drosophila* mobile dispersed genetic elements mdg3, *Genetika,* 22, 368, 1986.
144. **Jaenisch, R., Jahner, D., Nobis, P., Simon, I., Lohler, J., Harbers, K., and Grotkopp, D.,** Chromosomal position and activation of retroviral genomes inserted into the germ line of mice, *Cell,* 24, 519, 1981.
145. **Rose, M. and Winston, F.,** Identification of a Ty insertion within the coding sequence of the *Saccharomyces cerevisiae URA3* gene, *Mol. Gen. Genet.,* 193, 557, 1984.
146. **Natsoulis, G. and Boeke, J. D.,** unpublished data, 1987.
147. **O'Hare, K., Murphy, C., Levis, R., and Rubin, G. M.,** DNA sequence of the *white* locus of *Drosophila melanogaster, J. Mol. Biol.,* 180, 437, 1984.
148. **Johns, M.,** personal communication, 1986.
149. **Swaroop, A., Paco-Larson, M. L., and Garen, A.,** Molecular genetics of a transposon-induced dominant mutation in the *Drosophila* locus glued, *Proc. Natl. Acad. Sci. U.S.A.,* 82, 1751, 1985.
150. **Roeder, G. S., Farabaugh, P. J., Chaleff, D. T., and Fink, G. R.,** The origins of gene instability in yeast, *Science,* 209, 1375, 1980.
151. **Snyder, M. P., Kimbrell, D., Hunkapiller, M., Hill, R., Fristrom, J., and Davidson, N.,** A transposable element that splits the promoter region inactivates a *Drosophila* cuticle protein gene, *Proc. Natl. Acad. Sci. U.S.A.,* 79, 7430, 1982.
152. **Zachar, Z., Davison, D., Garza, D., and Bingham, P. M.,** A detailed developmental and structural study of the transcriptional effects of insertion of the *copia* transposon into the white locus of *Drosophila melanogaster, Genetics,* 111, 495, 1985.
153. **Levis, R., O'Hare, K., and Rubin, G. M.,** Effects of transposable element insertions on RNA encoded by the white gene of *Drosophila, Cell,* 38, 471, 1984.
154. **Kidd, S. and Young, M. W.,** Transposon dependent mutant phenotypes at the *Notch* locus of *Drosophila, Nature (London),* 323, 89, 1986.
155. **Scherer, S., Mann, C., and Davis, R. W.,** Reversion of a promoter deletion in yeast, *Nature (London),* 298, 815, 1982.
156. **Scott, M. P., Weiner, A. J., Hazelrigg, T. I., Polisky, B. A., Pirrotta, V., Scalenghe, F., and Kaufman, T. C.,** The molecular organization of the antennapedia locus of *Drosophila, Cell,* 35, 763, 1983.
157. **Campuzano, S., Balcells, L., Villares, R., Carramolino, L., Garcia-Alonso, L., and Modolell, J.,** Excess function hairy-wing mutations caused by gypsy and *copia* insertions within structural genes of the achaetescute locus of *Drosophila, Cell,* 44, 303, 1986.
158. **Eibel, H. and Philippsen, P.,** Preferential integration of yeast transposable element Ty into a promoter region, *Nature (London),* 307, 386, 1984.
159. **Simchen, G. Winston, F., Styles, C. A., and Fink, G. R.,** Ty-mediated expression of the *LYS2* and *HIS4* genes of *Saccharomyces cerevisiae* is controlled by the same *SPT* genes, *Proc. Natl. Acad. Sci. U.S.A.,* 81, 2431, 1984.
160. **Ikenaga, H. and Saigo, K.,** Insertion of a movable genetic element 297, into the T–A–T–A box for the H3 histone gene in *Drosophila melanogaster, Proc. Natl. Acad. Sci. U.S.A.,* 79, 4143, 1982.
161. **Inouye, S., Yuki, S., and Saigo, K.,** Sequence-specific insertion of the *Drosophila* transposable genetic element 17.6, *Nature (London),* 310, 332, 1984.
162. **Freund, R. and Meselson, M.,** Long terminal repeat nucleotide sequence and specific insertion of the gypsy transposon, *Proc. Natl. Acad. Sci. U.S.A.,* 81, 4462, 1984.
163. **Karlik, C. C. and Fyrberg, E. A.,** An insertion within a variably spliced *Drosophila* tropomyosin gene blocks accumulation of only one encoded isoform, *Cell,* 41, 57, 1985.

164. **Copeland, N. G., Hutchison, K. W., and Jenkins, N. A.,** Excision of the DBA ecotropic provirus in dilute coat-color revertants of mice occurs by homologous recombination involving the viral LTRs, *Cell,* 33, 379, 1983.
165. **Wirth, T., Gloggler, K., Baumruker, T., Schmidt, M., and Horak, I.,** Family of middle repetitive DNA sequences in the mouse genome with structural features of solitary retroviral long terminal repeats, *Proc. Natl. Acad. Sci. U.S.A.,* 80, 3327, 1983.
166. **Hutchison, K. W., Copeland, N. G., and Jenkins, N. A.,** Dilute-coat-color locus of mice: nucleotide sequence analysis of the $d+2J$ and $d+Ha$ revertant alleles, *Mol. Cell. Biol.,* 4, 2899, 1984.
167. **Rotman, G., Itin, A., and Keshet, E.,** "Solo" large terminal repeats (LTR) of an endogenous retrovirus-like gene family (VL30) in the mouse genome, *Nucleic Acids Res.,* 12, 2273, 1984.
168. **Mizrokhi, L. J., Obolenkova, L. A., Priimagi, A. F., Ilyin, Y. V., Gerasimova, T. I., and Georgiev, G. P.,** The nature of unstable insertion mutations and reversions in the locus *cut* of *Drosophila melanogaster:* molecular mechanism of transposition memory, *EMBO J.,* 4, 3781, 1985.
169. **Tschumper, G. and Carbon, J.,** High frequency excision of Ty elements during transformation of yeast, *Nucleic Acids Res.,* 14, 2989, 1986.
170. **Shepherd, B. M. and Finnegan, D. J.,** Structure of circular copies of the 412 transposable element present in *Drosophila melanogaster* tissue culture cells, and isolation of a free 412 long terminal repeat, *J. Mol. Biol.,* 180, 21, 1984.
171. **Roeder, G. S. and Fink, G. R.,** Movement of yeast transposable elements by gene conversion, *Proc. Natl. Acad. Sci. U.S.A.,* 79, 5621, 1982.
172. **Ciriacy, M. and Breilman, D.,** Delta sequences mediate DNA rearrangements in *Saccharomyces cerevisiae, Curr. Genet.,* 6, 55, 1982.
173. **Davis, P. S., Shen, M. W., and Judd, B. H.,** Insertion of the transposon *roo* into *copia* at the white locus of *Drosophila* creates a new allele and accounts for four classes of asymmetrical exchange products, *Proc. Natl. Acad. Sci. U.S.A.,* 84, 174, 1987.
174. **Gerasimova, T. I., Matjunina, L.V., Mizrokhi, L. J., and Georgiev, G. P.,** Successive transposition explosions in *Drosophila melanogaster* and reverse transpositions of mobile dispersed genetic elements, *EMBO J.,* 4, 3773, 1985.
175. **Gerasimova, T. I., Mizrokhi, L. J., and Georgiev, G. P.,** Transpositon bursts in genetically unstable *Drosophila melanogaster, Nature (London),* 309, 714, 1984.
176. **Jackson, I. J.,** Transposable elements and suppressor genes, *Nature (London),* 309, 751, 1984.
177. **Parkhurst, S. M. and Corces, V. G.,** Retroviral elements and suppressor genes in *Drosophila, Bioessays,* 5, 51, 1986.
178. **Parkhurst, S. M. and Corces, V. G.,** Mutations at the suppressor of forked locus increase the accumulation of gypsy-encoded transcripts in *Drosophila melanogaster, Mol. Cell. Biol.,* 6, 2271, 1986.
179. **Parkhurst, S. M. and Corces, V. G.,** Interactions among the gypsy transposable element and the yellow and the suppressor of hairy-wing loci in *Drosophila melanogaster, Mol. Cell. Biol.,* 6, 47, 1986.
180. **Farabaugh, P. J. and Fink, G. R.,** Insertion of the eukaryotic transposon Ty1 creates a 5-base pair duplication, *Nature (London),* 286, 352, 1980.
181. **Dunsmuir, P., Brorein, W. J., Jr., Simon, M. A., and Rubin, G. M.,** Insertion of the *Drosophila* transposable element *copia* generates a 5 base pair duplication, *Cell,* 21, 575, 1980.
182. **Levis, R., Dunsmuir, P., and Rubin, G. M.,** Terminal repeats of the *Drosophila* transposable element *copia:* nucleotide sequence and genomic organization, *Cell,* 21, 581, 1980.
183. **Inouye, S., Saigo, K., Yamada, K., and Kuchino, Y.,** Identification and nucleotide sequence determination of a potential primer tRNA for reverse transcription of a *Drosophila* retrotransposon, *297, Nucleic Acids Res.,* 14, 3031, 1986.
184. **Will, B. M., Bayev, A. A., and Finnegan, D. J.,** Nucleotide sequence of terminal repeats of 412 transposable elements of *Drosophila melanogaster, J. Mol. Biol.,* 153, 897, 1981.
185. **Toh, H., Kikuno, R., Hayashida, H., Miyata, T., Kugimiya, W., Inouye, S., Yuki, S., and Saigo, K.,** Close structural resemblance between putative polymerase of a *Drosophila* transposable genetic element 17.6 and *pol* gene product of Moloney murine leukaemia virus, *EMBO, J.,* 4, 1267, 1985.
186. **Miyata, T., Toh, H., Hayashida, H., Kikuno, R., Inokuchi, Y., and Saigo, K.,** Sequence homology among reverse transcriptase-containing viruses and transposable genetic elements: functional and evolutionary implications, in *Population Genetics and Molecular Evolution,* Ohia, T. and Aoki, K., Eds., Japan Scientific Society Press, Tokyo, 1985, 313.
187. **Scherer, G., Tschudi, C., Perera, J., Delius, H., and Pirrotta, V.,** B104, a new dispersed repeated gene family in *Drosophila melanogaster* and its analogies with retroviruses, *J. Mol. Biol.,* 157, 435, 1982.
188. **Meyerowitz, E. M. and Hogness, D. S.,** Molecular organization of a *Drosophila* puff site that responds to ecdysone, *Cell,* 28, 165, 1982.

188a. **Bayev, A. A., Jr., Lyubomirskaya, N. V., Dzhumagaliev, E. B., Ananiev, E. V., Amiantova, I. G., and Ilyin, Y. V.,** Structural organization of transposable element mdg4 from *Drosophila melanogaster* and a nucleotide sequence of its long terminal repeats, *Nucleic Acids Res.,* 12, 3707, 1984.

189. **Ilyin, Y. V., Chmeliauskaite, V. G., Ananiev, E. V., Lyubomirska, N. V., Bayev, A. A., and Georgiev, G. P.**, Mobile dispersed genetic element (3x) mdg1 of *Drosophila melanogaster, Nucleic Acids Res.*, 8, 5333, 1980.
190. **Kulguskin, V. V., Ilyin, Y. V., and Georgiev, G. P.**, Mobile dispersed genetic element (3x) mdg1 of *Drosophila melanogaster:* nucleotide sequence of long terminal repeats, *Nucleic Acids Res.*, 9, 3451, 1981.
191. **Ilyin, Y. V., Chmeliauskaite, V. G., Ananiev, E. V., and Georgiev, G. P.**, Isolation and characterization of a new family of mobile dispersed genetic elements, mdg3, in *Drosophila melanogaster, Chromosoma,* 81, 27, 1980.
192. **Bell, J. R., Bogardus, A. M., Schmidt, T., and Pellegrini, M.**, A new *copia*-like transposable element found in a *Drosophila* rDNA gene unit, *Nucleic Acids Res.*, 13, 3861, 1985.
193. **Friesen, P. D., Rice, W. C., Miller, D. W., and Miller, L. K.**, Bidirectional transcription from a solo long terminal repeat of the retrotransposon TED: symmetrical RNA start sites, *Mol. Cell. Biol.*, 6, 1599, 1986.
194. **Lueders, K. K. and Kuff, E. L.**, Intracisternal A-particle genes: Identification in the genome of *Mus musculus* and comparison of multiple isolates from a mouse gene library, *Proc. Natl. Acad. Sci. U.S.A.*, 77, 3571, 1980.
196. **Wirth, T., Schmidt, M., Baumruker, T., and Horak, I.**, Evidence for mobility of a new family of mouse middle repetitive DNA elements (LTR-IS), *Nucleic Acids Res.*, 12, 3603, 1984.
198. **Green, M. W.**, Mutant isoalleles at the vermilion locus in *Drosophila melanogaster, Proc. Natl. Acad. Sci. U.S.A.*, 38, 300, 1952.
199. **Searles, L. L. and Voelker, R. A.**, Molecular characterization of the *Drosophila* vermilion locus and its suppressible alleles, *Proc. Natl. Acad. Sci. U.S.A.*, 83, 404, 1986.
200. **Green, M. M.**, Spatial and functional properties of pseudoalleles at the *white* locus *Drosophila melanogaster, Heredity,* 13, 300, 1959.
201. **Jenkins, N. A., Copeland, N. G., Taylor, B. A., and Lee, B. K.**, Dilute (*d*) coat color mutation of DBA/2J mice is associated with the site of integration of an ecotropic MuLV genome, *Nature (London),* 293, 370, 1981.
202. **Varmus, H. and Swanstrom, R.**, Replication of Retroviruses, in *RNA Tumor Viruses,* 2nd ed., Weiss, R., Teich, N., Varmus, H.,and Coffin, J., Eds., Cold Spring Harbor Laboratory, Cold Spring Harbor, N.Y., 1984, chap. 5.
203. **Adams, S. E., Mellor, J., Gull, K., Sim, R. B., Tuite, M. F., Kingsman, S. M., and Kingsman, A. J.**, The functions and relationships of Ty-VLP proteins in yeast reflect those of mammalian retroviral proteins, *Cell,* 49, 111, 1987.
204. **Müller, F., Brühl, K.-H. Freidel, K., Kowallik, K. V., and Ciriacy, M.**, Processing of Tyl proteins and formation of Tyl virus-like particles in *Saccharomyces cerevisiae, Mol. Gen. Genet.*, 207, 421, 1987.
205. **Youngren, S., Boeke, J. D., Sanders, N., and Garfinkel, D. J.**, Functional organization of the retrotransposon Ty from *Saccharomyces cerevisiae:* the Ty protease is required for transposition, *Mol. Cell. Biol.*, submitted.
206. **Brown, P. O., Bowerman, B., Varmus, H. E., and Bishop, J. M.**, Correct integration of retroviral DNA in vitro, *Cell,* 49, 347, 1987.
207. **Yuki, S., Inouye, S., Ishimaru, S., and Saigo, K.**, Nucleotide sequence of characterization of a *Drosophila* retrotransposon, 412, *Eur. J. Biochem.*, 158, 403, 1986.
208. **Mietz, J., Grossman, Z., Lueders, K. K., and Kuff, E. L.**, Nucleotide sequence of a complete mouse intracisternal A-Particle genome, *J. Virol.*, 61, 3020, 1987.
209. **Kikuchi, Y., Ando, Y., and Shiba, T.**, Unusual mechanism of RNA-directed DNA synthesis in copia retrovirus-like particles of *Drosophila, Nature (London),* 323, 824, 1986.

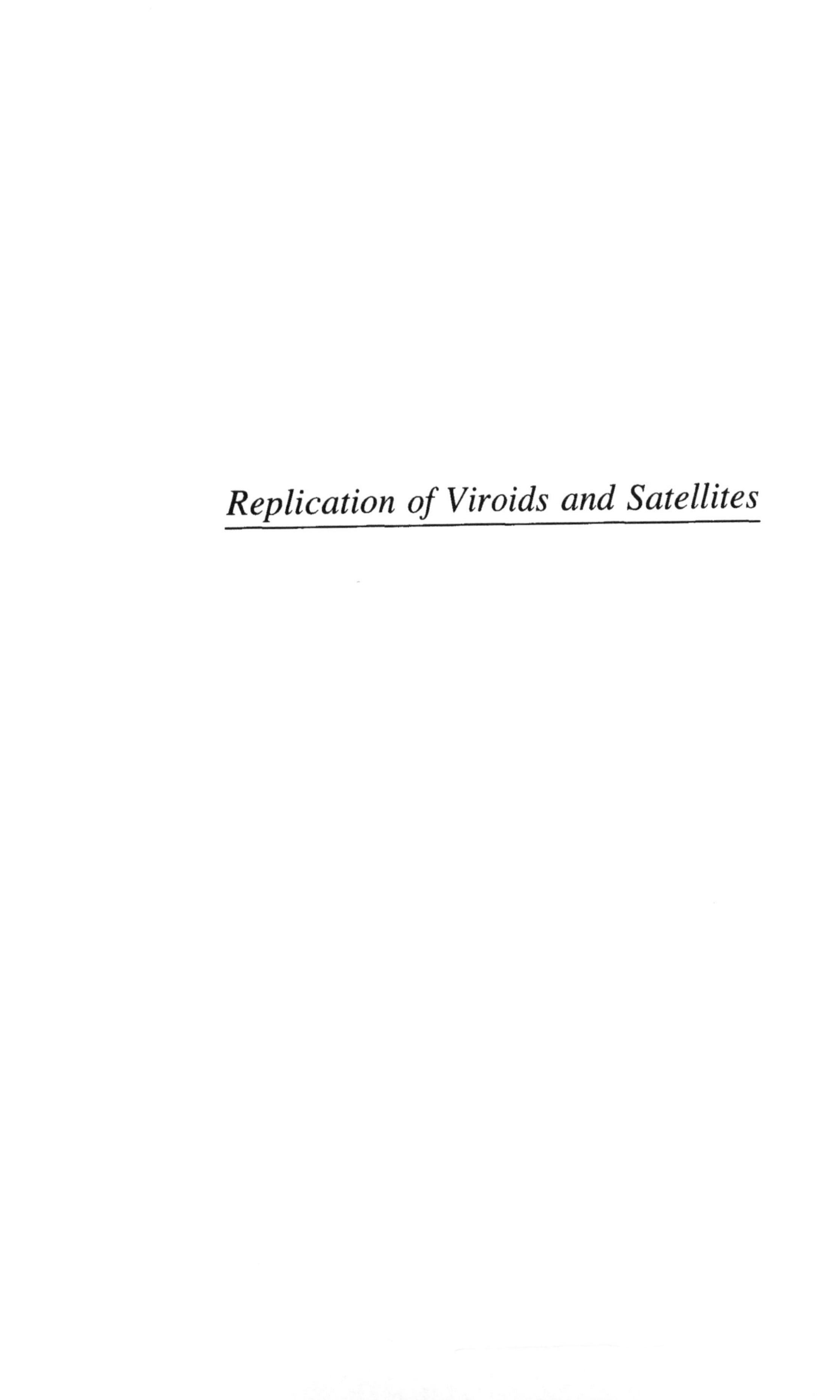

Replication of Viroids and Satellites

Chapter 5

STRUCTURE AND FUNCTION RELATIONSHIPS IN PLANT VIROID RNAs

Robert A. Owens and Rosemarie W. Hammond

TABLE OF CONTENTS

I. Introduction 108

II. Analysis of Relationships Between Viroid Structure and Function 108
- A. Primary and Secondary Structure 108
- B. Significance of Sequence Variants 110
- C. Mechanism of Replication 112
- D. Viroids as Potential mRNAs 112

III. Modification of Infectious Viroid cDNAs 112
- A. Factors Controlling Infectivity of Viroid cDNAs 113
- B. Site-Specific Mutagenesis 115
- C. Construction of Chimeric Viroid cDNAs 116
- D. Saturation Mutagenesis 118

IV. Emerging Approaches 118
- A. Development of Assays for Specific Viroid Functions 118
- B. Viroid Replication in Cells Transformed by *Agrobacterium* 119
- C. Viroid Infections as Analytical Tools 121

Acknowledgments 123

References 123

I. INTRODUCTION

Viroids are small (0.8 to 1.3 × 10^5 Mr), covalently circular RNAs that can be isolated from certain higher plant species afflicted with specific diseases. They replicate autonomously after introduction into healthy individuals of the same species and cause the appearance of the characteristic disease syndrome. Although many viroids were discovered because of their ability to cause readily recognizable disease symptoms in certain hosts, they may replicate in other species without causing obvious damage. In fact, it now appears that viroids are not restricted to diseased plants and may be more common in nature than previously believed.[1]

The discovery that viroids differ from conventional viruses in several fundamental respects[2] has been followed by a series of studies probing the nature of these unusual molecules. It is now clear that viroids constitute a group of replicating entities far simpler than viruses and that their replicative strategy seems to have evolved by adopting host pathways. They have become a model system for biochemical investigations, and the nature of their interaction with their hosts is being studied in several ways. The knowledge gained thus far suggests that what may appear to be a comparatively simple system for the study of host-pathogen interaction and RNA structure/function relationships is actually surprisingly complex.

II. ANALYSIS OF RELATIONSHIPS BETWEEN VIROID STRUCTURE AND FUNCTION

A. Primary and Secondary Structure

In 1978, Gross and collaborators reported the first complete viroid nucleotide sequence, the 359 nucleotide sequence of the type strain of PSTV.[3] Their secondary structure model for potato spindle tuber viroid (PSTV) (Figure 1) was derived by maximizing the number of potential base-pairs in the molecule, refined by considering the sites of preferential enzymatic cleavage and bisulfite modification, and confirmed by determining the binding pattern of specific tRNA anticodons to certain loop regions of PSTV.[4] Features of particular interest include a stretch of 18 purines (mostly adenosines) in positions 48 to 65, the absence of both modified nucleotides and AUG initiator triplets,[3] and the presence of 1 to 3 pairs of inverted repeat sequences containing 9 to 10 nucleotides and capable of forming hairpin-like structures. The locations of two such pairs of inverted repeats in PSTV are shown in the lower portion of Figure 1. These hairpins are not part of the thermodynamically preferred configurations present at room temperature, but they are transiently formed during thermal denaturation.[5,6] At least one pair of inverted repeats is present in all viroids, suggesting that they are essential to viroid function in vivo.

Over the last 8 years, complete nucleotide sequences for a number of other viroids have been determined (Table 1). Overall sequence homologies between different pairs of viroids range from 35 to 76%,[21] and their most probable secondary structures are very similar to that of PSTV shown in Figure 1. Although their lengths vary from about 250 to more than 370 nucleotides, all viroids contain extensive regions of intramolecular complementarity which results in the formation of many short base-paired regions interrupted by small internal loops. Because unbranched structures are thermodynamically preferred over branched structures, the molecules assume a rod-like, quasi-double-stranded conformation.

Comparative analysis of viroid sequences has allowed Keese and Symons[21] to identify additional structural regularities and to divide the native structure into the five domains shown in Figure 1: a central conserved region; flanking pathogenicity and variable regions; and two highly homologous terminal loops. The sequences which comprise the upper and lower portions of the central conserved region are particularly well conserved among all viroids except avocado sunblotch viroid (ASBV) and appear to be involved in one or more essential viroid functions (see below). Dramatic differences in the degree of sequence ho-

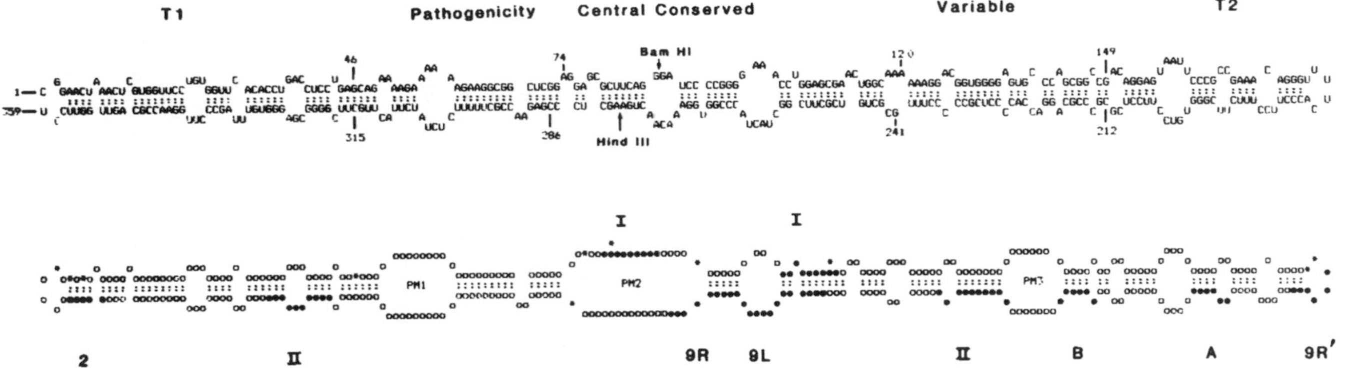

FIGURE 1. Relative locations of PSTV nucleotides important for its structure in vitro and its function in vivo. (Top) The five structural and functional domains proposed by Keese and Symons[21] are indicated above the native structure of PSTV proposed by Sänger and co-workers.[3] Nearby numbers indicate exact boundary positions. (Bottom) Nucleotides comprising the stems of secondary hairpins I and II and regions homologous to the Group I intron "boxes" (2, 9R, 9L, B, A, and 9R′, right to left)[28] are indicated by the filled circles; *, locations of various site-specific mutations in PSTV cDNA discussed in the text; PM1 to PM3, premelting loops defined in Reference 19.

Table 1
SEQUENCE ANALYSIS OF VIROIDS AND VIROID ISOLATES

Viroid	Nucleotide number	Ref.
Avocado sunblotch (ASBV)	247	7
Chrysanthemum stunt (CSV)	353—356	8—10
Citrus exocortis (CEV)	370—375	9,11,12
Coconut cadang-cadang (CCCV)	246, 247	13
Columnea latent	354	14
Hop stunt (HSV)	297	15
Cucumber pale fruit (CPFV)	303	16
Grapevine	297	17
Potato spindle tuber (PSTV)	359	3,18,19
Tomato apical stunt (TASV)	360	20
Indonesian tomato	363	10
Tomato "planta macho" (TPMV)	360	20

mology between the various domains of different viroids suggests that viroid evolution has involved repeated RNA rearrangements.[21]

B. Significance of Sequence Variants

Several different types of viroid sequence heterogeneity have been detected during sequence analysis studies. In the simplest case, a single "lethal" isolate of PSTV was found to contain approximately equal amounts of two sequence variants that differ by only a single nucleotide substitution at position 309.[19] Each variant contains 359 nucleotides, and theoretical calculations suggest that both should induce "lethal" symptoms. More extensive sequence heterogeneity has been reported to exist within a number of citrus exocortis viroid (CEV) isolates.[11,12] Although certain isolates appear to contain only a single major sequence variant, an Australian isolate inducing severe symptoms on tomato (isolate "J") is a complex mixture of nine variants. Visvader and Symons[12] classified seven of these variants as "severe", while two were classified as "mild". A third example of sequence heterogeneity has been observed among different cadang-cadang coconut viroid (CCCV) isolates.[13] The smaller or "fast" forms of both RNA 1 and 2 can be heterogeneous as a result of the presence or absence of a single cytosine residue at position 198. The various "slow" forms of CCCV RNAs 1 and 2, generated by insertion of different size classes of repeated and internally base-paired sequences between nucleotides 123 and 124, are homogeneous and appear to be the result of a single, rare sequence duplication event.

Sequence analysis of naturally occurring PSTV isolates that vary in the severity of symptoms produced in tomato[18,19] has shown that minor sequence variations can have profound biological effects. Most of the nucleotide exchanges, insertions, or deletions are clustered within a small portion of the PSTV pathogenicity region (see Figure 2). Schnölzer and coworkers[19] have proposed that the additional adenosine found in the upper portion of the variable region of some isolates (see Figure 1) is required to maintain a nucleotide number of 359, but the sanctity of this number is difficult to understand. Characterized strains of CEV and chrysanthemum stunt vivoid (CSV) contain 370-375 and 353-356 nucleotides, respectively, and closely related viroids such as tomato "planta macho" viroid (TPMV) and tomato apical stunt viroid (TASV) contain 360 nucleotides (see Table 1).

A potential connection between symptom severity and viroid secondary structure has been noted by Schnölzer et al.,[19] who recognized that increasing symptom severity can be correlated with the decreasing thermodynamic stability of a portion of the PSTV pathogenicity

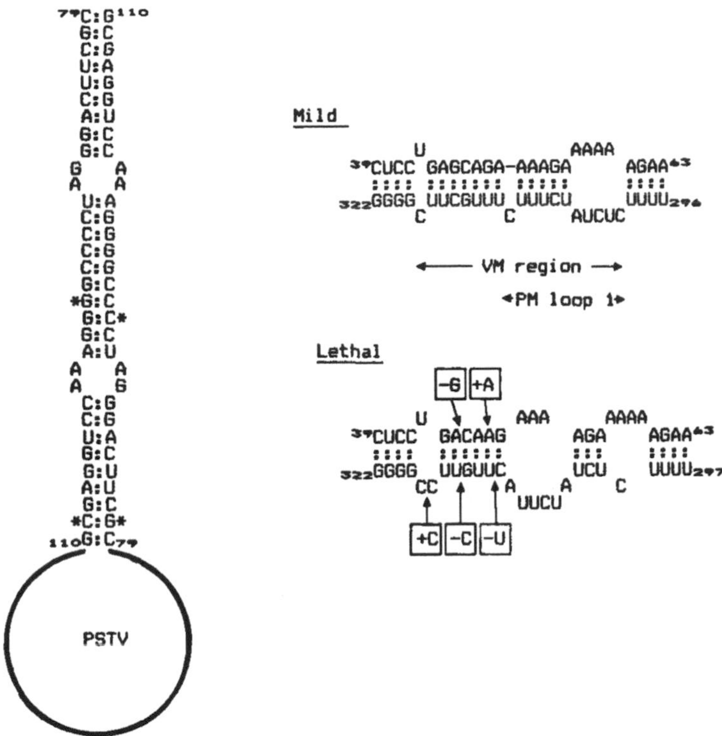

FIGURE 2. PSTV regions particularly amenable to site-specific mutagenesis. (Left) The thermodynamically extremely stable base-paired configuration that partial or complete viroid oligomers can assume has been redrawn from Diener.[47] The locations of various site-specific mutations are discussed in the text. (Right) Portions of the native structures of two naturally occurring PSTV isolates have been redrawn from Schnölzer et al.[19] Sequence changes which distinguish the lethal strain of PSTV from the mild strain have been boxed, and definitions of the terms "virulence modulating region" (VM) and "premelting loop 1" (PM) can be found in Reference 19.

domain. Lowest free-energy structures for the "virulence-modulating" regions of mild and lethal strains of PSTV are shown in Figure 2. Although structure/function correlations of this type provide a valuable conceptual framework for site-specific mutagenesis of infectious viroid cDNAs, it is important to remember that their validity is usually open to question. For example, even though the majority of sequence differences among various CEV isolates occur within an analogous region, a similar relationship between symptom severity and structural stability of the "virulence modulating" region was not discernible.[12] The structures of many sequence variants and their possible relation to symptoms and disease progression are further discussed by Keese, Visvader, and Symons in Volume III, Chapter 4.

Finally, the structure of viroids in vivo may actually be quite different from their well-characterized in vitro structures. Although purified viroids are composed solely of RNA and virion-like nucleoprotein particles have never been detected in infected plants, the possibility that viroids are complexed with cellular constituents in vivo has often been discussed.[22,23] Recently, evidence has been obtained that substantiates the existence of such complexes. Wolff et al.[24] have demonstrated the in vivo association of PSTV with proteins of 41,000 and 31,000 Mr, as well as histones, by isolating and characterizing such complexes and demonstrating in vitro interaction between the separated components. The predominant viroid-containing species present in a nucleosomal fraction was a 12- to 15-S complex.

C. Mechanism of Replication

Much of the mystery that had previously surrounded viroid replication now appears to have been dispelled, and recent results (see Reference 25 for review) have produced a unified concept of the molecular mechanisms involved. In brief, viroids are transcribed from complementary RNA (not DNA) templates, and these templates (as well as progeny viroids) are synthesized by preexisting host enzymes — possibly RNA polymerase I and/or RNA polymerase II functioning as RNA-dependent RNA polymerases. Double-stranded replicative intermediates are synthesized from infecting circular viroids by a rolling circle-type mechanism that results in the formation of oligomers of opposite polarity from which progeny viroids are in turn transcribed. With ASBV, the infecting monomeric ASBV may serve as template for synthesis of a circular (−) strand from which the oligomeric (+) strand viroids are transcribed.[26] Although a specific mechanism appears essential for the precise cleavage of oligomeric replicative intermediates into linear monomers and their subsequent ligation to form circular progeny, the molecular details of such a series of reactions remain to be established.

Comparative sequence analysis of viroids has identified features characteristic of the group I introns found in nuclear rRNA and certain mitochondrial mRNA and rRNA introns.[27,28] These features (see Figure 1) include a 16-nucleotide consensus sequence and three pairs of short complementary sequences (boxes 9L and 2; 9R and 9R′; A and B). When compared with random sequences of equal length, the occurrence of these group I intron-like features in viroids is statistically highly significant,[28] but whether or not they are actually involved in viroid cleavage and ligation is presently unknown.

D. Viroids as Potential mRNAs

Viroids are of sufficient chain length to code for a polypeptide of about 10,000 daltons, and the uneven nucleotide number of viroids such as PSTV could theoretically permit up to three rounds of translation with a frameshift each time. Neither PSTV nor CEV, however, functions as a messenger RNA in a variety of cell-free protein synthesis systems,[29,30] and both lack AUG initiation codons. Complementary RNAs synthesized during viroid replication provide a second potential source of viroid mRNA,[31] but not all such RNAs contain AUG codons. It is unlikely that the GUG codons present in all sequenced viroids function as translation initiators because AUG has been shown to be the only initiation codon used by eukaryotic cells.[32] Major differences between the possible translation products of PSTV and CSV[8] and conservation of only a single polypeptide containing 15 amino acids and initiated by GUG among CEV variants[12] have led Symons and co-workers to conclude that viroids do not encode functional polypeptides. Although synthesis of at least two proteins was enhanced in PSTV- or CEV-infected plant tissue,[33,34] subsequent studies indicated that these proteins are host and not viroid specific.[35,36] If more sensitive methods of analysis fail to disclose the presence of viroid-encoded polypeptides in infected cells, one must conclude that viroids have dispensed altogether with the need for translation and have become completely dependent upon host transcriptional systems.

III. MODIFICATION OF INFECTIOUS VIROID cDNAs

The preceding sections have shown that viroids provide a minimal genetic and biological system ideally suited to detailed analysis of the interrelationship between RNA structure and function. Viruses are entirely dependent upon the translational capacity of their hosts, but they do code for certain essential enzymes and structural proteins. Although obligate parasites of the translational machinery of their hosts, conventional viruses are either independent or only partially dependent upon their transcriptional systems. Viroids, on the other hand, are the only autonomously replicating pathogens that do not code for pathogen-specific proteins and may be regarded as obligate parasites of the transcriptional machinery of their hosts.

Unlike conventional viral RNAs, the sole theoretical requirement of a viroid is the presence of appropriate signals to trigger its replication in susceptible cells. The potential significance of these unusual pathogens to plant molecular biology can be defined as a series of questions: what are the molecular signals which viroids possess (and cellular RNAs evidently lack) that allow host enzyme(s) to accept them as templates? Are the molecular mechanisms responsible for viroid replication also operative in uninfected cells? How do viroids induce disease? In the absence of viroid-specified proteins, these processes must result from a direct interaction between viroid-specific RNA and certain host constituents. Localization of viroid-specific RNAs within the nucleus suggests a direct interaction with the host genome.[2] Although speculative mechanisms for such interactions have been proposed, very little is known about the molecular mechanisms of such interactions.

Two independent approaches, physical-chemical studies of purified viroids (see Reference 37 for review) and comparative sequence analysis of different viroids and viroid strains,[19,21] suggest that defined portions of the native structure control such essential viroid functions as replication and pathogenesis. The recent demonstration that certain cloned viroid cDNAs (and their corresponding in vitro RNA transcripts) are infectious permits the use of a third approach, the introduction of desired mutations into an infectious viroid cDNA and subsequent bioassay to detect phenotypic variation. This approach has been widely used to analyze biological and structural functions encoded by a variety of DNA genomes, but it is equally applicable to RNA genomes that can be copied into biologically active cDNAs. Because naturally occurring variants are unlikely to contain alterations that severely affect essential functions, site-specific mutagenesis techniques complement comparative sequence analysis of naturally occurring variants.

A. Factors Controlling Infectivity of Viroid cDNAs

Evidence from several laboratories has demonstrated that appropriately constructed, viroid-specific cDNAs (as well as their respective RNA transcripts) are infectious when introduced into viroid-susceptible plants.[38-44] Inoculated plants develop the characteristic symptoms of viroid infection, and progeny viroids of the predicted sequence are synthesized. Evidently, the infecting cDNAs must be transcribed into viroid-specific RNAs from which the initial viroid progeny are synthesized. Although the exact mechanisms responsible for this process are unknown, there are certain specific requirements that must be met in order for the cDNAs or their RNA transcripts to be infectious (Table 2).

Although naturally occurring linear viroid monomers[45] and artificially nicked viroids with 2′,3′-cyclic phosphate termini[46] are as infectious as circular molecules, the presumed rolling circle mechanism of viroid replication suggests that repeated units of the viroid sequence may be required for transcription/processing of an entire sequence using linear DNA templates. Such sites are statistically unlikely to occur at the exact beginning of a monomeric insert, and the very low infectivity of several monomeric viroid cDNAs or their respective in vitro RNA transcripts (see Table 2) was not unexpected. Transcripts from multimeric inserts that have same polarity as the infectious viroid, on the other hand, are highly infectious. Experiments involving the insertion of double-stranded PSTV cDNA monomers into various plasmid vectors have provided important clues about the exact amount of sequence duplication required for infectivity.

pBR322 recombinants containing a PSTV cDNA BamHI monomer inserted in the BamHI site are either noninfectious[38] or only marginally infectious,[40] but insertion of the same cDNA into the BamHI site of either bacteriophage M13mp9[40] or plasmids pUC9 or pSP64[41,44] produces constructs that exhibit relatively high levels of infectivity. Examination of the DNA sequences adjacent to the junction between the vector DNA and the viroid-specific insert reveals that pBR322 recombinants contain the 359 nucleotides of the monomeric PSTV sequence, plus 6 PSTV-specific nucleotides derived from the vector. Insertion into vectors

Table 2
INFECTIVITIES OF CLONED VIROID cDNAs AND RNA TRANSCRIPTS

Insert size	Inoculum	Transcript polarity[a]	Infectivity[b]	Ref.
Monomer[c]	ssDNA	+	− or ±	40
		−	−	40
	dsDNA			
	intact plasmid	+	− or ±	38, 40
		−	− or ±	38, 40
	excised fragments	+	+ or + +	40, 42
	RNA			
	E. coli	+	±	38
	in vitro transcripts	+	−	41, 43
		−	−	41, 43
Multimer	ssDNA	+	+ + +	40
		−	+ + +	41
	dsDNA			
	intact plasmid	+	+ + +	38, 40
		−	+ + +	38, 40
	excised fragment	+	+ + +	38, 40
	RNA			
	E. coli	+	+ + +	38, 40
		−	−	38
	in vitro transcripts	+	+ + +	41, 43
		−	−	41, 43

[a] +, polarity of infectious viroid; −, polarity of viroid complement.
[b] −, no infectivity; ±, trace; +, low; + +, medium; + + +, high levels of infectivity.
[c] Six or fewer terminal nucleotide duplications.

Adapted from Diener, T. O., *Proc. Natl. Acad. Sci. U.S.A.*, 83, 58, 1986.

such as pUC9 or pSP64 produces clones containing the 359 nucleotides of PSTV plus 11 PSTV-specific nucleotides (GGATCCCCGGG) derived from the multiple cloning site. This sequence is part of the central conserved region, and Tabler and Sänger[40] have suggested that this difference of five nucleotides seems to be essential for the infectivity of cloned PSTV cDNA. The infectivity of cDNAs with an 11-nucleotide repeat is still, however, less than that of clones containing a complete duplication of viroid sequence (see Table 2).

Diener[47] has proposed a model for viroid processing that identifies a thermodynamically extremely stable, base-paired configuration which may explain this requirement for sequence duplication. The structure postulated to be essential for precise cleavage and ligation (see Figure 2) can be assumed by certain partial viroid oligomers and involves structural features common to all viroids, i.e., the central conserved region and secondary hairpin I. It also explains why vector-derived sequences on either or both sides of the viroid sequence present in a DNA restriction fragment do not appear in the viroid progeny. Although it is compatible with recent evidence implicating the upper portion of the central conserved region as the site of RNA cleavage-ligation during viroid replication,[48,49] as well as with results from site-specific mutagenesis of infectious viroid cDNAs,[44,49] other pathways for the production of circular viroid progeny have not yet been ruled out.

The infectivity of certain "monomeric" viroid cDNAs and their corresponding in vitro RNA transcripts allows the use of site-specific DNA mutagenesis techniques to dissect structure/function relationships within the viroid molecule. Several different techniques can be used to introduce mutations into DNA copies of viroid RNA before bioassay to determine the effect of the alteration. Two of these techniques, site-specific mutagenesis and construction of chimeric viroid cDNAs, appear particularly promising.

B. Site-Specific Mutagenesis

Cleavage/ligation of greater-than-unit-length RNA transcripts was the first step in viroid replication to be studied by site-specific mutagenesis techniques. As part of an investigation of a possible processing site within the 11-nucleotide repeat discussed above, Visvader et al.[49] used oligonucleotide-directed mutagenesis to introduce G → A/T substitutions at position 97 within the central conserved region of cloned CEV strain A cDNA. Although the mutant CEV cDNAs were infectious when the intact pSP64 plasmid DNA or RNA transcripts were used an inoculum, the excised mutant CEV cDNA monomer was noninfectious. This result suggests that the 11-nucleotide duplication created by the plasmid vector is required for RNA processing, i.e., in vivo processing of greater-than-unit-length RNAs may occur at positions adjacent to the mutation at position 97. The end result of such a cleavage is the excision of a viable viroid RNA monomer that contains no mutations.

Owens et al.[44] utilized a combination of chemical and oligonucleotide-directed mutagenesis techniques to introduce alterations into two different regions of cloned PSTV cDNA. Directed chemical mutagenesis using bisulfite produced C → T transitions at positions 92 and 103 within the central conserved region (see Figure 1). These mutations rendered the viroid cDNA significantly less infectious than the parental cDNA and suggests that cleavage and ligation of multimeric PSTV RNAs may occur between positions 86 and 92. This result is consistent with the observations made by Visvader et al.[49] and the model proposed by Diener.[47]

Introduction of a single C → T transition at position 284 in PSTV cDNA creates a unique HindIII restriction endonuclease site that can be used to construct chimeric viroid cDNAs (see Figure 1). The corresponding C → U transition in PSTV RNA results in the replacement of a single G:C base-pair by a G:U base-pair, but this mutation seemed unlikely to significantly destabilize the native structure of PSTV. We were, therefore, initially surprised to find that neither tandem dimers of the mutant cDNA nor their in vitro RNA transcripts were infectious.[44] Position 284 lies within one of three ''premelting regions'' identified by Schnölzer et al.,[19] but assignment of a precise role for this region is not yet possible. Construction of pseudorevertants in which the base-pairing disrupted by the lethal mutation has been restored is discussed below.

A variety of small insertions and deletions have been introduced into infectious viroid cDNAs in order to determine the effect of altered thermodynamic stability on viability. Ishikawa et al.[50] used the presence of unique restriction endonuclease sites within the terminal loops of HSV to create a four-nucleotide insertion in the left terminal region and a four-nucleotide deletion in the right terminal region. Both alterations were lethal, and coinoculation of mutant cDNAs or their RNA transcripts produced no evidence for either DNA or RNA recombination in vivo.

Recent studies in our laboratory[69] have shown that single nucleotide substitutions and small insertions/deletions in several regions of PSTV also result in a loss of infectivity (see Figure 1 for precise locations). Introduction of either nucleotide substitutions/small insertions into the terminal loops or a single G → C substitution in the ''pathogenesis'' domain of PSTV was lethal. Because comparative sequence analysis had shown that sequence variation among PSTV isolates is largely confined to the pathogenesis region (see Figure 2), this region appeared particularly amenable to mutation. Suitably chosen multiple mutations in this region should, however, reveal how PSTV virulence is modulated.[19]

The dramatic and deleterious effect of apparently minor sequence alterations suggests the existence of previously unsuspected interactions within viroids, and necessitates a more

Viroid	Homology PSTV	Homology CEV	Hairpin I	Hairpin II	Box 9L • Box 2	Box 9R • Box 9R′
PSTV	100	55	^{79}CGCUUCAGG^{87} ::::::::: $_{110}$GCGAGGUCC$_{102}$	^{228}ACCCCUCGCCCC^{236} :::::::::: $_{330}$GUGGGAGCGGGG$_{319}$	^{257}UACUACCCGGUGGA^{270} :::: $_{356}$UUGGU	:::::: CCACUU$_{179}$
TPMV	76	60	^{81}CGCUUCAGG^{89} ::::::::: $_{112}$GCGAGGUCC$_{104}$	^{219}CACCCUCGCCCG^{230} ::::::::::: $_{328}$GUGGGAGCGGGG$_{317}$	^{254}AGACUACCCGGUGGA^{268} ::::: $_{358}$CCUGGU	:::::: CCACUU$_{178}$
TASV	64	73	^{82}UCCUUCAGG^{90} ::::::::: $_{113}$AGGAGGUCC$_{105}$	^{230}CCCUCGCCCGGAG^{242} ::::::::::::: $_{330}$GGGAGCGGGCCUC$_{318}$	^{254}AGACUACCCGGUGGA^{268} :::: $_{358}$CCUGGG	:::::: CCACUU$_{180}$
CEV	55	100	^{81}UCCUUCAGG^{89} ::::::::: $_{112}$AGGAGGUCC$_{104}$	^{239}CCCUCGCCCGGAG^{251} ::::::::::::: $_{339}$GGGAGCGGGCCUC$_{327}$	^{263}AGACUACCCGGUGGA^{277} :: ::: $_{363}$UUAGAU	:: ::: CUUCCU$_{187}$

FIGURE 3. Potential structural interactions within viroids. Three specific secondary interactions are illustrated. Values for overall sequence homology with PSTV and CEV have been taken from Keese and Symons;[21] Riesner[37] has reviewed formation of secondary hairpins I and II during thermal denaturation. The hypothetical Box 9L:Box 2 and Box 9R:Box 9R′ interactions are those discussed by Hadidi.[28]

sophisticated approach to studies of potential nucleotide interactions. As suggested by Schnölzer et al.,[19] it is possible that most insertions and deletions destabilize the viroid secondary structure unless corresponding deletions and insertions are also made. One such approach involves the construction of pseudorevertants, in which introduction of lethal single nucleotide substitutions is followed by the introduction of selected secondary mutations. The purpose of the secondary mutations is to restore certain of the potential base-pairings disrupted by the primary mutations. If these pseudorevertants are viable, considerable support is obtained for the functional importance of that particular potential interaction or conformation.

Hammond and Owens[69] have used this approach to assess the relative importance of secondary hairpin I and the native structure for PSTV viability. For example, the G → A (position 80) and C → U (position 109) substitutions in PSTV cDNA (see Figure 1) should have only minimal effects upon the native structure, but at least the position 80 G → A change should disrupt the stem of secondary hairpin I (Figure 3). The simultaneous presence of both substitutions should restore this secondary interaction. Neither cDNAs containing the individual mutations nor pseudorevertant containing both mutations was infectious, however, suggesting that the ability to form secondary hairpin I is not the only essential function which involves the upper portion of the central conserved region. When, however, a G → A substitution was introduced at position 76 in a noninfectious PSTV cDNA containing a C → T transition at positon 284, the resulting psuedorevertant regained its ability to replicate. The net effect of these two changes is to replace a G:C base-pair in the native structure with an A:U base-pair. Other applications of this strategy will undoubtedly appear.

C. Construction of Chimeric Viroid cDNAs

The various oligonucleotide-directed mutagenesis strategies discussed above are most appropriate for dissection of structure/function relationships suggested by previous comparative sequence analysis or physical-chemical studies. In essence, the experimental protocol asks the inoculated plant to either accept (replicate) or reject (fail to replicate) the altered viroid RNA sequence. Because mutations must be tested individually, extensive studies using site-specific mutagenesis are laborious. If, however, host plants are inoculated with viroid cDNAs that can give rise to a number of potentially viable RNA transcripts, individual infected plants might contain viroid clones whose sequence would vary from plant to plant. Characterization of these clones would identify nucleotide substitutions that do not destroy viability.

One method to offer the host such a choice is inoculation with chimeric viroid cDNAs, i.e., full-length or longer cDNAs constructed from fragments derived from either different viroids or viroid strains. Simple construction of such cDNA chimeras depends upon the presence of homologous restriction sites in individual viroid cDNAs. Owens et al.[44] have constructed infectious mixed cDNA dimers from full-length PSTV and TASV cDNAs with BamHI termini. Individual infected tomatoes contained either PSTV or TASV, but no evidence for either mixed infections or viroid chimeras could be obtained by nucleic acid hybridization analyses using full-length viroid cDNA probes. Such a result is consistent with the infectivity of the individual viroid cDNAs (see Table 2), the probable location of the site for processing of oligomeric viroid replication intermediates within the overlapping BamHI-SmaI site which joins the two full-length cDNA,[47] and the well-known phenomenon of viroid cross protection.[51] However, because every inoculated cell simultaneously receives a copy of both infectious viroid cDNAs, such mixed tandem cDNA dimers should be useful in future studies of the molecular basis of cross protection.

The unique recognition sequences for BamHI and HindIII within or near the central conserved regions of several viroids (including CEV and TASV) allow their full-length, infectious cDNAs to be divided almost precisely in half (see Figure 1). Visvader et al.[49] have suggested that the relative roles of the "pathogenicity" and "variable" domains in CEV symptom expression can be assessed by constructing full-length chimeric cDNAs containing BamHI-HindIII fragments derived from mild and severe CEV variants. Unfortunately, the same strategy cannot be used for PSTV because all known isolates lack the required HindIII recognition site, and creation of a HindIII site in PSTV cDNA by introducing a single C $\rightarrow$ T transition at position 284 abolished its infectivity.[44]

Owens et al.[44] have used BamHI-HindIII fragments from TASV and this mutant PSTV cDNA to construct chimeric viroid cDNAs that contain the thermodynamically less stable left half of one viroid and the more stable right half of the second viroid. Our goal in such studies was to develop a method to assess the importance of interactions not predicted by the native structure of viroids. Although the potential secondary structures of their central conserved regions and the ability to form a group I intron "core" sequence appear unaltered, sequence divergence outside the central conserved region of the PSTV-TASV chimeras would prevent formation of two of the nine base-pairs in the stem of secondary hairpin I and several base-pairs in the stem of secondary hairpin II (see Figure 3). When recombinant plasmid DNA was used as inoculum, these chimeric PSTV-TASV cDNAs were noninfectious.

Chimeric cDNA monomers can also be constructed from viroids more closely related than PSTV and TASV, and data summarized in Figure 3 suggest that certain of these constructions may well be infectious. For example, several of the potential structural interactions discussed above would not be disturbed in CEV-TASV or TPMV-PSTV chimeras. Because neither PSTV nor TPMV contain a HindIII site in the lower portion of their central conserved regions, alternative sites would have to be found for chimera construction.

Even chimeric cDNAs that are noninfectious when recombinant plasmid DNA is used as inoculum may eventually prove useful. Because sequence reversion at the RNA level can "rescue" apparently lethal mutations in PSTV cDNA when the infection is mediated by the

Ti plasmid of *Agrobacterium tumefaciens,*[44] it may be possible to isolate viable pseudo-revertant viroids from plants inoculated with *A. tumefaciens* strains whose Ti plasmids contain nonviable chimeric cDNAs. Such an approach would certainly be less laborious than construction and bioassay of multiple point mutations in the individual cDNAs, but an important limiting factor will be the number of individual mutations required to restore viability. At the present time, no data are available concerning the infectivity of recombinant Ti plasmids whose viroid cDNA inserts contain multiple point mutations.

D. Saturation Mutagenesis

Use of the mutagenesis strategies discussed above should lead to rapid refinements in our understanding of the molecular mechanisms involved in viroid replication and symptom expression. As the questions posed become more detailed, however, rapid and efficient methods for generating and isolating large numbers of random, single-base substitutions will be required. Maniatis and co-workers have recently described a series of protocols for random chemical mutagenesis of single-stranded DNA that allows the isolation and characterization of large numbers of such mutations in the absence of phenotypic selection.[52,53] In these methods, double-stranded DNA fragments containing base substitutions are physically separated from wild-type fragments by denaturing gradient gel electrophoresis. The procedures are applicable to DNA fragments containing 30 to 600 base-pairs, but fragments smaller than 200 base-pairs are usually chosen to facilitate DNA sequence analysis of the mutants obtained. For larger DNAs (e.g., viroid cDNAs containing 250 to 375 nucleotides), mutations can be mapped by pancreatic RNase cleavage at single-base mismatches present in RNA:DNA duplexes formed by annealing [^{32}P]-labeled RNA probe complementary to the wild-type sequence and unlabeled single-stranded mutant DNAs. The location of the mismatch is determined by gel electrophoretic analysis of the resulting RNA cleavage products. More than 50% of all possible single-base substitutions can be detected by this procedure.[53]

If the particular region of interest is sufficiently small and flanked by restriction sites, a second approach is also possible. Hutchison et al.[54] have shown that it is possible to control the number of mutations in synthetic DNA molecules by adjusting the concentration of "dopant" nucleoside phosphoramidate added to each of the four monomer reservoirs of the DNA synthesizer during automated synthesis of the target sequence. No matter what method is used to prepare libraries of point mutations in infectious viroid cDNAs, however, selection of appropriate assays to screen the altered DNAs for functional changes will be extremely important.

IV. EMERGING APPROACHES

A. Development of Assays for Specific Viroid Functions

Conventional bioassay techniques for cDNA infectivity are based upon the ability of viroid infections to rapidly become systemic and may be unable to differentiate mutations that are actually lethal from those that severely inhibit replication or cell-to-cell spread. The selective pressure in a systemic bioassay also favors the propagation of wild-type viroid generated by sequence reversion at the RNA level[44] rather than propagation of the presumably less-fit mutant. Bioassay systems where the Ti plasmid of *Agrobacterium tumefaciens* is used to mediate viroid infection[55] can provide large amounts of tissue in which viroid infection need not be systemic to be detectable. This characteristic will be advantageous in certain situations to be discussed later, but *Agrobacterium*-based assays also have certain potential disadvantages, i.e., the possible persistence of viroid-related RNAs transcribed from the integrated cDNA in the absence of viroid replication and the ability of sequence reversion at the RNA level to "rescue" apparently lethal mutations.[44] Quantitative in vivo and in vitro assay systems for specific viroid functions are required for the detailed analysis of viroid structure/function relationships by site-specific mutagenesis.

Specific assays for in vivo RNA splicing have been reported for the rRNA intron of *Tetrahymena thermophila*.[56,57] A DNA fragment containing the self-splicing intron sequence was cloned into plasmid or phage vectors in such a way that expression of the α-fragment of β-galactosidase is dependent upon intron excision from the mRNA. Using this approach, a simple color assay can be used to screen for splicing-deficient mutants generated by either random or directed mutagenesis techniques. Analysis of structure/function relationships in RNA enzymes without the influence of preexisting models is possible, and long-range tertiary interactions between nucleotides that cannot be predicted by secondary structure calculations can be identified.

Specific in vitro assays are also valuable for characterizing the effects of certain mutations. Hall and collaborators[58] have used the ability of mutant brome mosaic virus (BMV) RNAs synthesized in vitro to act as substrates for both tyrosyl-tRNA synthetase and BMV RNA replicase in order to study RNA structure/function relationships in the 3′-terminal region. A potential first step toward similar in vitro assays for individual viroid functions have been reported by Robertson et al.[59] They reported that dimeric PSTV RNA synthesized in vitro can undergo spontaneous cleavage to form full-length linear PSTV molecules. Unfortunately, subsequent results from other laboratories suggest that progress in this area may not be as rapid as one might hope.[70]

B. Viroid Replication in Cells Transformed by *Agrobacterium*

Gardner and co-workers[55,60] have shown that *A. tumefaciens* strains containing recombinant Ti plasmids can initiate viroid infection by transferring PSTV cDNAs to plant cells at wound sites on tomato stems. Subsequent PSTV accumulation in the leaves of inoculated plants indicates that such infections rapidly become systemic. The process utilizes the T-DNA transfer mechanisms of *Agrobacterium* because PSTV infection required four of the five virulence (*vir*) genes and at least one of the border sequences which flank the T-DNA.[60] Because this approach is proving quite useful for a number of different applications, important characteristics of the experimental system are summarized below.

In one set of experiments, Gardner and Knauf[60] used a binary vector system to introduce the viroid cDNA into *A. tumefaciens*. A wide-host-range plasmid containing a PSTV cDNA trimer between the left and right borders of the T-DNA was conjugated into either a virulent *A. tumefaciens* strain A722 containing an octopine-type Ti plasmid or into a nontumorigenic strain that contains a Ti plasmid with *vir* loci, but no T-DNA. Although only the wild-type strain caused tumors on tomato plants, plants inoculated with either strain showed typical symptoms of systemic PSTV infection after 10 to 12 days. Mutations in *vir* loci, A, B, C, and D, blocked the ability of *Agrobacterium* to promote infection with PSTV. Thus, genes corresponding to these loci must be required for transfer of T-DNA into the plant cell. Because the two *vir* E mutants tested gave rise to systemic PSTV infections, this locus appears nonessential for transfer of the PSTV-containing T-DNA into the plant.

The role of the T-DNA borders in the infection process was examined by inserting the PSTV cDNA-containing plasmid into wild-host-range plasmids containing different numbers and arrangements of borders. All plasmids with at least one border were infectious, and infectivity was independent of the location of the PSTV cDNA with respect to the borders. PSTV cDNA located outside the T-DNA borders could also be transferred to the plant, and the consistent appearance of PSTV symptoms showed that this assay is a sensitive indicator of DNA transfer. T-DNA transfer from nontumorigenic *Agrobacterium* strains must be limited to a small number of cells around the wound site. The reliability of the assay is indicated by the fact that all plants inoculated with a given strain develop symptoms within a few days of each other.

In a related series of experiments, the influence of the cauliflower mosaic virus (CaMV) 35 S RNA promotor on expression of PSTV cDNA was determined.[55] PSTV cDNAs with

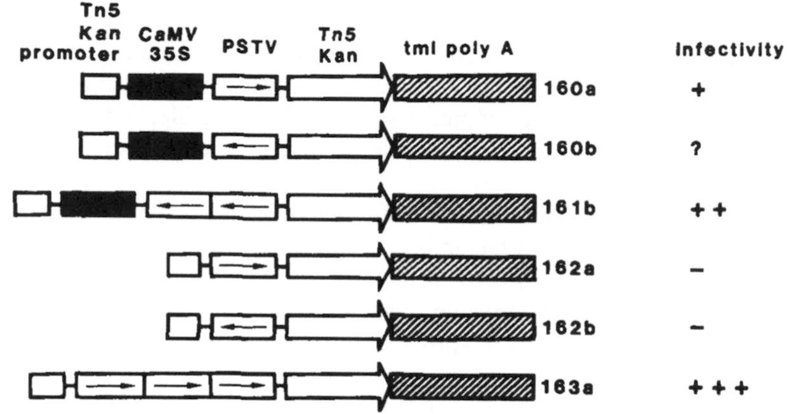

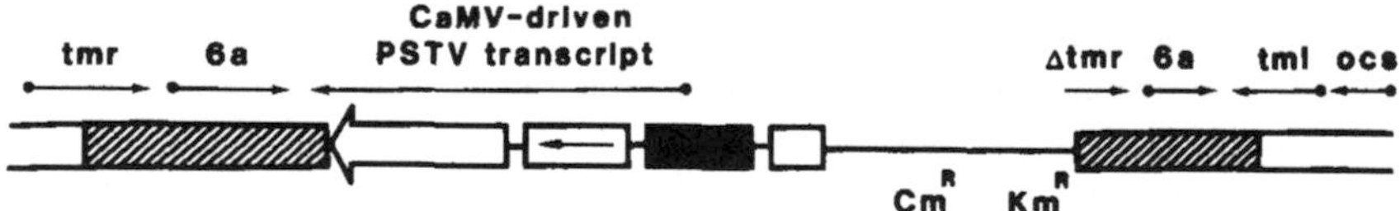

FIGURE 4. PSTV cDNA constructs in the Ti plasmid of *Agrobacterium tumefaciens*. (A) The structures of six PSTV cDNA-containing constructs are shown. The upper three constructs are insertions into pCGN149a, a vector which contains the CaMV 35 S promoter, while the lower three involve the promoterless vector pCGN157. The arrows within the PSTV cDNAs indicate the direction of transcription giving rise to PSTV (+) strand RNAs. (B) The structure of one of the constructs (pCGN160a) following recombination into the T-DNA of pTiA6 (not to scale). Transcripts predicted to occur in transformed cells are calculated from published nucleotide sequence data and restriction endonuclease maps (see Reference 54 for details). The duplicated segment of the T-DNA (shaded) corresponds to the fragment from pCGN157 containing nucleotides 9,062 to 11,207.

BamHI termini were inserted into the unique BglII sites of two *Agrobacterium* shuttle vectors and transferred into the T-DNA of the octopine plasmid pTiA6 of *Agrobacterium* strain A722 using a single crossover procedure and selection for kanamycin resistance. The constructions obtained are diagrammed in Figure 4. When the ability of these six constructs to give rise to systemic PSTV infection was tested by inducing gall formation on tomato plants, three were consistently infectious (Figure 4). Infectivity results were generally consistent with those obtained by inoculation of cotyledons with purified DNA[38,40,42] or RNA transcripts.[41,43,49] Furthermore, they provide genetic evidence that one mechanism of infection by viroid DNA involves excision of viroid RNA from a longer RNA polymerase II-derived transcript.

Preliminary results[55,61] suggest that this experimental system may be useful in the analysis of viroid replication. *A. tumefaciens* induces crown gall formation on a wide variety of dicotyledonous plants, only some of which are known for hosts for PSTV. Further analysis of tissue from galls or transformed plants might allow identification of the step at which replication is blocked in nonhost species, because they provide large numbers of transformed cells possibly containing RNA replicative intermediates. Similar analyses can be performed with nonviable mutants of PSTV introduced into a host that is permissive for replication. Together these approaches may help elucidate individual steps in the viroid infection cycle.

The fact that clone 160a (Figure 4) was infectious may have important implications for the mechanism by which multimeric viroid RNAs are cleaved and ligated to form circular progeny. RNA transcripts from this clone contain only a 4-nucleotide (GATC) sequence duplication rather than the 6- to 11-nucleotide duplication shown to be essential for infectivity when intact plasmid DNAs are used as inoculum (see Figure 2 and Table 2). It is not difficult to devise possible explanations for this apparent increase in specific infectivity when the Ti plasmid is used to deliver the viroid cDNA to susceptible cells, but additional experimentation will be required to identify the correct explanation.

C. Viroid Infections as Analytical Tools

Apparent similarities between viroid replication and the mRNA splicing machinery of eukaryotic cells have led to the proposal that viroids and RNA viruses may have originated from a novel class of eukaryotic RNA whose primary function is the intercellular exchange of genetic information.[62] The origin, amplification, and extracellular exchange of these hypothetical regulatory RNAs would be dependent upon intermolecular RNA recombination, a poorly characterized phenomenon first detected in studies of poliovirus[63] and foot-and-mouth disease virus mutants.[64] More recently, Bujarski and Kaesberg[65] have reported the first molecular evidence for genetic recombination between the RNA components of BMV, a tripartite plant virus.

Figure 5 indicates how noninfectious viroid cDNAs could be used to detect RNA recombination and assess its role in viroid evolution. Although cDNAs containing nucleotide insertions/deletions should be unable to serve as template for the synthesis of wild-type PSTV transcripts in vivo, intermolecular RNA recombination might be able to generate wild-type PSTV following *Agrobacterium*-mediated inoculation with a mixture of two defective cDNAs. The major difference between the pathways illustrated in Figures 5A and B is the relative timing of the intermolecular RNA cleavage/ligation (RNA recombination) reaction. Both direct RNA cleavage/ligation and discontinuous transcription provide plausible mechanisms for the RNA rearrangements diagrammed.[21]

This type of assay system to detect RNA recombination should be extremely sensitive because only the desired product is capable of autonomous replication, but generation of wild-type PSTV cDNA templates by DNA rearrangements in vivo[66,67] must be excluded by appropriate control experiments. If RNA recombination between viroids can be detected, its rate will probably exhibit a strong dependence upon both the distance between the individual mutations and their specific locations. In the case of RNA recombination involving brome mosaic virus RNAs 1 to 3, the mutant RNA 3 to be "repaired" was able to replicate at a reduced rate. It is not known whether or not RNA genomes need to replicate in order to undergo RNA recombination, but carefully chosen experiments should provide a great deal of useful information about the molecular mechanism of viroid replication.

Finally, Gardner and Knauf[60] have used an *Agrobacterium*-mediated assay for PSTV infection to subdivide the process of plant transformation by *A. tumefaciens* into two stages, i.e., DNA transfer from the bacterium to the host cell in the absence of *vir* E followed by integration of the T-DNA into the host chromosome. There are many potential applications for assays of this type which measure the transient expression of foreign DNAs in plant cells. *Agrobacterium* mutants that efficiently transfer DNA into plant cells, but do not permit its integration may provide a means to isolate autonomously replicating plant DNA sequences. A second potential application involves the construction of "suicide" vectors capable of introducing transposable genetic elements into heterologous cells. Similar applications have also been proposed by Grimsley et al.[68] for what they term "agroinfection" (*Agrobacterium*-mediated infections involving cloned DNA viral genomes).

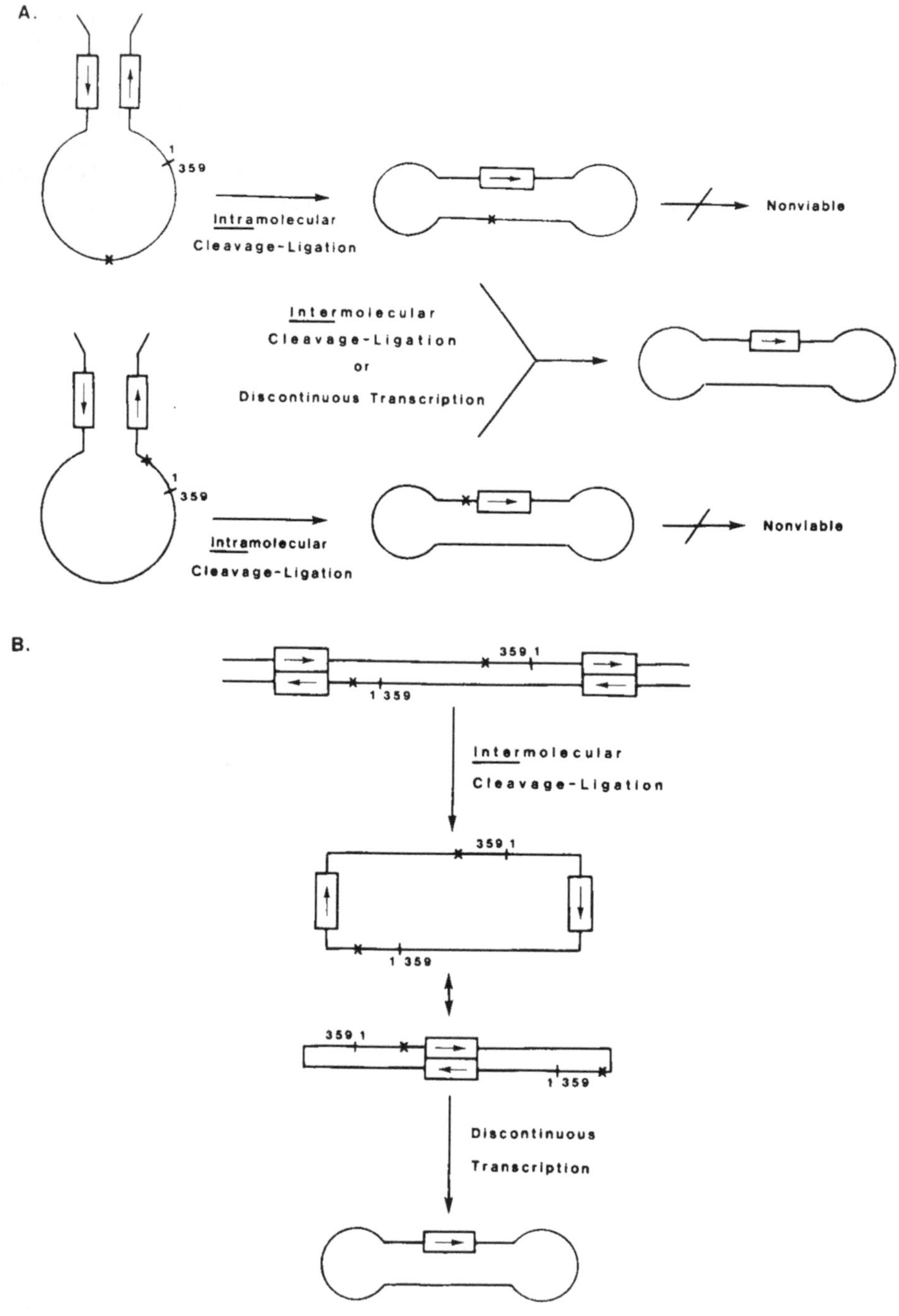

FIGURE 5. Viroid RNA transcripts as substrates for RNA recombination. Two partially overlapping pathways (A and B) are depicted for the generation of wild-type viroid progeny from two defective viroid RNA transcripts. X, location of inactivating mutations; boxed arrow, nucleotides forming the extremely stable, base-paired configuration which may be essential for precise cleavage and ligation of oligomeric viroid RNAs (see Figure 2).

ACKNOWLEDGMENTS

We thank Dr. T. O. Diener for continuing discussion of experimental results, Drs. T. Candresse and L. F. Salazar for permission to cite unpublished results, and Drs. J. M. Kaper and L. Vodkin for constructive criticism of this manuscript. R. W. H. has been supported by funds provided by U.S. Department of Agriculture Competitive Research Grants 81-CRCR-1-0719 and 85-CRCR-1-1738.

REFERENCES

1. **Diener, T. O.,** Biological properties, in *The Viroids*, Diener, T. O., Ed., Plenum Press, New York, 1987, 9.
2. **Diener, T. O.,** Potato spindle tuber "virus". IV. A replicating, low molecular weight RNA, *Virology*, 45, 411, 1971.
3. **Gross, H. J., Domdey, H., Lossow, C., Jank, P., Raba, M., Alberty, H., and Sänger, H. L.,** Nucleotide sequence and secondary structure of potato spindle tuber viroid, *Nature (London)*, 273, 203, 1978.
4. **Wild, U., Ramm, K., Sänger, H. L., and Riesner, D.,** Loops in viroids. Accessibility to tRNA anticodon binding, *Eur. J. Biochem.*, 103, 227, 1980.
5. **Henco, K., Sänger, H. L., and Riesner, D.,** Fine structure melting of viroids as studied by kinetic methods, *Nucleic Acids Res.*, 6, 3041, 1979.
6. **Riesner, D. and Gross, H. J.,** Viroids, in *Annu. Rev. Biochem.*, 54, 531, 1985.
7. **Symons, R. H.,** Avocado sunblotch viroid: primary sequence and proposed secondary structure, *Nucleic Acids Res.*, 9, 6527, 1981.
8. **Haseloff, J. and Symons, R. H.,** Chrysanthemum stunt viroid: primary sequence and secondary structure, *Nucleic Acids Res.*, 9, 2741, 1981.
9. **Gross, H. J., Krupp, G., Domdey, H., Raba, M., Alberty, H., Lossow, C. H., Ramm, K., and Sänger, H. L.,** Nucleotide sequence and secondary structure of citrus exocortis and chrysanthemum stunt viroid, *Eur. J. Biochem.*, 121, 249, 1982.
10. **Candresse, T.,** unpublished data.
11. **Visvader, J. E., Gould, A. R., Bruening, G. E., and Symons, R. H.,** Citrus exocortis viroid: nucleotide sequence and secondary structure of an Australian isolate, *FEBS Lett.*, 137, 288, 1982.
12. **Visvader, J. E. and Symons, R. H.,** Eleven new sequence variants of citrus exocortis viroid and the correlation of sequence with pathogenicity, *Nucleic Acids Res.*, 13, 2907, 1985.
13. **Haseloff, J., Mohamed, N. A., and Symons, R. H.,** Viroid RNA of cadang-cadang disease of coconuts, *Nature (London)*, 299, 316, 1982.
14. **Hammond, R. W.,** unpublished data.
15. **Ohno, T., Takamatsu, N., Meshi, T., and Okada, Y.,** Hop stunt viroid: molecular cloning and nucleotide sequence of the complete DNA copy, *Nucleic Acids Res.*, 11, 509, 1983.
16. **Sano, T., Uyeda, I., Shikata, E., Ohno, T., and Okada, Y.,** Nucleotide sequence of cucumber pale fruit viroid: homology to hop stunt viroid, *Nucleic Acids Res.*, 12, 3427, 1984.
17. **Sano, T., Ohshima, K., Uyeda, I., Shikata, E., Meshi, T., and Okada, Y.,** Nucleotide sequence of grapevine viroid: a grapevine isolate of hopstunt viroid, *Proc. Jpn. Acad.*, 61(B), 265, 1985.
18. **Gross, H. J., Liebl, U., Alberty, H., Krupp, G., Domdey, H., Ramm, K., and Sänger, H. L.,** A severe and mild potato spindle tuber viroid isolate differ in 3 nucleotide exchanges only, *Biosci. Rep.*, 1, 235, 1981.
19. **Schnölzer, M., Haas, B., Ramm, K., Hofmann, H.,and Sänger, H. L.,** Correlation between structure and pathogenicity of potato spindle tuber viroid (PSTV), *EMBO J.*, 4, 2181, 1985.
20. **Kiefer, M. C., Owens, R. A., and Diener, T. O.,** Structural similarities between viroids and transposable genetic elements, *Proc. Natl. Acad. Sci. U.S.A.*, 80, 6234, 1983.
21. **Keese, P. and Symons, R. H.,** Domains in viroids: evidence of intermolecular RNA rearrangements and their contribution to viroid evolution, *Proc. Natl. Acad. Sci. U.S.A.*, 82, 4582, 1985.
22. **Diener, T. O.,** *Viroids and Viroid Diseases*, Wiley-Interscience, New York, 1979.
23. **Sänger, H. L.,** Biology, structure, functions and possible origin of viroids, in *Encyclopedia of Plant Physiology*, New Series, Vol 14B, Parthier, B. and Boulter, D., Eds., Springer-Verlag, Berlin, 1982, 368.
24. **Wolff, P., Gilz, R., Schumacher, J., and Riesner, D.,** Complexes of viroids with histones and other proteins, *Nucleic Acids Res.*, 13, 355, 1985.

25. **Sänger, H. L.,** Viroid replication, in *The Viroids,* Diener, T. O., Ed., Plenum Press, New York, 1987, 117.
26. **Bruening, G., Gould, A. R., Murphy, P. J., and Symons, R. H.,** Oligomers of avocado sunblotch viroid are found in infected avocado leaves, *FEBS Lett.,* 148, 71, 1982.
27. **Dinter-Gottlieb, G.,** Viroids and virusoids are related to group I introns, in *The Viroids,* Diener, T. O., Ed., Plenum Press, New York, 1987, 189.
28. **Hadidi, A.,** Relationship of viroids and certain other plant pathogenic nucleic acids to group I and II introns, *Plant Mol. Biol.,* 7, 129, 1986.
29. **Davies, J. W., Kaesberg, P., and Diener, T. O.,** Potato spindle tuber viroid. XII. An investigation of viroid RNA as a messenger for protein synthesis, *Virology,* 61, 281, 1974.
30. **Hall, T. C., Wepprich, R. K., Davies, J. W., Weathers, L. G., and Semancik, J. S.,** Functional distinctions between the ribonucleic acids from citrus exocortis viroid and plant viruses: cell-free translation and aminoacylation reactions, *Virology,* 61, 486, 1974.
31. **Matthews, R. E. F.,** Are viroids negative-strand viruses?, *Nature (London),* 276, 850, 1978.
32. **Sherman, F., McKnight, G., and Stewart, J. W.,** AUG is the only initiation codon in eukaryotes, *Biochem. Biophys. Acta,* 609, 343, 1980.
33. **Zaitlin, M. and Hariharasubramanian, V.,** A gel electrophoretic analysis of proteins from plants infected with tobacco mosaic and potato spindle tuber viruses, *Virology,* 47, 296, 1972.
34. **Conejero, V. and Semancik, J. S.,** Exocortis viroid: alteration in the proteins of *Gynura aurantiaca* accompanying viroid infection, *Virology,* 77, 221, 1977.
35. **Flores, R., Chroboczek, J., and Semancik, J. S.,** Some properties of the CEV-P_1 protein from citrus exocortis viroid-infected *Gynura aurantiaca* DC, *Physiol. Plant Pathol.,* 13, 193, 1978.
36. **Conejero, V., Picazo, I., and Segado, P.,** Citrus exocortis viroid (CEV): protein alterations in different hosts following viroid infection, *Virology,* 97, 454, 1979.
37. **Riesner, D.,** Structure formation, in *The Viroids,* Diener, T. O., Ed., Plenum Press, New York, 1987, 63.
38. **Cress, D. E., Kiefer, M. C., and Owens, R. A.,** Construction of infectious potato spindle tuber viroid cDNA clones, *Nucleic Acids Res.,* 11, 6821, 1983.
39. **Ohno, T., Ishikawa, M., Takamatsu, N., Meshi, T., Okada, Y., Sano, T., and Shikata, E.,** *In vitro* synthesis of infectious RNA molecules from cloned hop stunt viroid complementary DNA, *Proc. Jpn. Acad.,* 59B, 251, 1983.
40. **Tabler, M. and Sänger, H. L.,** Cloned single- and double-stranded DNA copies of potato spindle tuber viroid (PSTV) RNA and co-inoculated subgenomic DNA fragments are infectious, *EMBO J.,* 3, 3055, 1984.
41. **Tabler, M. and Sänger, H. L.,** Infectivity studies on different potato spindle tuber viroid (PSTV) RNAs synthesized *in vitro* with the SP6 transcription system, *EMBO J.,* 4, 2191, 1985.
42. **Meshi, T., Ishikawa, M., Ohno, T., Okada, Y., Sano, T., Ueda, I., and Shikata, E.,** Double-stranded cDNAs of hop stunt viroid are infectious, *J. Biochem.,* 95, 1521, 1984.
43. **Ishikawa, M., Meshi, T., Ohno, T., Okada, Y., Sano, T., Ueda, I., and Shikata, M.,** A revised replication cycle for viroids. The role of longer than unit length RNA in viroid replication, *Mol. Gen. Genet.,* 196, 421, 1984.
44. **Owens, R. A., Hammond, R. W., Gardner, R. C., Kiefer, M. C., Thompson, S. M., and Cress, D. E.,** Site-specific mutagenesis of potato spindle tuber viroid cDNA: alterations within premelting region 2 that abolish infectivity, *Plant Mol. Biol.,* 6, 179, 1986.
45. **Owens, R. A., Erbe, E., Hadidi, A., Steere, R. L., and Diener, T. O.,** Separation and infectivity of circular and linear forms of potato spindle tuber viroid, *Proc. Natl. Acad. Sci. U.S.A.,* 74, 3859, 1977.
46. **Hashimoto, J., Suzuki, K., and Uchida, T.,** Infectivity of artificially nicked viroid molecules, *J. Gen. Virol.,* 66, 1545, 1985.
47. **Diener, T. O.,** Viroid processing: a model involving the central conserved region and hairpin I, *Proc. Natl. Acad. Sci. U.S.A.,* 83, 58, 1986.
48. **Meshi, T., Ishikawa, M., Watanabe, Y., Yamaya, J., Okada, Y., Sano, T., and Shikata, E.,** The sequence necessary for the infectivity of hop stunt viroid cDNA clones, *Mol. Gen. Genet.,* 200, 199, 1985.
49. **Visvader, J. E., Forster, A. C., and Symons, R. H.,** Infectivity and *in vitro* mutagenesis of monomeric cDNA clones of citrus exocortis viroid indicates the site of processing of viroid precursors, *Nucleic Acids Res.,* 13, 5843, 1985.
50. **Ishikawa, M., Meshi, T., Okada, Y., Sano, T., and Shikata, E.,** *In vitro* mutagenesis of infectious viroid cDNA clone, *J. Biochem.,* 98, 1615, 1985.
51. **Niblett, C. L., Dickson, E., Fernow, K. H., Horst, R. K., and Zaitlin, M.,** Cross protection among four viroids, *Virology,* 91, 198, 1978.
52. **Myers, R. M., Lerman, L. S., and Maniatis, T.,** A general method for saturation mutagenesis of cloned DNA fragments, *Science,* 229, 242, 1985.

53. **Myers, R. M., Larin, Z., and Maniatis, T.,** Detection of single base substitutions by ribonuclease cleavage at mismatches in RNA:DNA duplexes, *Science,* 230, 1242, 1985.
54. **Hutchison, C. A., III, Nordeen, S. K., Vogt, K., and Edgell, M. H.,** A complete library of point substitution mutations in the glucocorticoid response element of mouse mammary tumor virus, *Proc. Natl. Acad. Sci. U.S.A.,* 83, 710, 1986.
55. **Gardner, R. C., Chonoles, K., and Owens, R. A.,** Potato spindle tuber viroid infections mediated by the Ti plasmid of *Agrobacterium tumefaciens, Plant Mol. Biol.,* 6, 221, 1986.
56. **Waring, R. B., Ray, J. A., Edwards, S. W., Scazzocchio, C., and Davies, R. W.,** The Tetrahymena rRNA intron self-splices in *E. coli: in vivo* evidence for the importance of key base-paired regions of RNA for RNA enzyme function, *Cell,* 40, 371, 1985.
57. **Price, J. V. and Cech, T. R.,** Coupling of Tetrahymena ribosomal RNA splicing to b-galactosidase expression in *Escherichia coli, Science,* 228, 719, 1985.
58. **Dreher, T. W., Bujarski, J. J., and Hall, T. C.,** Mutant viral RNAs synthesized *in vitro* show altered aminoacylation and replicase activities, *Nature (London),* 311, 171, 1984.
59. **Robertson, H. D., Rosen, D. L., and Branch, A. D.,** Cell free synthesis and processing of an infectious dimeric transcript of potato spindle tuber viroid RNA, *Virology,* 142, 441, 1985.
60. **Gardner, R. C. and Knauf, V. C.,** Transfer of Agrobacterium DNA to plants requires a T-DNA border but not the virE locus, *Science,* 231, 725, 1986.
61. **Salazar, L.,** unpublished results.
62. **Zimmern, D.,** Do viroids and RNA viruses derive from a system that exchanges genetic information between eukaryotic cells?, *Trends Biochem. Sci.,* 7, 205, 1982.
63. **Cooper, P. D.,** Genetics of picornaviruses, in *Comprehensive Virology,* Vol. 9, Fraenkel-Conrat, H. and Wagner, R. R., Eds., Plenum Press, New York, 1977, 133.
64. **King, A. M. Q., McCahon, D., Slade, W. R., and Newman, J. W. I.,** Recombination in RNA, *Cell,* 29, 921, 1982.
65. **Bujarski, J. J. and Kaesberg, P.,** Genetic recombination between RNA components of a multipartite plant virus, *Nature (London),* 321, 528, 1986.
66. **Calos, M. P., Lebkowski, J. S., and Botchan, M. R.,** High mutation frequency in DNA transfected into mammalian cells, *Proc. Natl. Acad. Sci. U.S.A.,* 80, 3015, 1983.
67. **Dixon, L. K. and Hohn, T.,** Cloning and manipulating cauliflower mosaic virus, in *Recombinant DNA Research and Viruses,* Becker, Y., Ed., Martinus Nijhoff, Netherlands, 1985, 247.
68. **Grimsley, N., Hohn, B., Hohn, T., and Walden, R.,** "Agroinfection", an alternative route for viral infection of plants by using the Ti plasmid, *Proc. Natl. Acad. Sci. U.S.A.,* 83, 3282, 1986.
69. **Hammond, R. W. and Owens, R. A.,** Mutational analysis of potato spindle tuber viroid reveals complex relationships between structure and infectivity, *Proc. Natl. Acad. Sci. U.S.A.,* 84, 3967, 1987.
70. **Tsagris, M., Tabler, M., and Sänger, H. L.,** Oligomeric potato spindle tuber viroid (PSTV) RNA does not process autocatalytically under conditions where other RNAs do, *Virology,* 157, 227, 1987.

Chapter 6

REPLICATION OF SMALL SATELLITE RNAs AND VIROIDS: POSSIBLE PARTICIPATION OF NONENZYMIC REACTIONS

George Bruening, Jamal M., Buzayan, Arnold Hampel, and Wayne L. Gerlach

TABLE OF CONTENTS

I. Introduction 128

II. Autolytic Reactions of STobRV RNA 129
- A. Autolytic Processing of STobRV (+) RNA 129
- B. Autolytic Processing of STobRV (−) RNA 129

III. Sobemovirus Satellite RNAs 131

IV. Small Satellite RNA of Cucumber Mosaic Virus 132

V. Nonenzymic Ligation Reactions of STobRV RNA 133
- A. Ligation of STobRV (−)RNA 133
- B. The Bond Formed in Ligated STobRV(−) RNA 133
- C. Ligation of STobRV (+)RNA 138

VI. A Model for the Replication of STobRV RNA 139

VII. Enzymic Ligation of Sobemovirus Satellite RNAs and Other RNAs 140

VIII. RNA Processing in Viroid Replication 141

Acknowledgments 142

References 142

I. INTRODUCTION

The subjects of this review may be considered to be "RNA-level" infectious agents, RNAs that replicate in plants, but apparently neither engender any encoded protein nor appear in any DNA form during such replication. The agents are viroids and certain small satellite RNAs, such as those of the nepoviruses[1-3] and of sobemoviruses.[4] Each of these has fewer than 400 nucleotide residues in the most abundant form. Each occurs partly or even predominantly as circles. The small satellite RNA of cucumber mosaic virus is similar to the other satellite RNAs, but may encode a small polypeptide and has not been detected in circular form (Section IV and Chapter 10, Volume III).

Viroids replicate in inoculated plants that are free of any other known infectious agent[5] and, after increase, appear in extracts of infected plants in an unencapsidated form[5,6] (also see Owens and Hammond, Volume II, and Keese, Visvader and Symons, Volume III). Diener[7] demonstrated that the infectivity associated with extracts of tomato plants infected with the potato spindle tuber agent is the effect of a specific, unencapsidated RNA species, the potato spindle tuber viroid (PSTV). Tissues that are infected with PSTV or other viroids contain RNAs that are complementary to the abundant, relatively easily isolated form of the viroid,[8] but viroid sequences are not found in the DNA of such tissues. There is no indication that either PSTV or its complementary RNA serves as a messenger RNA (reviewed by Diener[9] and by Symons et al.[10]).

In contrast to viroids, small satellite RNAs[4,11-13] (also see Kaper and Collmer, Volume III) of plant viruses exhibit *dependence*. They replicate extensively and become encapsidated only in mixed infections in which the host plant has been inoculated with both the satellite RNA and any of set of specific strains of a plant virus, which serves as the supporting virus. The satellite RNA becomes encapsidated in virus-like particles, protected by the coat protein of the supporting virus. The satellite RNA may be 80 to 95%, or even more, of the encapsidated RNA in some infections (see examples by Schneider,[1] Randles et al.,[14] and Gould et al.[15]). However, the phenomenon of dependence dictates that in every such mixed infection some of the encapsidated RNA will be virus genomic RNA.

The close functional association of the small satellite RNA and the genomic RNA(s) of the supporting virus is not reflected in a corresponding degree of nucleotide sequence relationship: there is *no extensive nucleotide sequence homology* between satellite RNA and supporting virus genomic RNAs. It is this lack of nucleotide sequence homology that distinguishes plant virus satellite RNAs from the RNAs of the defective-interfering particles that have been found in association with certain animal viruses.[16] The nucleotide sequence of a defective-interfering RNA is derived entirely or almost entirely[17] from the genomic RNA of the virus upon which it depends for replication, by a series of deletions and rearrangements of the helper virus RNA. Recently, Hillman et al.[18] and Simon and Howell[19] have reported small "satellite-like" RNAs of plant viruses that *do* have extensive homology with the supporting virus.

The 359-nucleotide residue satellite RNA of tobacco ringspot virus (STobRV RNA) has been studied in this laboratory and is the satellite RNA about which we here present the most detailed information. STobRV RNA propagates in the presence of tobacco ringspot virus (TobRV). TobRV has two genomic RNAs, separately encapsidated. These have a 3′-polyadenylate sequence and a 5′-linked protein, designated VPg. Neither structure is characteristic of STobRV RNA.[20,21] Nevertheless, STobRV RNA acts effectively as a parasite of TobRV by reducing its accumulation and ameliorating the symptoms that TobRV alone induces.[1] There is no evidence for a translation product of STobRV RNA[21,22] nor for transcription of STobRV RNA sequences into DNA in the infected plant. Thus STobRV RNA effects control over TobRV without messenger activity or apparent structural similarity to the RNAs of its supporting virus.

That small RNAs can replicate without the participation of DNA templates and without specifying a protein implies the existence of multiple functions for the RNA. Observations on nonreplicating RNAs have revealed a considerable repertoire. Guerrier-Takada et al.[23] discovered a naturally occurring, truly catalytic RNA molecule, the active subunit of ribonuclease P of *Escherichia coli*. Krueger et al.[24] found that the splicing of an intron from the precursor to the 26 S ribosomal RNA of a protozoan is accomplished in vitro entirely by the RNA itself, without enzymic intervention. These and other observations have greatly increased the interest in the chemistry of RNA molecules, including the replicating RNAs reviewed here.

II. AUTOLYTIC REACTIONS OF STobRV RNA

Almost no information about the replication of STobRV RNA is available from in vivo experiments. We describe some chemical reactions of STobRV RNA and related RNAs that have been observed in vitro. We postulate that these unusual reactions have a role in replication of STobRV RNA.

The dissimilar nucleotide sequence and end groups of STobRV RNA and the genomic RNAs of TobRV raises the possibility of distinct replication mechanisms for the TobRV RNAs and STobRV RNA. Tissue infected with STobRV RNA and TobRV accumulates multimeric forms of STobRV RNA of both polarities.[20,21] Extracts of such tissues also have circular STobRV RNAs of the same polarity as those found encapsidated in TobRV protein,[25] arbitrarily specified as the (+) polarity. We know of no evidence for either circular or multimeric forms of the TobRV genomic RNAs, supporting the notion of distinct replication mechanisms for the satellite and genomic RNAs. In addition to the bulk of monomeric STobRV (+)RNA, virus-like particles also have small amounts of multimeric STob RV (+)RNA, with diminishing quantities of each succeedingly larger multimer.[20] No circular STobRV RNA was recovered from virus-like particles.[25]

A. Autolytic Processing of STobRV (+)RNA

The dimeric STobRV RNA from virus-like particles has a nucleotide sequence that is a simple, direct repeat of the monomeric STobRV RNA sequence. The bond between monomeric units does not appear to be branched or unusual in any way except for its lability, since reverse transcriptase is able to use dimeric STobRV (+)RNA from virus-like particles as a template.[26-28] Presumably the dimers and higher-order multimeric forms from virus-like particles are composed of direct repeats of the STobRV RNA monomeric form, connected by ordinary 3′-to-5′ phosphodiester bonds. It is unlikely that the dimeric and higher-order STobRV (+)RNAs in preparations from virus-like particles could have formed by nonenzymic ligation (see Section V), as the RNA was isolated and maintained[20] under conditions that are not favorable to such reactions.

Multimeric forms of STobRV (+)RNA might be the precursors of the monomeric form or, via ligation reactions, the products of the monomeric form. The precursor hypothesis predicts a processing reaction in which cleavage of the multimers will occur at a specific bond (termed a junction) within a characteristic nucleotide sequence to generate RNA molecules of precise monomeric length and with the characteristic end groups of monomeric STobRV (+)RNA. The initial attempts to obtain purified dimeric RNA from virus-like particles, to serve as a substrate for a presumed enzyme-catalyzed processing reaction, were frustrated by the instability of the dimeric RNA. Electrophoretically purified dimeric RNA preparations gradually accumulated RNA with the mobility of monomeric form, implying nonenzymic, autolytic processing.

The in vitro production of monomeric STobRV (+)RNA in preparations of dimeric RNA was not mimicked by limited nuclease digestions of the dimeric RNA and was not inhibited by proteinases or detergents, separately or together.[29,30] That is, the reaction occurred under conditions which a folded protein molecule is unlikely to survive. The only reaction requirements are a buffered solution in the neutral pH range and any of several divalent or trivalent cations. The production of monomeric STobRV RNA from dimeric RNA in vitro thus is an autolytic processing reaction with likely in vivo significance: the cleavage is at a single bond, and the monomeric STobRV (+) RNA product has the terminal groups and biological activity that are characteristic of STobRV RNA from virus-like particles.[21,30]

Cleavage at a specific site within a specific sequence, i.e., at an autolytic junction, is an important criterion for distinguishing authentic autolytic processing from less precise nuclease-catalyzed, or chemical, degradation of an RNA molecule. Cleavage at a junction is indicated by detection of products that have a single pair of new end groups and are of specific sizes. Less specific cleavage, possibly guided to a general region of the RNA molecule by its conformation, would not be expected to generate only one detectable new end group for each product RNA.

cDNA transcribed from STobRV (+)RNA was cloned in a circularly permuted, dimeric form. The RNA transcript of such a cloned sequence, formed by the action of bacteriophage SP6 RNA polymerase, has two junction sites. Autolytic processing occurred at both of these junctions[26] to generate biologically active, monomeric STobRV RNA. The autolytic processing activity of the circularly permuted, dimeric RNA shows that no (tightly bound) protein from the virus-like particles or from the host plant are necessary for processing. Apparently the RNA itself is the only macromolecule required for the autolytic cleavage at a specific CpA bond, a reaction that generates a new 5′-terminal adenosine-5′-hydroxyl group and a new 3′-terminal cytidine-2′:3′-cyclic phosphodiester. In fact, less than one fourth of the STobRV (+)RNA sequence is sufficient for the autolytic processing reaction.[31]

An autolytic processing reaction that is at least superficially similar to that of STobRV (+)RNA is exhibited by the bacteriophage T4 RNA p2Sp1. Although the conditions for efficient processing of p2Sp1 RNA are different from those that are most suitable for STobRV RNA, the bond that is cleaved is in the sequence CpA, and the products have a 5′-hydroxyl group and are 3′ phosphorylated.[32]

B. Autolytic Processing of STobRV (−)RNA

When the circularly-permuted, dimeric STobRV RNA cDNA sequence was inserted in reverse orientation relative to the SP6 promoter, bacteriophage SP6 polymerase generated, in vitro, a new autolytically processing transcript, a transcript of the (−) polarity (Figure 1B). Processing occurred at two, identical ApG bonds (Figure 1A and B), thus forming three RNAs: the monomeric STobRV (−)RNA (M) and two flanking fragments. One of the flanking fragments, RNA P, is proximal to the SP6 promoter and the other, RNA D, is distal to the promoter. The *only* new end groups detected were a 5′-terminal guanosine-5′-hydroxyl of RNA M and RNA D and a 3′-terminal adenosine-2′:3′-cyclic phosphodiester of RNA P and RNA M. Processing was not dependent upon the proteins that are present in the SP6 transcription reaction mixture. Residual transcript, designated P-M-D, with hyphens representing junctions, as well as partial processing products P-M and M-D, were purified and incubated in a buffered solution of magnesium ions and spermidine; each generated the expected processing products.

The cleaved ApG sequence corresponds to residues 49 and 48 of the STobRV (+)RNA sequence (Figure 1A and Reference 33). Thus, the cleavage site for STobRV (+)RNA, between residues 359 and 1, and the site for STobRV (−)RNA lie 48 residues apart in the STobRV RNA sequence, being separated by almost one seventh of the sequence. A permuted monomeric STobRV (−)RNA, with a single junction site, also processed autolytically (Figure 1C). An implication of this observation is that the linear, double-stranded RNA that was isolated from infected tissue and was found to have multimeric STobRV RNAs of both

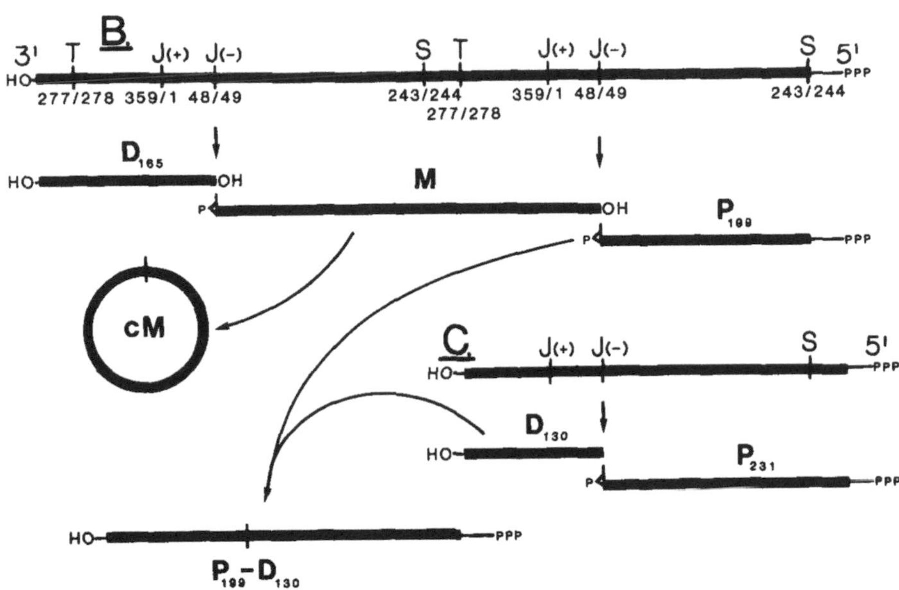

FIGURE 1. Autolysis and ligation reactions of STobRV (−)RNA. Sequences and diagrams of (−)RNA are shown in the 3′-to-5′, left-to-right, orientation to emphasize their identity as (−) polarity sequences. Nucleotide residue numbering, based on the STobRV (+)RNA sequence, increases from left to right. (A) The top line displays the nucleotide sequences in the region of the autolytic cleavage junction, J(−), located between residues 49 and 48 of STobRV(−)RNA.[33] The second line shows the nucleotide sequence of an oligodeoxyribonucleotide that is designated d35-68)(+). (B) A diagram of the transcript of SmaI-linearized plasmid pSP641,[33] which has a circularly permuted dimeric STobRV (−)RNA sequence. The two J(−) junctions separate the central unpermuted monomeric RNA sequence from two bordering sequences, P and D. J(+) is the corresponding junction of STobRV (+)RNA, between residues 1 and 359. Sites in the template for restriction endonucleases Sau3AI (S, between residues 244 and 243) and TaqI (T, between residues 278 and 277) are indicated. Staight arrows show autolytic cleavage reactions that generate monomeric STobRV(−)RNA and the two bordering RNA fragments, D_{165} and P_{199}. The subscripts indicate the number of STobRV RNA-derived nucleotide residues in the sequence of the RNA. The terminal triphosphate (ppp), cyclic phosphodiester (>p) and hydroxyl groups are marked. (C) The primary transcript of BamHI-linearized plasmid P231J130, that has a circularly permuted monomeric STobRV RNA insert. The single J(−) site generates two RNA fragments, P_{231} and D_{130}. Curved arrows define ligation reactions that were used to produce RNAs cM and P_{199}-D_{130}, for analysis of the junction sequences.

polarities[20,34] should have staggered rather than blunt ends. This possibility remains to be tested.

Although autolytic processing of the circularly permuted, dimeric STobRV (−)RNA transcript precisely generated monomeric STobRV (−)RNA, such RNA was not biologically active,[26] failing to give rise to STobRV RNA when inoculated to plants that were infected by TobRV. The double-stranded DNA clones of the circularly permuted, dimeric STobRV RNA were biologically active, although the corresponding single-stranded DNA of either polarity was not.

III. SOBEMOVIRUS SATELLITE RNAs

The sobemoviruses have a single genomic RNA, designated RNA 1. However, isolates of the sobemoviruses lucerne transient streak virus (LTSV), *Solanum nodiflorum* mottle virus (SNMV), subterraneum clover mottle virus (SCMoV), and velvet tobacco mottle virus (VTMoV) each are associated with RNAs that are less than one tenth the size of the genomic RNA and are designated RNA 2 and RNA 3. RNA 2 is circular. The less abundant RNA 3 is the linear form of RNA 2. These RNAs have from 327 to 388 nucleotide residues and often are referred to as "virusoids" because of their physical resemblance to viroids.[4] The sobemovirus satellite RNAs were discovered in field isolates of the corresponding sobemoviruses. In contrast, STobRV RNA[1] has thus far been detected only in association with laboratory isolates of TobRV. The satellite nature of RNA 2 is shown by its inability to replicate when inoculated alone and its ability to replicate efficiently when coinoculated with an RNA 1 from heterologous sobemovirus.[35]

The sobemovirus satellite RNAs produce multimeric forms of both polarities in infected tissue,[10,36] and transcripts of both polarities from cDNA clones of LTSV RNA 2 exhibited autolytic processing.[36a] Although the nepoviruses and the sobemoviruses are distinct virus groups, and their members have only a few features in common, RNA 2 of SNMV and RNA 2 of VTMoV show extensive nucleotide sequence homology with STobRV (+)RNA in the region of the autolytic cleavage site.[10,21,37] Lesser but presumably significant homology exists between this region of STobRV (+)RNA and the RNA 2 of LTSV and of SCMoV.[10,37]

These homologies indicate a possible autolytic processing site in each of the sequences of the sobemovirus satellite RNAs. The circular sobemovirus satellite RNAs have, at the dinucleoside phosphate sequence at which autolytic processing is predicted to occur, a 2′-phosphate group[28] (see Section VII). This structure is anticipated at the junction if the precursor of the circular RNA 2 is a linear RNA with a 2′:3′-cyclic phosphodiester and the circularization reaction is catalyzed by a plant RNA ligase. The linear monomeric precursor to RNA 2 is, in turn, presumed to be derived from a multimeric, linear precursor.

IV. SMALL SATELLITE RNA OF CUCUMBER MOSAIC VIRUS

An extensively studied (reviewed by Kaper and Tousingant[12] and by Francki;[11] also see Kaper and Collmer, Volume III) small satellite RNA not discussed so far is the satellite RNA of cucumber mosaic virus (CMV). CMV is a member of the cucumovirus group, the members of which share fewer properties with the nepoviruses and sobemoviruses than members of these two groups share with each other. Some satellite RNAs of CMV have been designated cucumber mosaic virus-associated RNA 5, or CARNA 5. CMV satellite RNA has very limited nucleotide sequence homologies with CMV genomic RNA[38] and the complement of CMV genomic RNA.[39] The encapsidated CMV satellite RNAs, like the CMV genomic RNAs, have a 5′ cap structure and a 3′ hydroxyl group.[40,41] It is possible that some CMV satellite RNAs code for small peptides.[11,42]

Although multimeric forms of the CMV satellite RNAs have been detected in infected tissues and in virus-like particles,[43,44] circular forms of the satellite were not detected.[25] Tobacco plants were transformed with a circularly permuted, slightly greater-than-dimer cDNA sequence of a CMV satellite RNA. Transcripts detected in the regenerated plants were not processed.[44] However, upon infection by CMV, abundant quantities of monomeric and lesser amounts of dimeric CMV satellite RNA accumulated and continued to replicate upon transfer with CMV to other hosts.

The similarities in terminal residues of the genomic and satellite RNAs and the lack of circular forms of the CMV satellite RNAs are consistent with replication of the satellite and genomic RNAs by similar mechanisms. The presence of multimeric forms, and the ability of multimeric forms in transformed plants to initiate the replication of CMV satellite RNA are consistent with distinct mechanisms for the replication of the satellite and genomic RNAs. As suggested by Linthorst and Kaper,[43] it is conceivable that CMV satellite RNAs replicate by more than one mechanism.

V. NONENZYMIC LIGATION REACTIONS OF STobRV RNA

The autolysis of (−) polarity RNA transcribed from a circularly permuted, dimeric clone of the STobRV RNA sequence occurs during transcription, but the autolysis is incomplete. Thus, upon analysis by gel electrophoresis, the reaction mixtures generated six zones through the action of enzymic transcription and nonenzymic processing: P-M-D, P-M, M-D, P, M, and D.[33] However, the analyses also showed two other species, one of them minor in abundance.

A. Ligation of STobRV (−)RNA

The minor component ultimately was identified as P-D and the more prominent, as cM. P-D and cM are derived from two reactions — respectively, the joining of two flanking STobRV (−)RNA fragments, P and D, to form P-D, and the circularization of monomeric STobRV (−)RNA to form RNA species cM.[33] The reactions demonstrate the reversibility of the autolytic cleavage of STobRV (−)RNA under the conditions of the transcription reaction. They are diagrammed in Figure 1B and C.

The formation of P-D does not require the entire STobRV RNA sequence since, as indicated in Figure 1, an RNA P and an RNA D from different constructions will join to produce a P-D that lacks 30 nucleotide residues from the central portion of the STobRV RNA sequence. Circularization of M is an efficient reaction. When M was incubated in a buffered solution of 6 m*M* magnesium ions and 2 m*M* spermidine for 1 hr at 37°C, cM amounted to about 40% of the product, with trace amounts of linear dimer also being formed. Unlike the reaction of P and D to form P-D, the circularization of M did not require magnesium ions or spermidine and occurred in the presence of EDTA. Thus, the circularization reaction may be less dependent than the joining of P and D, or the joining of two monomeric RNAs, on stabilization induced by divalent or trivalent ions.

Presumably, these ligation reactions require an attack of the 5′-hydroxyl group on a 2′:3′-cyclic phosphodiester bond to form a linear phosphodiester bond. The reactivity of P, D and M were unaffected by treatment with phosphomonoesterase. These are the expected results if reactive RNA P and RNA M each is terminated with a phosphomonoesterase-resistant, 2′:3′-cyclic phosphodiester group (Figure 1B and C). 5′-Phosphorylation of either D or M prevented their participation in subsequent ligation reactions.[33,45]

B. The Bond Formed in Ligated STobRV (−)RNA

Species cM and P-D survived heating in formamide and urea solutions and electrophoresis through hot, urea-permeated polyacrylamide gel. These observations suggest that covalent bonds were formed when P-D and cM were generated. Two forms of RNA P-D were prepared. One was the uncleaved, residual P-D STobRV (−)RNA obtained directly from the transcription of a permuted monomer clone (Figure 1C). The other was obtained by incubating RNAs P and D, purified from the same transcription reaction mixture, under ligation conditions. Each RNA was hybridized to a primer, and the primer was extended by the action of reverse transcriptase. Nucleotide sequence analyses of the transcripts in Figure 2 show that the uncleaved RNA and the ligation product have the same nucleotide sequence. The same primer was extended on a cM template, and the nucleotide sequence analysis of the transcript demonstrated that the circularization of M to cM also has the sequence expected from a simple ligation reaction.

A 34-nucleotide-residue oligodeoxyribonucleotide, designated d35-68(+) to signify its correspondence to residues 35 through 68 of the STobRV (+)RNA sequence (Figure 1A), is expected to protect the junction region of STobRV (−)RNA from the action of ribonuclease T_1 under high-salt conditions.[45] The largest oligoribonucleotide expected to survive digestion of protected cM or P-D is an oligoribonucleotide 39-mer. This 39-mer should correspond

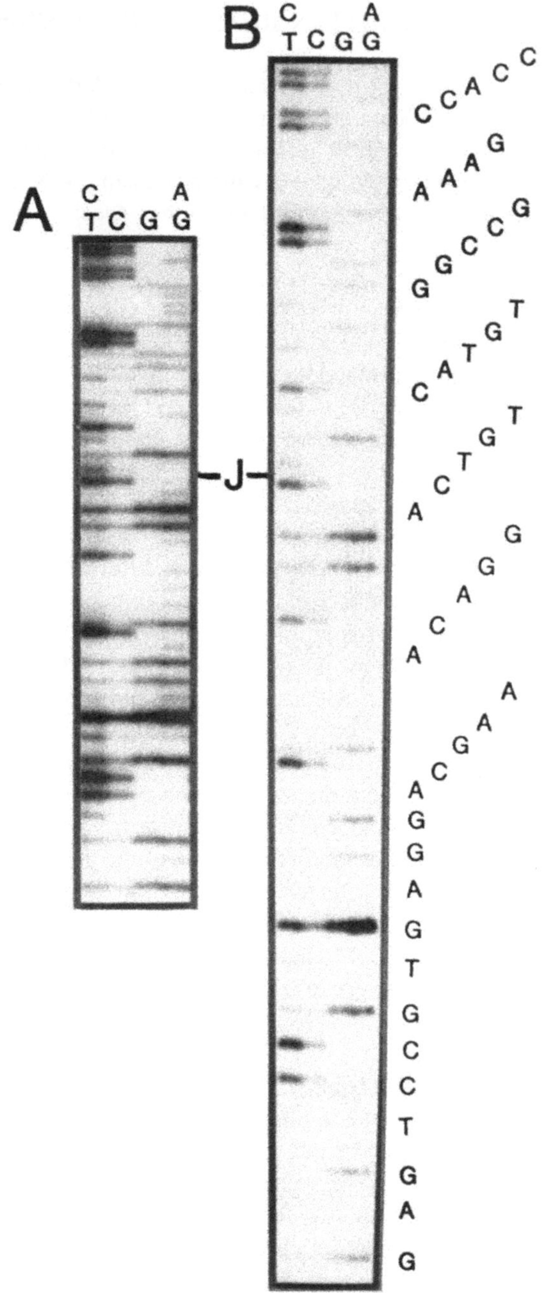

FIGURE 2. Determination of the nucleotide sequences near an uncleaved and a newly ligated junction of STobRV (−)RNA by analysis of cDNA transcripts. (A) Uncleaved P_{231}-D_{130} was recovered directly from a transcription reaction mixture by preparative gel electrophoresis. The primed cDNA transcript of this RNA was electrophoretically purified before sequence analysis by partial, base-specific chemical cleavage. J indicates the J(−) junction site. (B) P_{231}-D_{130} was prepared by spontaneous ligation of P_{231} and D_{130}, and the partial sequence was determined as for the uncleaved RNA. The two panels reveal the same sequence in the same region of the STobRV (−)RNA.

to nucleotide residues 69 through 31 of the STobRV (−)RNA sequence, which includes the junction ApG (Figure 1A). The time course of digestion by ribonuclease T_1 (Figure 3) shows that an apparent 39-mer resisted the ribonuclease T_1 digestion of both cM and P-D when these were hybridized to d35-68(+). The 39-mer was presumptively identified from its mobility relative to that of 5′-phosphorylated d35-68(+), and its absence when d35-68(+) was omitted from the reaction mixture. It was the most slowly migrating resistant oligoribonucleotide (Figure 3). The set of oligoribonucleotides obtained from the RNA/DNA hybrid can be expected to include not only the 39-mer and the limit digest oligoribonucleotides, but also partial digestion products and the d35-68(+)-protected terminal regions of linear monomeric STobRV(−) RNAs.

The 39-mers derived from P-D and from cM have the anticipated nucleotide sequences, as revealed by partial digestion by ribonucleases T_1 and U_2 (Figure 4). The partial sequence from P-D, 5′-GGYYAYYYGAYAGYYYYGYYYYG-3′ (Y = pyrimidine), and from cM, 5′-GGYYAYAYGAYAGYYYYGYYYYG-3′, are as predicted from the nucleotide sequences of the cDNA clones[21] from which the RNAs were transcribed. cM corresponds to the less common sequence and P-D to the more common sequence observed at position 54 of STobRV (+)RNA. The junction containing dinucleoside phosphate sequence is underlined.

Figure 4 shows an unexpected band in the T_1 lane at the position of the junction adenylate residue. This adenylate was identified not only by results in Figure 2, but also from the nucleotide sequence of STobRV RNA.[21] Possible explanations for this zone are comigration of the phosphomonoester form of a guanylate-terminated oligoribonucleotide in the T_1 lane with the adenylate-cyclic-phosphodiester-terminated oligoribonucleotide in the U_2 lane, or the presence of cleaved oligoribonucleotide in the preparation (lane C of the cM analysis), or to both effects. It is conceivable that the junction sequence of oligoribonucleotide has an unusually labile phosphodiester bond since it is this bond that is cleaved in the autolysis reaction of STobRV (−)RNA.

The ability of the newly ligated sequences of cM and P-D to serve as templates for reverse transcriptase implies, but does not prove, that the ligation has produced a 3′-to-5′ phosphodiester bond. That a 3′-to-5′ ApG bond was formed is shown by its susceptibility to digestion by ribonuclease U2.[46] As expected, ribonuclease U_2 failed to hydrolyze a (2′ → 5′)ApG standard.[45] As a further test of the structure of the junction ApG sequence, ApG was isolated after complete digestion of the 39-mer, derived from cM, by the combined action of the guanylate-specific ribonuclease T_1, the pyrimidine-specific pancreatic ribonuclease A, and calf intestinal alkaline phosphatase. The product (Figure 5) comigrated with (3′ → 5′)ApG and not with the 2′-to-5′ isomer. The ApG from d35-68(+)-protected cM was digested readily by ribonuclease T_2, whereas ribonuclease T_2 did not digest (2′ → 5′)ApG.[46] The 39-mer from P-D gave similar results. As an independent analysis of the ApG phosphodiester bond, unprotected cM and P-D also were digested by treatment with ribonucleases and phosphatase to release all ApGs (23 ApGs in circular STobRV (−)RNA). There was no indication of a 2′-to-5′ phosphodiester bond in the recovered ApG.[45]

The reversibility of the autolysis and ligation reactions[33] implies that there is no gain or loss of a nucleotide or nucleoside residue during autolysis and ligation. This conclusion is supported by nucleotide sequence analyses of cDNA transcripts and of the d35-68(+)-protected, junction-containing RNA fragment. Since the ligated STobRV (−)RNA sequences serve as a template for reverse transcriptase, the ligated RNA must have no branch point[27] or 2′-phosphorylation site[28] at the newly ligated junction. Thus, the net ligation reaction must be the exact reverse of the autolysis of the original 3′-to-5′, ApG phosphodiester bond created during polymerization of the polyribonucleotide chain.

Phosphodiester bond formation during ligation of P and D apparently is not *directly* catalyzed either by a divalent or trivalent cation. Either magnesium ions or spermidine can be omitted individually without preventing ligation.[45] The formation of cM occurs even in

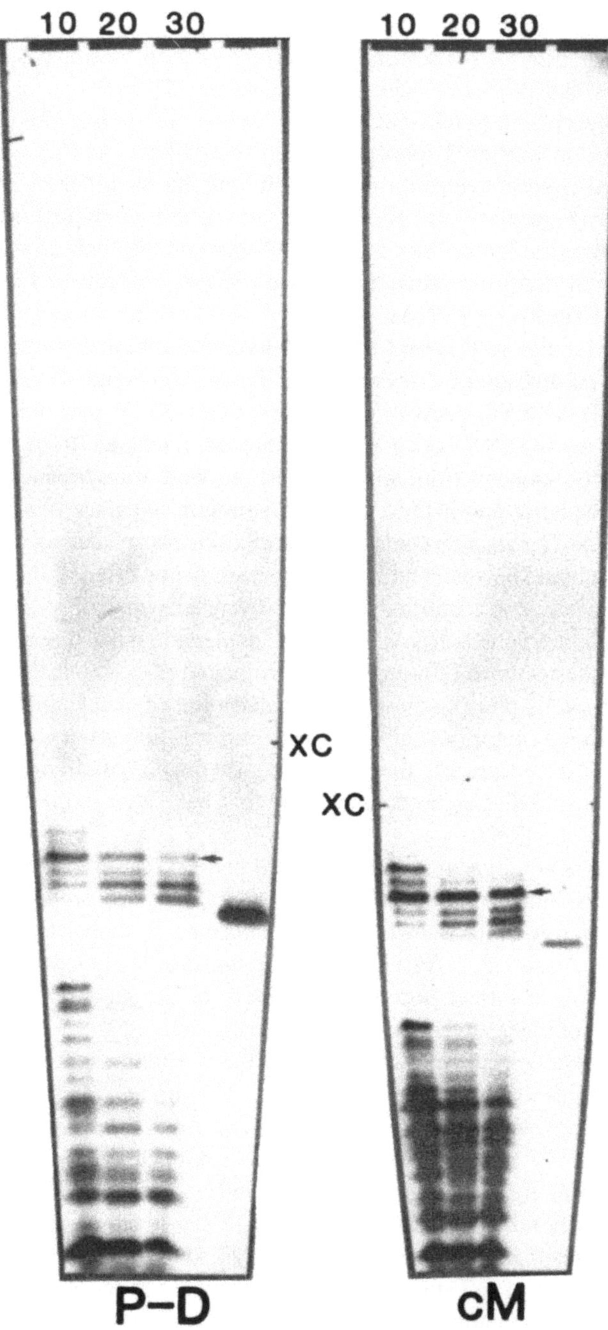

FIGURE 3. Isolation of an RNA fragment from the junction region of P-D and cM RNAs formed in nonenzymic ligation reactions. Preparations of cM (right-hand lane) and P_{199}-D_{130} (left-hand panel) were hybridized to deoxyoligoribonucleotide d35-68(+), and aliquots were incubated with ribonuclease T_1 for 10, 20, and 30 min, respectively, before electrophoresis through a 12% polyacrylamide gel in 7M urea. The standard in the fourth lane is 5′-[^{32}P]d35-68)+). The arrow marks the expected location of an oligoribonucleotide of 39 residues if it is terminated by a 2′:3′-cyclic phosphodiester group. The position of the xylene cyanol FF tracking dye, XC, is indicated.

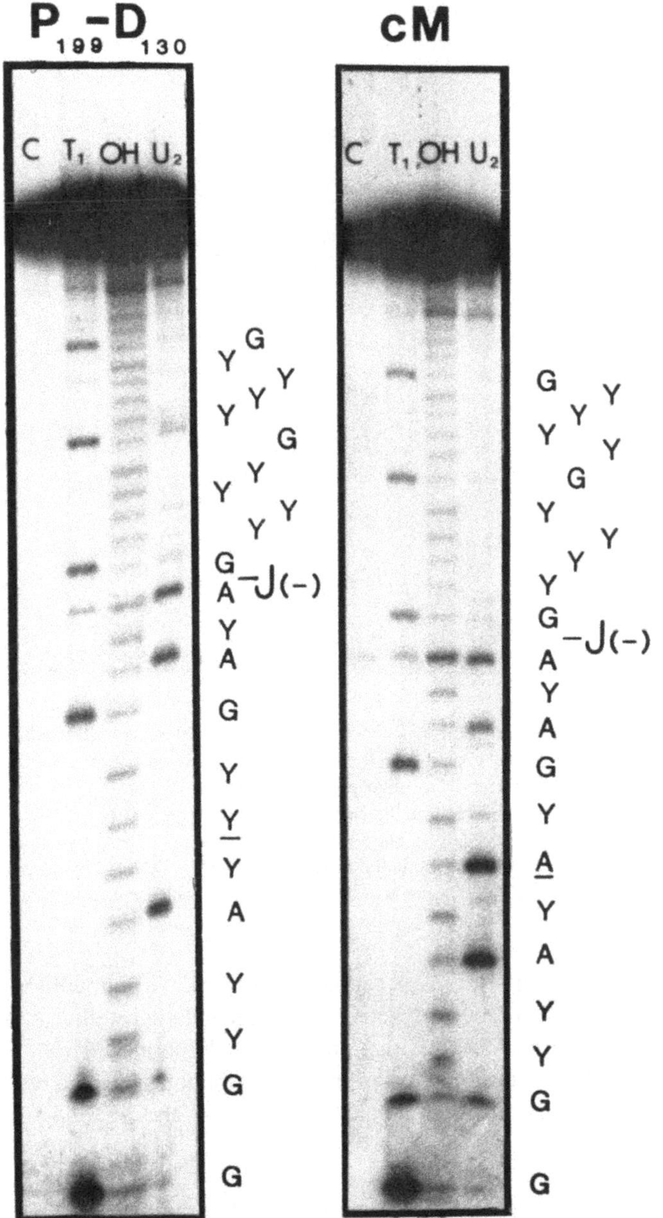

FIGURE 4. Partial nucleotide sequence determination for related oligoribonucleotides derived from the junction region of P_{199}-D_{130} and of cM. The 39-mer indicated by the arrow in Figure 3 was 5′-terminally labeled by phosphorylation and was applied directly to the gel, lane C, or incubated with ribonuclease T_1, base (lane OH), or ribonuclease U_2 before application to the second through fourth lanes, respectively. The derived partial nucleotide sequence (Y = pyrimidine residue) is indicated to the right of each panel, including the location of the junction, J(−). The residues that correspond to the single difference in the nucleotide sequences of these two RNA fragments, an adenylate in cM and a pyrimidine (cytidylate) in P_{199}-D_{130}, are underlined.

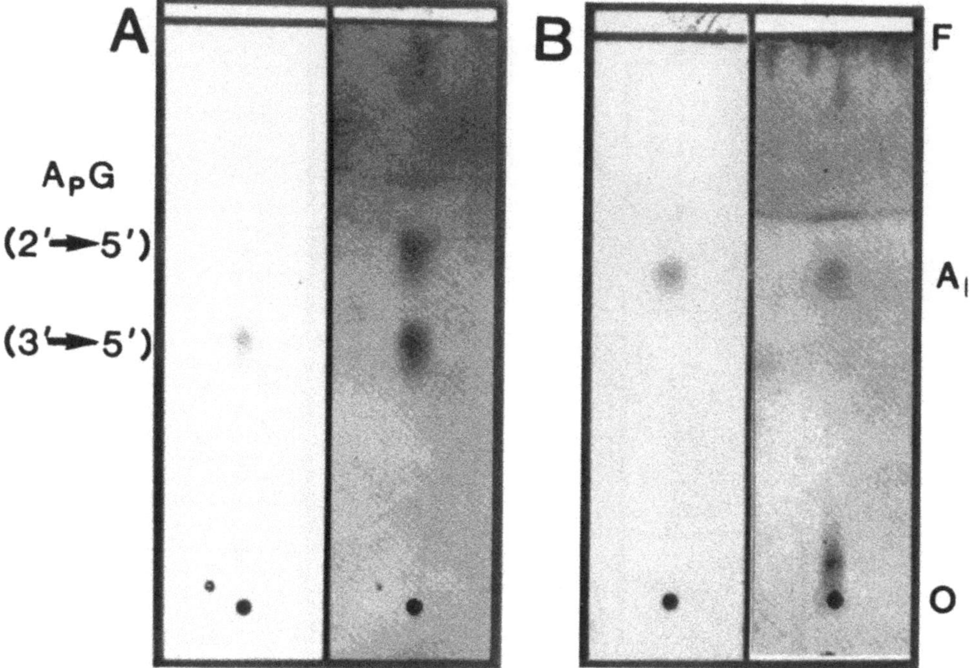

FIGURE 5. A (3′ → 5′)ApG bond in the junction region of cM. A 39-nucleotide residue oligoribonucleotide (arrow, right-hand panel, Figure 3) that contains the J(−) junction was recovered from a preparative-scale digestion of cM that was protected by hybridization to oligodeoxyribonucleotide d35-68(+). (A) Digestion with T_1 and pancreatic ribonucleases and phosphatase and preliminary chromatography generated a ^{32}P-labeled species (left half of panel) that comigrated with (3′ → 5′)ApG, but not with (2′ → 5′)ApG. These two standards were added to the sample before chromatography on PEI CTL in 0.5 *M* KH_2PO_4 and were detected by quenching of fluorescence (right half of panel). (B) The ^{32}P-labeled species was digested with ribonuclease T_2 and combined with Ap before chromatography on PEI CTL.

the absence of added divalent or trivalent cation (Buzayan, J. M., unpublished observations). Probably magnesium ions or, more effectively, protonated spermidine ions, stabilize conformations of the RNA P and RNA D that promote the ligation reaction. Presumably, this is accomplished by favoring a trigonal bipyramid conformation about the reactive phosphorus atom in the expected transition state complex for autolysis and ligation.[47]

C. Ligation of STobRV (+)RNA

Prody et al.[30] reported the nonenzymic ligation of monomeric STobRV (+)RNA to generate dimers in a yield of less than 1%. Unpublished observations of Buzayan and Bruening revealed the circularization of monomeric STobRV (+)RNA. Typical reaction conditions for circularization are 4°C overnight in solution buffered at from pH 6[30] to 7.5. The yield of circles is only about 1%. However, the circularization of monomeric STobRV (+)RNA, like that of STobRV (−)RNA, occurred even in the presence of EDTA and no added divalent metal ion or spermidine. For RNAs of both polarities, the spontaneous formation of the circles of the monomeric RNA implies a close proximity of the 5′ and 3′ ends of each of these RNAs in at least one conformation. Such proximity is in agreement with other observations on monomeric STobRV (+)RNA. STobRV (+)RNA is labeled only very inefficiently in a polynucleotide kinase-catalyzed reaction. However, removal of about one fourth of the 3′-sequence created a molecule that was very efficiently labeled at the 5′-end in such a reaction.[21]

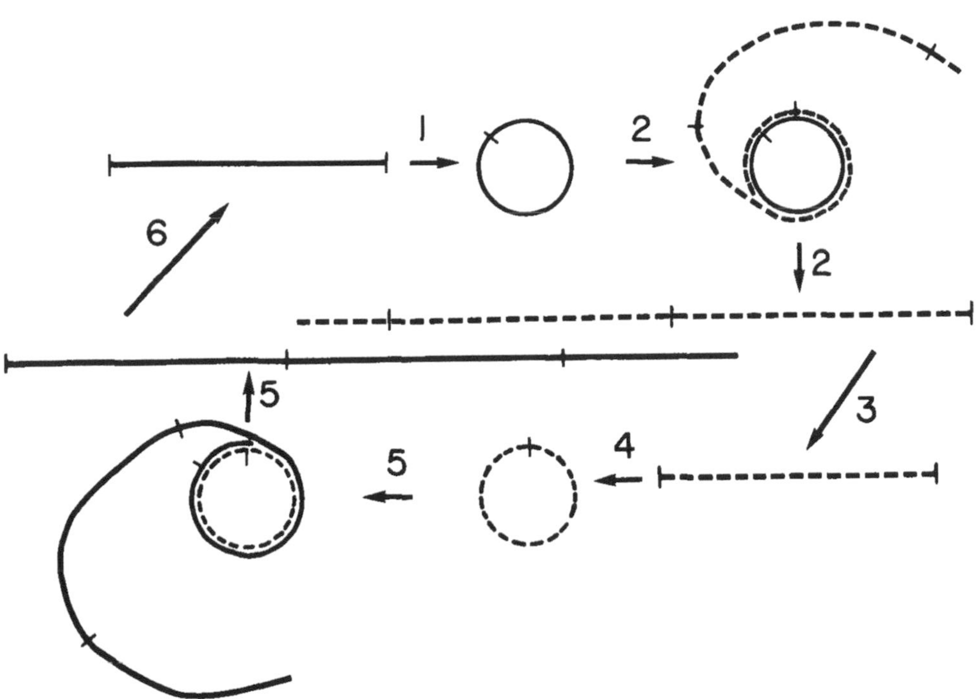

FIGURE 6. A model for the replication of STobRV RNA. Six reactions are represented. The circularization of STobRV (+)RNA and StobRV (−)RNA are reactions 1 and 4, respectively. The rolling circle transcription of the resulting templates are drawn as two steps each for reactions 2 and 5. Two autolytic processing reactions, 3 and 6, complete the cycle. The autolytic processing junctions, which are 48 nucleotide residues apart in the STobRV RNA sequence, are located on the diagram by marks perpendicular to the representations of the (+) polarity (solid line) and the (−) polarity (dashed line) polynucleotide chains.

As has been noted, the extent of the autolysis and ligation reactions differs greatly for STobRV (+)RNA and STobRV (−)RNA,[33] for both linear[30] and circular forms. Autolysis is favored for STobRV (+)RNA, whereas STobRV (−)RNA exhibits a significant ligation reaction, showing how strongly the nucleotide sequence can influence RNA reactivity. STobRV RNA has provided the first examples of a spontaneous, *nonenzymic* RNA ligation reaction in which a 2′:3′-cyclic phosphodiester is attacked by a 5′-hydroxyl group.

VI. A MODEL FOR THE REPLICATION OF STobRV RNA

Our observations on the in vitro reactions of STobRV RNAs and the presence in infected tissues of multimeric and circular STobRV RNAs[20,25,34] suggest a replication model for STobRV RNA that has just six principal steps. These steps require ligation, rolling circle transcription, and autolytic processing. Reviews by Branch and Robertson,[48] Branch et al.,[49] and Symons et al.[10] discuss rolling circle models for small, replicating RNAs in general terms. The first three steps of the replication scheme as proposed for STobRV RNA are (1) circularization of (+)RNA, (2) transcription of circular (+)RNA to generate multimeric (−)RNA, and (3) autolytic processing of multimeric (−)RNA. The remaining three steps are similar, beginning with monomeric STobRV (−)RNA. They are diagrammed in Figure 6.

The distinct locations of the junctions for the (+) and (−) polarities of satellite RNA are indicated by marks that are perpendicular to the polynucleotide chains of the (+)RNA (solid line) and the (−)RNA (dashed line). As is summarized above, four (numbers 1, 3, 4, and 6) of these six reactions proceed to at least a limited extent in Eppendorf® tubes in buffered solutions of spermidine and/or divalent metal ion. The model in Figure 6 is attractive because of its symmetry and simplicity. Conceivably, the contributions to this scheme from the reactions of the STobRV RNA molecules themselves could limit the necessary contribution of enzymes from host and/or supporting virus to that of a single polymerase. Such a polymerase would accept circular STobRV (+)RNA or circular STobRV (−)RNA as template.

We speculate that the greater propensity of STobRV (−)RNA, as opposed to STobRV (+)RNA, to form circles may be part of a mechanism for controlling the relative extents of synthesis of STobRV (+)RNA and STobRV (−)RNA. Only linear STobRV (+)RNA molecules have been found in TobRV capsids.[25] If the principal function of (−)RNA molecules is to serve as a template for (+)RNA synthesis, production of STobRV (−)RNA circles should be facile. If most (+)RNA molecules are to be encapsidated, STobRV(+)RNA should occur principally in the linear form.

We favor the model shown in Figure 6, with four nonenzymic steps, on the basis of our observations of in vitro reactions. However, the circularization of STobRV (+)RNA is such a weak reaction that it is easy to imagine that in vivo it is assisted by the action of some protein. Of course, enzymic ligation reactions and alternative models with linear transcription (e.g., Symons et al.[10]), or even completely different mechanisms, have not been ruled out by in vivo data.

Linthorst and Kaper[25] observed dimeric STobRV (+)RNA circles in infected tissue. Because one seventh of the STobRV RNA sequence lies between the (+) polarity and (−) polarity junctions, transcription of a dimeric circle that is composed of two different nucleotide sequence variants of STobRV RNA could generate, after autolytic processing of the transcribed STobRV (−)RNA, a recombinant. In this scheme, the nucleotide sequences of the recombinant that are located in the one seventh of the molecule between the two junctions will be derived from a different parental variant than the variant from which the remainder of the molecule was derived.

VII. ENZYMIC LIGATION OF SOBEMOVIRUS SATELLITE RNAs AND OTHER RNAs

As was indicated in Section III, a 2′-phosphate group is found at a specific site in the circular satellite RNAs of VTMoV and SNMV,[28] and possibly is present in the circular satellite RNAs of LTSV and SCMoV as well (Haseloff, J., personal communication, as cited by Keese and Symons[37]). Kiberstis et al.[28] consider the 2′-phosphate to be a possible remnant from circularization reactions catalyzed by a host enzyme(s). It is found at the site proposed for processing of multimeric forms of the sobemovirus satellite RNAs.[10,37] An RNA with a 2′:3′-cyclic phosphodiester terminal group is the substrate for wheat germ RNA ligase.[50] The other reactant has a 5′-phosphate group, and the product has a junction with a 3′ → 5′-phosphodiester bond and a 2′-phosphate group. The 2′-phosphate at the junction is characteristic of yeast[51] and plant[52] tRNA splicing, as well. Thus, the replication model of Figure 6 can be proposed for the sobemovirus satellite RNAs, with enzymic ligation as the most likely mode of circularization for the monomeric (+)RNA.

The tRNA half-molecules, that are ligated by extracts from HeLa cells[53] in an ATP-dependent reaction, have the 2′:3′-cyclic phosphodiester and 5′-hydroxyl termini. These also are the groups that participate in the nonenzymic ligations of STobRV RNAs. Filipowicz et al.[54] and Reinberg et al.[55] describe an enzymic activity that converts 3′-phosphates to 2′:3′-cyclic phosphodiesters, so such termini may occur in several types of RNAs.

Other enzymic and nonenzymic RNA ligation reactions, including those associated with splicing reactions, have, as reactants or intermediates, RNA molecules with 3′-hydroxyl and 5′-phosphoryl groups. Examples are bacteriophage T4 RNA ligase,[56] self-splicing Group I introns,[57] self-splicing Group II introns,[58,59] and nuclear mRNA precursor splicing.[60,61] Because of the differences in terminal phosphate groups, these reactions must be considered to be in classes that are distinct from the classes of reactions that the nepovirus and sobemovirus satellite RNAs undergo.

VIII. RNA PROCESSING IN VIROID REPLICATION

Multimeric viroid RNAs of one or both polarities appear in tissues infected with any of several viroids.[36,62-65] Hutchins et al.[66] transcribed cloned avocado sunblotch viroid (ASBV) sequences and found that dimeric transcripts of both polarities exhibited efficient and precise autolytic cleavage. Each cleavage reaction was at a specific bond, between residues 55 and 56 (CpU) of the (+) polarity (most abundant form) ASBV RNA and residues 70 and 69 (CpG) of the (−) polarity ASBV RNA. The new ends of the products have a 5′-hydroxyl group and a 2′:3′-cyclic phosphodiester. Wheat germ RNA ligase catalyzed the cyclization of both naturally occurring linear ASBV (+)RNA and the autolytically released, in vitro synthesized monomeric RNAs of both polarities,[66] confirming the stated end groups for the linear ASBV monomeric RNAs. However, neither ASBV nor other viroid circular RNAs from infected tissues have been found to bear a 2′-phosphate.[37]

Hutchins et al.[66] electrophoretically purified residual, unprocessed dimeric and monomeric ASBV RNAs from transcription reaction mixtures. These junction-containing RNAs autolytically cleaved in solutions of divalent cation(s), showing no requirement for a specific cation. In the most efficient reaction reported, the purified, dimeric ASBV (−)RNA transcript cleaved completely when incubated in 50 m*M* magnesium ions for 1 hr at 40°C.

Hutchins et al.[66] propose a consensus secondary structure for the nucleotide sequences about the junction sites of ASBV (+)RNA and ASBV (−)RNA, elements of which also are found in the proposed and experimentally determined autolytically processing sites of the sobemovirus satellite RNAs and STobRV (+)RNA.[10,37] The junctions represented are ApU, CpA, CpG, and CpU. The determined and proposed (+) and (−) polarity junctions of ASBV and the sobemovirus satellite RNAs lie closer together in their respective nucleotide sequences than the (+) and (−) junctions of STobRV RNA. In addition, it is difficult to reconcile the ApG junction and surrounding sequence of STobRV (−)RNA with the consensus deduced from STobRV (+)RNA, ASBV, and the sobemovirus satellite RNAs.

The nucleotide sequence of ASBV[67] clearly distinguishes it from other viroids of known structure, which are recognizably related to each other.[68] The autolytically processing sites of ASBV do not have corresponding structures[66] in other viroids, such as PSTV. There is no evidence at present for precise autolytic cleavage of PSTV[69] or any of the related viroids that have been tested.

Indirect evidence for a (presumably enzymic) processing site in non-ABSV viroids comes from observations on infectious viroid cDNA. Cress et al.[70] found that dimeric, but not monomeric, clones of PSTV, when embedded in the plasmid DNA, are infectious. However, addition of a small terminal repeat of the central conserved region[68] of the viroid to a monomeric sequence is sufficient to make the DNA infectious.[71-74] The stimulatory effect of a specific short terminal repeat on the infectivity of viroid cDNAs suggests a requirement for two specific processing sites in RNA transcripts. However, the infectivity of such transcripts may have a less specific origin: a regional cleavage (for example, see Reference 75), not necessarily bond specific, possibly guided by RNA conformation. If the 3′ end bears a 2′:3′-cyclic phosphodiester group, plant ligases can generate circular forms, some of which will have exactly the monomeric RNA sequence. Such cleavage also could contribute to the generation of the monomeric CMV satellite RNA that appears in transformed tobacco plants after inoculation with CMV[44] (see Section IV.).

Tsagris et al.[76] report an enzymic activity from nuclei of cultured potato cells that is capable of cutting transcripts of cloned oligomeric PSTV sequences and of forming RNA circles from the monomeric PSTV so released. Two sites for circularization were identified; the most active of which is in the same highly conserved central domain of PSTV[76a,76b] and other viroids that was found in the infectivity studies. The transcripts of one clone, with a complete PSTV sequence, plus a terminal repeat of just six residues from the conserved central domain, both were substrates for the processing and the circularizing activities and were infectious. The correlation of these in vitro results with the requirements for highly infectious PSTV cDNA, if the correlation reflects conditions in vivo, strongly supports PSTV replication mechanisms in which junction-specific processing has a role.

Thus, the ability of a (+)RNA sequence to be converted to a circular form under the influence of a plant cell nuclear extract is correlated with infectivity. This, in turn, implicates rolling circle transcription, which often has been postulated to have a role in viroid replication (see examples by Branch et al.[77] and Owens and Diener[65]). Present data are limited, however, and the structures found in viroid-infected tissues must be considered at this point to be consistent with several variations of rolling circle replication or even with replication that does not require rolling circle transcription.

ACKNOWLEDGMENTS

Research in this laboratory on STobRV RNA has been supported by the Competitive Grants Program of the U.S. Department of Agriculture Science and Education Administration, Grant 84-CRCR-1-1449 and by the Agricultural Experiment Station of the University of California. J. M. Buzayan was a McKnight Training Grant Fellow. A. Hampel received a Presidential Award from Northern Illinois University. W. L. Gerlach was the recipient of a Harkness Fellowship from the Commonwealth Fund, Harkness House, New York, New York. We thank Stephen D. Daubert for his comments on this manuscript and Paul A. Feldstein for the construction of some of the plasmids used in research reported here.

REFERENCES

1. **Schneider, I. R.,** Defective plant viruses, in *Beltsville Symposia in Agricultural Research. I. Virology in Agriculture,* Romberger, J. A., Ed., Allenheld, Montclair, N.J., 1977, 201.
2. **Davies, D. L. and Clark, M. F.,** A satellite-like nucleic acid of Arabis mosaic virus associated with hop nettle disease, *Ann. Appl. Biol.*, 103, 439, 1983.
3. **Piazzolla, P. and Rubino, L.,** Evidence that the low-molecular weight RNA associated with chicory yellow mottle virus is a satellite, *Phytopathol. Z.*, 111, 199, 1984.
4. **Francki, R. I. B., Randles, J. W., Chu, P. W. G., Rohozinski, J., and Hatta, T.,** Viroid-like RNAs incorporated in conventional virus capsids, in *Subviral Pathogens of Plants and Animals: Viroids and Prions,* Maramorosch, K. and McKelvey, J. J., Jr., Eds., Academic Press, New York, 1985, 265.
5. **Diener, T. O.,** Potato spindle tuber viroid. IV. A replicating low molecular weight RNA, *Virology,* 45, 411, 1971.
6. **Semancik, J. S. and Weathers, L. G.,** Exocortis virus: an infectious free-nucleic acid plant virus with unusual properties, *Virology,* 47, 456, 1972.
7. **Diener, T. O.,** Potato spindle tuber viroid. VIII. Correlation of infectivity with UV-absorbing component and thermal denaturation properties of the RNA, *Virology,* 50, 606, 1972.
8. **Grill, L. K. and Semancik, J. S.,** RNA sequences complementary to citrus exocortis viroid in nucleic acid preparations from infected *Gynura aurantiaca, Proc. Natl. Acad. Sci. U.S.A.*, 75, 896, 1978.
9. **Diener, T. O.,** Portraits of viruses: the viroid, *Intervirology,* 22, 1, 1984.
10. **Symons, R. H., Haseloff, J., Visvader, J. E., Keese, P., Murphy, P. J., Gordon, K. H. J., and Bruening, G.,** On the mechanism of replication of viroids, virusoids, and satellite RNAs, in *Subviral Pathogens of Plants and Animals: Viroids and Prions,* Maramorosch, K. and McKelvey, J. J., Jr., Eds., Academic Press, New York, 1985, 235.

11. **Francki, R. I. B.,** Plant virus satellites, *Annu. Rev. Microbiol.*, 39, 151, 1985.
12. **Kaper, J. M. and Tousignant, M. E.,** Viral satellites: parasitic nucleic acids capable of modulating disease expression, *Endeavour New Ser.*, 8, 194, 1984.
13. **Murant, A. F. and Mayo, M. A.,** Satellites of plant viruses, *Annu. Rev. Phytopathol.*, 20, 49, 1982.
14. **Randles, J. W., Davies, C., Hatta, T., Gould, A. R., and Francki, R. I. B.,** Studies on encapsidated viroid-like RNA. I. Characterization of velvet tobacco mottle virus, *Virology*, 108, 111, 1981.
15. **Gould, A. R., Francki, R. I. B., Hatta, T., and Hollings, M.,** The bipartite genome of red clover necrotic mosaic virus, *Virology*, 108, 499, 1981.
16. **Perrault, J.,** Origin and replication of defective interfering particles, *Curr. Top. Microbiol. Immunol.*, 93, 151, 1981.
17. **Monroe, S. S. and Schlesinger, S.,** RNAs from two independently isolated defective interfering particles of Sindbis virus contain a cellular tRNA sequence at their 5′ ends, *Proc. Natl. Acad. Sci. U.S.A.*, 80, 3279, 1983.
18. **Hillman, B. I., Carrington, J. C., and Morris, T. J.,** A defective interfering RNA that contains a mosaic of a plant virus genome, *Cell*, 51, 427, 1987.
19. **Simon A. E. and Howell, S. H.,** The virulent satellite RNA of turnip crinkle virus has a major domain homologous to the 3′ end of the helper genome, *EMBO J.*, 5, 3423, 1986.
20. **Kiefer, M. C., Daubert, S. D., Schneider, I. R., and Bruening, G.,** Multimeric forms of satellite tobacco ringspot virus RNA, *Virology*, 137, 371, 1982.
21. **Buzayan, J. M., Gerlach, W. L., Bruening, G., Keese, P., and Gould, A.,** Nucleotide sequence of satellite tobacco ringspot virus RNA and its relationship to multimeric forms, *Virology*, 150, 186, 1986.
22. **Owens, R. A. and Schneider, I. R.,** Satellite of tobacco ringspot virus RNA lacks detectable mRNA activity, *Virology*, 80, 222, 1977.
23. **Guerrier-Takada, C., Gardiner, K., Marsh, T., Pace, N., and Altman, S.,** The RNA moiety of ribonuclease P is the catalytic subunit of the enzyme, *Cell*, 35, 849, 1983.
24. **Krueger, K., Grabowski, P. J., Zaug, A. J., Sands, J., Gottschung, D. E., and Cech, T. R.,** Self-splicing RNA: autoexcision and autocyclization of the ribosomal RNA intervening sequence of Tetrahymena, *Cell*, 31, 147, 1982.
25. **Linthorst, H. J. M. and Kaper, J. M.,** Circular satellite-RNA molecules in satellite of tobacco ringspot virus-infected tissue, *Virology*, 137, 206, 1984.
26. **Gerlach, W. L., Buzayan, J. M., Schneider, I. R., and Bruening, G.,** Satellite tobacco ringspot virus RNA: biological activity of DNA clones and their *in vitro* transcripts, *Virology*, 150, 172, 1986.
27. **Rushkin, B., Krainer, A. R., Maniatis, T., and Green, M. R.,** Excision of an intact intron as a novel lariat structure during pre-mRNA splicing in vitro, *Cell*, 38, 317, 1984.
28. **Kiberstis, P. A., Haseloff, J., and Zimmern, D.,** 2′ Phosphomonoester, 3′—5′ phosphodiester bond at a unique site in a circular viral RNA, *EMBO J.*, 4, 817, 1985.
29. **Prody, G. A., Bakos, J. T., Buzayan, J. M., Schneider, I. R., and Bruening, G.,** Self-processing of multimeric forms of satellite tobacco ringspot virus RNA, in *Abstr. 3rd Cold Spring Harbor RNA Processing Meet.*, Cold Spring Harbor Laboratory, Cold Spring Harbor, N.Y., 1984, 8.
30. **Prody, G. A., Bakos, J. T., Buzayan, J. L., Schneider, I. R., and Bruening, G.,** Autolytic processing of dimeric plant virus satellite RNA, *Science*, 231, 1577, 1986.
31. **Buzayan, J. M., Gerlach, W. L., and Bruening, G.,** Satellite tobacco ringspot virus RNA: A subset of the RNA sequence is sufficient for autolytic processing, *Proc. Natl. Acad. Sci. U.S.A.*, 83, 8859, 1986.
32. **Watson, N., Gurevitz, M., Ford, J., and Apirion, D.,** Self cleavage of a precursor RNA from bacteriophage T4, *J. Mol. Biol.*, 172, 301, 1984.
33. **Buzayan, J. M., Gerlach, W. L., and Bruening, G.,** Spontaneous ligation of RNA fragments with sequences that are complementary to a plant virus satellite RNA, *Nature (London)*, 323, 349, 1986.
34. **Sogo, J. M. and Schneider, I. R.,** Electron microscopy of double-stranded nucleic acids found in tissue infected with the satellite of tobacco ringspot virus, *Virology*, 117, 401, 1982.
35. **Francki, R. I. B., Grivell, C. J., and Gibb, K. S.,** Isolation of velvet tobacco mottle virus capable of replication with and without a viroid-like RNA, *Virology*, 148, 381, 1986.
36. **Hutchins, C. J., Keese, P., Visvader, J. E., Rathjen, P. D., McInnes, J. L., and Symons, R. H.,** Comparison of multimeric plus and minus forms of viroids and virusoids, *Plant Mol. Biol.*, 4, 293, 1985.

36a. **Forster, A. C. and Symons, R. H.,** Self-cleavage of plus and minus RNAs of a virusoid and a structural model for the active sites, *Cell*, 49, 211, 1987.

37. **Keese, P. and Symons, R. H.,** The structure of viroids and virusoids, in *Viroids and Viroid-Like Pathogens*, Semancik, J. S., Ed., CRC Press, Boca Raton, Fla., 1986.
38. **Gordon, K. H. J. and Symons, R. H.,** Satellite RNA of cucumber mosaic virus forms a secondary structure with partial 3′-terminal homology to genomal RNAs, *Nucleic Acids Res.*, 11, 947, 1983.
39. **Rezaian, M. A., Williams, R. H., and Symons, R. H.,** Nucleotide sequence of cucumber mosaic virus RNA 1. Presence of a sequence complementary to part of the viral satellite RNA and homologies with other viral RNAs, *Eur. J. Biochem.*, 150, 331, 1985.

40. **Richards, K. E., Jonard, G., Jacquemond, M., and Lot, H.,** Nucleotide sequence of cucumber mosaic virus-associated RNA 5, *Virology,* 89, 395, 1978.
41. **Collmer, C. W. and Kaper, J. M.,** Double-stranded RNAs of cucumber mosaic virus and its satellite contain an unpaired terminal guanosine: implications for replication, *Virology,* 145, 249, 1985.
42. **Owens, R. A. and Kaper, J. M.,** Cucumber mosaic virus associated RNA 5. II. *In vitro* translation in a wheat germ protein-synthesis system, *Virology,* 80, 196, 1977.
43. **Linthorst, H. J. M. and Kaper, J. M.,** Replication of peanut stunt virus and its associated RNA 5 in cowpea protoplasts, *Virology,* 139, 317, 1984.
44. **Baulcombe, D. C., Saunders, G. R., Bevan, M. W., Mayo, M. A., and Harrison, B. D.,** Expression of biologically active viral satellite RNA from the nuclear genome of transformed plants, *Nature (London),* 321, 446, 1986.
45. **Buzayan, J. M., Hampel, A., and Bruening, G.,** Nucleotide sequence and newly formed phosphodiester bond of spontaneously ligated satellite tobacco ringspot virus RNA, *Nucleic Acids Res.,* 14, 9729, 1986.
46. **Ball, L. A.,** 2′,5′-Oligoadenylate synthetase, in *The Enzymes,* Vol. 15, 3rd ed., Boyer, P. D., Ed., Academic Press, New York, 1982, 281.
47. **Westheimer, F. H.,** Pseudo-rotation in the hydrolysis of phosphate esters, *Acc. Chem. Res.,* 1, 70, 1968.
48. **Branch, A. D. and Robertson, H. D.,** A replication cycle for viroids and other small infectious RNAs, *Science,* 223, 450, 1984.
49. **Branch, A. D., Willis, K. K., Davatelis, G., and Robertson, H.,** *In vivo* intermediates and the rolling circle mechanism in viroid replication, in *Subviral Pathogens of Plants and Animals: Viroids and Prions,* Maramorosch, K. and McKelvey, J. J., Jr., Eds., Academic Press, New York, 1985, 201.
50. **Konarska, M., Filipowicz, W., Domdey, H., and Gross, H.,** Formation of a 2′-phosphomonoester, 3′,5′-phosphodiester linkage by a novel RNA ligase in wheat germ, *Nature (London),* 293, 112, 1981.
51. **Greer, C., Peebles, C., Gegenheimer, P., and Abelson, J.,** Mechanism of action of a yeast RNA ligase in tRNA splicing, *Cell,* 32, 537, 1983.
52. **Tyc, K., Kikuchi, Y., Konarska, M., Filipowicz, W., and Gross, H.,** Ligation of endogenous tRNA 3′ half molecules to their corresponding 5′ halves via 2′-phosphomonoester, 3′,5′-phosphodiester bonds in extracts of Chlamydomonas, *EMBO J.,* 2, 605, 1983.
53. **Filipowicz, W. and Shatkin, A.,** Origin of splice junction phosphate in tRNAs processed by HeLa cell extract, *Cell,* 32, 547, 1983.
54. **Filipowicz, W., Strugala, K., Konarska, M., and Shatkin, A.,** Cyclization of RNA 3′-terminal phosphate by cyclase from HeLa cells proceeds via formation of N(3′)pp(5′)A activated intermediate, *Proc. Natl. Acad. Sci. U.S.A.,* 82, 1316, 1985.
55. **Reinberg, D., Arenas, J., and Hurwitz, J.,** The enzymatic conversion of 3′-phosphate terminated RNA chains to 2′,3′-cyclic phosphate derivatives, *J. Biol. Chem.,* 260, 6088, 1985.
56. **Romaniuk, P. and Uhlenbeck, O.,** Joining of RNA molecules with RNA ligase, *Methods Enzymol.,* 100, 52, 1983.
57. **Cech, T.,** The generality of self-splicing RNA: relationship to nuclear mRNA splicing, *Cell,* 44, 207, 1986.
58. **Peebles, C., Perlman, P., Mecklenburg, K., Petrillo, J., Tabor, J., Jarrell, K., and Cheng, H.,** A self-splicing RNA excises an intron lariat, *Cell,* 44, 213, 1986.
59. **Van der veen, R., Arnberg, A., Van der Horst, G., Bonen, L., Tabak, H., and Grivell, L.,** Excised group II introns in yeast mitochondria are lariats and can be formed by self-splicing in vitro, *Cell,* 44, 225, 1986.
60. **Konarska, M., Grabowski, P., Padgett, R., and Sharp, C.,** Characterization of the branch site in lariat RNAs produced by splicing of mRNA precursors, *Nature (London),* 313, 552, 1985.
61. **Sharp, P.,** On the origin of RNA splicing and introns, *Cell,* 42, 397, 1985.
62. **Branch, A. D., Robertson, H. D., and Dickson, E.,** Longer-than-unit-length viroid minus strands are present in RNA from infected plants, *Proc. Natl. Acad. Sci. U.S.A.,* 78, 6381, 1981.
63. **Rohde, W. and Sänger, H. L.,** Detection of complementary RNA intermediates of viroid replication by northern blot hybridization, *Biosci. Rep.,* 1, 327, 1981.
64. **Bruening, G., Gould, A. R., Murphy, P. J., and Symons, R. H.,** Oligomers of avocado sunblotch viroid are found in infected avocado leaves, *FEBS Lett.,* 148, 71, 1982.
65. **Owens, R. A. and Diener, T. O.,** RNA intermediates in potato spindle tuber viroid replication, *Proc. Natl. Acad. Sci. U.S.A.,* 79, 113, 1982.
66. **Hutchins, C. J., Rathjen, P. D., Forster, A. C., and Symons, R. H.,** Self-cleavage of plus and minus RNA transcripts of avocado sunblotch viroid, *Nucleic Acids Res.,* 14, 3627, 1986.
67. **Symons, R. H.,** Avocado sunblotch viroid: primary sequence and proposed secondary structure, *Nucleic Acids Res.,* 9, 6527, 1981.
68. **Keese, P. and Symons, R. H.,** Domains in viroids: evidence of intermolecular RNA rearrangements and their contribution to viroid evolution, *Proc. Natl. Acad. Sci. U.S.A.,* 82, 4582, 1985.

69. **Tsagris, M., Tabler, M., and Sänger, H. L.,** Oligomeric potato spindle tuber viroid (PSTV) RNA does not process autocatalytically under conditions where other RNAs do, *Virology*, 157, 227, 1987.
70. **Cress, D. E., Kiefer, M. C., and Owens, R. A.,** Construction of infectious potato spindle tuber viroid cDNA clones, *Nucleic Acids Res.*, 11, 6821, 1983.
71. **Tabler, M. and Sänger, H. L.,** Cloned single- and double-stranded DNA copies of potato spindle tuber viroid (PSTV) RNA and co-inoculated subgenomic DNA fragments are infectious, *EMBO J.*, 3, 3055, 1984.
72. **Visvader, J. E., Forster, A. C., and Symons, R. H.,** Infectivity and *in vitro* mutagenesis of monomeric cDNA clones of citrus exocortis viroid indicates the site of processing of viroid precursors, *Nucleic Acids Res.*, 13, 5843, 1985.
73. **Hashimoto, J. and Machida, Y.,** The sequence in the potato spindle tuber viroid required for its cDNA to be infective: a putative processing site for viroid replication, *J. Gen. Appl. Microbiol.*, 31, 551, 1985.
74. **Meshi, T., Ishikawa, M., Watanabe, Y., Yamaya, J., Okada, Y., Sano, T., and Shikata, E.,** The sequence necessary for the infectivity of hop stunt viroid cDNA clones, *Mol. Gen. Genet*, 200, 199, 1985.
75. **Robertson, H. D., Rosen, D. L., and Branch, A. D.,** Cell-free synthesis and processing of an infectious dimeric transcript of potato spindle tuber viroid RNA, *Virology*, 142, 441, 1985.
76. **Tsagris, M., Tabler, M., Muhlbach, H. P., and Sänger, H. L.,** Linear oligomeric potato spindle tuber viroid (PSTV) RNAs are accurately processed *in vitro* to the monomeric circular viroid proper when incubated with a nuclear extract from healthy potato cells, *EMBO J.*, 6, 2173, 1987.

76a. **Diener, T. O.,** Viroid processing: a model involving the central conserved region and hairpin I, *Proc. Natl. Acad. Sci. U.S.A.*, 83, 58, 1986.

76b. **Steger, G., Tabler, M., Brüggemann, W., Colpan, M., Sänger, H. L., and Riesner, D.,** Structure of viroid replicative intermediates: physico-chemical studies on SP6 transcripts of cloned oligomeric potato spindle tuber viroid, *Nucleic Acids Res.*, 14, 9613, 1986.

77. **Branch, A. D., Robertson, H. D., Greer, C., Geggenheimer, P., Peebles, C., and Abelson, J.,** Cell-free circularization of viroid progeny RNA by an RNA ligase from wheat germ, *Science*, 217, 1147, 1982.

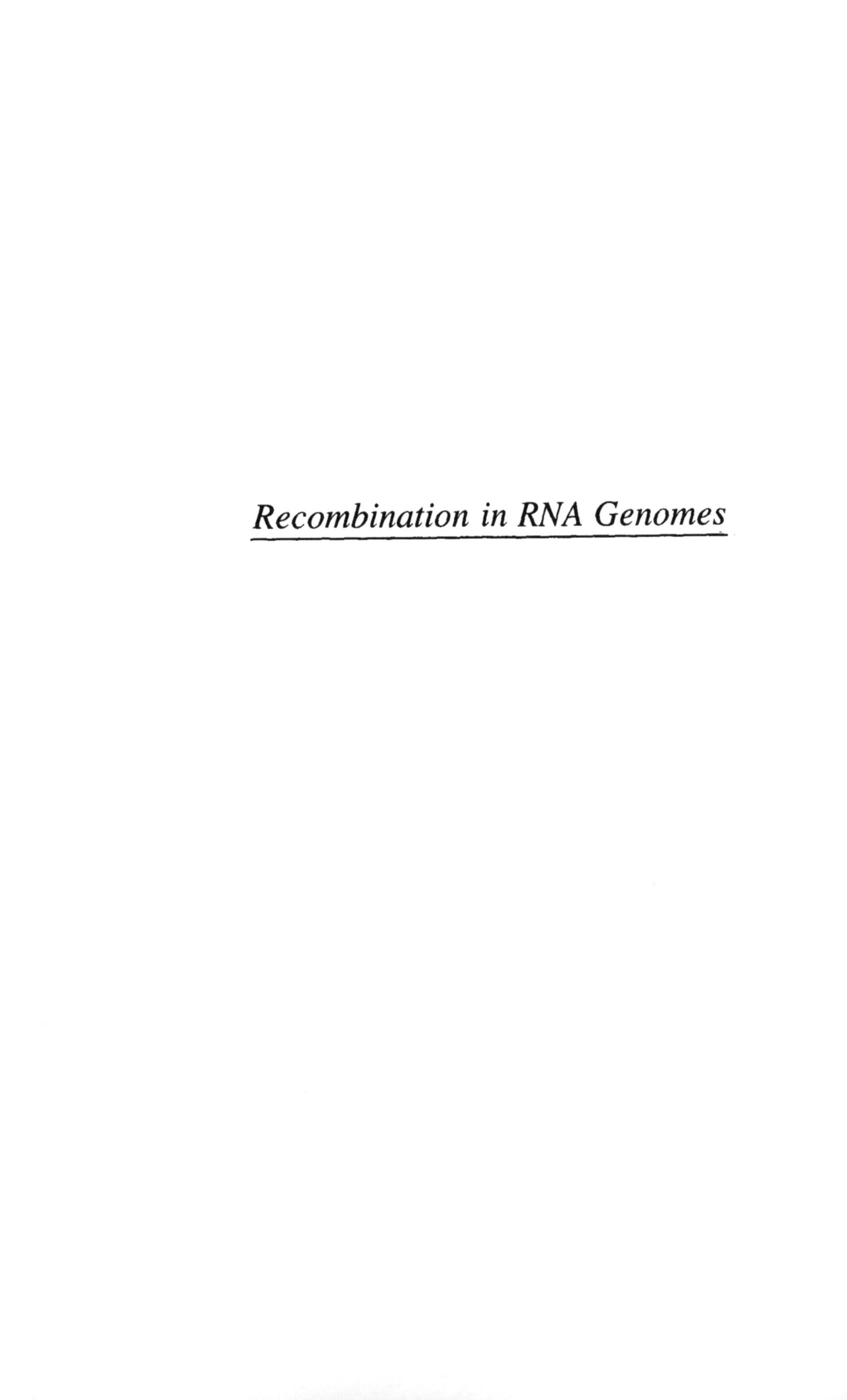

Recombination in RNA Genomes

Chapter 7

GENETIC RECOMBINATION IN POSITIVE STRAND RNA VIRUSES

Andrew M. Q. King

TABLE OF CONTENTS

I. Introduction 150

II. Homologous Recombination 150
- A. Homologous Recombination Among Picornaviruses 150
  - 1. Detection and Isolation 150
  - 2. Genetic Recombination Maps of Picornaviruses 151
  - 3. Frequency of Recombination 151
    - a. Between Isogenic Parents 151
    - b. Between Allogenic Parents 152
  - 4. Recombination Occurs at Many Sites in the Picornavirus Genome 152
  - 5. Genetic Exchanges Between Distantly Related Picornaviruses 154
    - a. Recombination in Tissue Culture 154
    - b. Recombination in Poliovirus Vaccinees 156
  - 6. Is a Minimum Length of Sequence Homology Required at the Crossover Site? 157
  - 7. Recombination is a General Phenomenon 157
- B. Homologous Recombination in Coronaviruses 158
- C. Homologous Recombination Among Other RNA Viruses 158
  - 1. Introduction 158
  - 2. Tobraviruses 158
  - 3. Reversiviruses 159
- D. Mechanism of Homologous Recombination 159
- E. Do RNA Viruses Recombine in Nature? 160
- F. Why do RNA Viruses Recombine? 160
  - 1. Recombination Generates New Virus Variants 161
  - 2. Recombination is an RNA Repair Mechanism 161
  - 3. Recombination Eliminates Deleterious Mutations 161

III. Nonhomologous Recombination 162
- A. Sindbis Virus 162
- B. Brome Mosaic Virus 162
- C. Mechanism of Nonhomologous Recombination 162
- D. Nonhomologous Recombination in Evolution 163

Acknowledgments 163

References 163

I. INTRODUCTION

Genetic recombination in RNA viruses is defined as any process involving the exchange of information between genomic RNA molecules. We are not concerned here with the analogous process of reassortment, which occurs between segmented RNA viruses, nor with internal rearrangements seen the genomes of defective interfering particles, which are reviewed in Chapter 8 of this volume.

RNA recombination is of interest for several reasons. First, it offers a means of manipulating RNA genomes, e.g., for genetic mapping. Second, the ability to exchange genetic information may confer a selective advantage on the virus and thus be a significant factor in its evolution. Third, although homologous recombination was reported in poliovirus more than 20 years ago, only recently has the process been studied in molecular terms.

Recombinational processes, as defined above, are of two kinds: homologous and nonhomologous. In the former, the parental RNAs are related to each other and the location of the genetic crossover is the same in both sequences, so preserving any reading frame and producing a potentially functional recombinant molecule. In nonhomologous recombination neither of these restrictions applies. Of the two processes, homologous recombination has been the most extensively studied. Most of this paper will be devoted to reviewing the progress that has been made towards answering the following questions: can homologous recombination occur anywhere in the genome? To what extent can distantly related RNA viruses exchange genetic information? Must there be homology between the parental nucleotide sequences at the crossover point, and is there a preferred sequence or secondary structure at which recombination occurs?

II. HOMOLOGOUS RECOMBINATION

A. Homologous Recombination Among Picornaviruses

1. Detection and Isolation

Homologous recombinants may be selected by assaying the progeny of a mixed infection under conditions in which the growth of both parental viruses is independently suppressed. An enhancement in the yield of resistant virus, compared with singly infected controls, is indicative of genetic recombination. In the extensive work of Cooper's group,[1] and in our laboratory,[2] this was achieved by growing mixtures of *ts* mutants at the restrictive temperature. Alternative selective agents are neutralizing antibody or the antiviral compound, guanidine, which have been used either together[3,4] or in combination with restrictive temperature.[5]

Homologous recombination was first demonstrated in poliovirus by Hirst[6] and Ledinko[7] in the early 1960s. Soon after, foot-and-mouth disease virus (FMDV or "aphthovirus") was also shown to have the ability to recombine.[5] These two picornaviruses remained the only RNA viruses in which genetic recombination had been shown until the discovery recently of recombination in coronaviruses.[8]

The standard procedure for crossing viruses, known as the yield test, is to grow the mixture of parents under permissive conditions and assay their progeny under restrictive conditions. An advantage of temperature selection is that it is possible to begin applying the selective pressure during the initial mixed infection. This is the basis of the more sensitive "infectious center" method, introduced by McCahon and Slade.[9] In this method, the mixedly infected cells are resuspended before they have had time to lyse, and assayed at the restrictive temperature. In crosses between mutants of the same strain, up to 30% of infected cells can be shown to contain ts^+ recombinants. We have used this method in all of our recent experiments with different FMDV types and subtypes. A similar method was used for isolating coronavirus recombinants.[8]

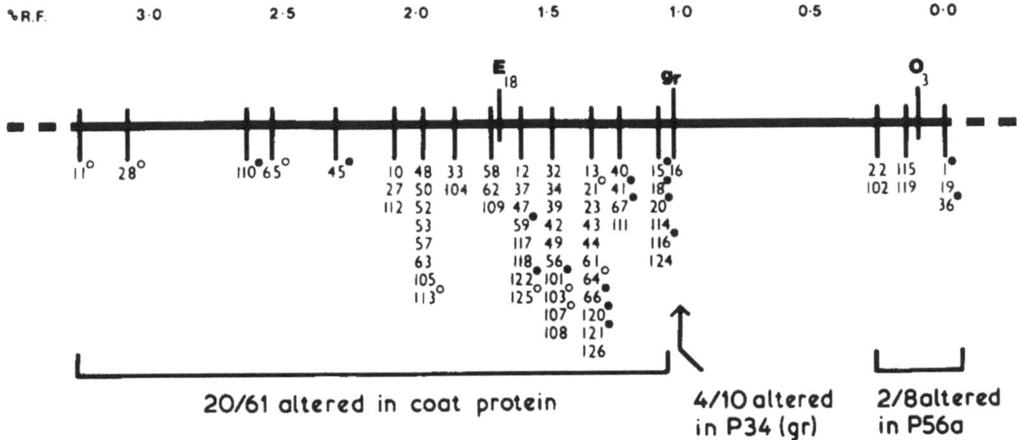

FIGURE 1. The physical distribution of mutations of the FMDV genetic recombination map[2] as determined by electrofocusing virus-induced proteins. All mutations are *ts* except for those designated *gr*. (From King, A. M. Q., McCahon, D., Newman, J. W. I., Crowther, J., and Carpenter, W. C., *Curr. Topics Microbiol.*, 104, 219, 1983. With permission.)

2. Genetic Recombination Maps of Picornaviruses

Frequencies of recombination, as measured by the yield test, tend to be variable and small in comparison to the background reversion rate. Despite these problems, Cooper[10] succeeded in constructing a map, based on recombination frequencies between *ts* mutations of a strain of poliovirus, that was linear and approximately additive. This map featured a single, centrally located, guanidine-resistance locus. A similar map was constructed for FMDV,[11] to be followed later by an extended version,[2] which is shown in Figure 1.

Unfortunately, with a limited range of biological markers, it was difficult to distinguish recombinants from revertants. These attempts at mapping the picornavirus genome rested on two unproven assumptions, (1) that the recombination assays do actually measure recombination and not some other process like complementation-enhanced reversion, and (2) that recombination frequency is proportional to genomic distance (i.e., that crossovers are randomly distributed).

As we shall see, the weight of biochemical evidence now supports both assumptions. Moreover, the FMDV recombination map does seem to be a real representation of the physical genome. We have determined the locations of many of the FMDV *ts* mutations by correlating their *ts* character with a change in the isoelectric point (pI) of a viral protein. The bottom part of Figure 1 summarizes all the pI changes, detected by gel electrofocusing, that coreverted with a *ts* mutation. Of the mutations that we were able to map physically, all 20 *ts* mutations on the left of the *gr* locus are carried by the coat proteins,[13,14] which are encoded within the 5′ half of the genome. The four *gr* mutations are all in the central P34-coding region,[15] and the two *ts* mutations at the extreme right of the map are polymerase mutants[16] and hence near the 3′ end.

Recombination mapping is only approximate, however. The standard error, estimated from the loci of a mutant that was mapped independently 15 times, is 0.2%, a distance on the map roughly equivalent to one of the major coat protein genes. For this reason, the relation between *ts* loci and individual coat protein genes is somewhat blurred.[14]

3. Frequency of Recombination

a. Between Isogenic Parents

We can use the maps to estimate the fraction of genomes in the virus yield that are derived from recombination events. The standard cross, *gr* × *ts*03, included by McCahon et al. in all their recombination experiments,[2] gave a recombination frequency of 0.92%. This was the mean value of 15 replicates. Allowing for the fact that the selection only detected recombination (1) between markers now known to be separated by 1.5 to 3 kilobases, (2) in one direction, and (3) between different parents, leads us to conclude that the overall recombination frequency is between 10 and 20%; i.e., 10 to 20% of the viral genomes undergo recombination in a single growth cycle. Recently, Kirkegaard and Baltimore[17] observed a recombination frequency of 0.1% in a cross between *gr* and *ts* markers separated by only 190 bases in the poliovirus genome. Making the same assumptions as before, this leads to a similarly high estimate of the overall recombination frequency of 15% for poliovirus. It is also comparable with a conservative estimate of about 5% based on the length of the poliovirus map. All three figures should be regarded as underestimates, since they assume that in every cell, of each cross, the ratio of parental genomes never fluctuated from the optimum value of 1:1.

b. Between Allogenic Parents

As we have just seen, the rate of genetic exchange within populations of isogenic viruses is extremely rapid. However, the frequency dropped sharply when different subtypes of FMDV were recombined,[18] and attempts to cross different serotypes of poliovirus or FMDV produced insignificant or very low numbers of recombinants.[19] In our laboratory, the problem was solved by using the infectious center assay, but this does not provide a meaningful estimate of the recombination frequency. Tolskaya et al.[3] succeeded in isolating recombinants between poliovirus types 1 and 3 at the very low frequency of 1×10^{-5} to 2×10^{-4}. For selectable markers separated by approximately 2 kilobases (Figure 3A), this frequency is about 100-fold lower than expected for crosses between isogenic parents. The same large reduction in recombination frequency was observed by Kirkegaard and Baltimore when their *gr/ts* mutant (Figure 3A) of poliovirus type 1 was crossed with wild-type type 2, instead of the isogenic wild-type strain.

Thus, recombination appears to be exquisitely sensitive to parental strain differences. Whether this is due to a reduction in the number of potential recombination sites in the genome, or to the fact that the same sites are used less efficiently, will be discussed in Section II.A.7.

4. Recombination Occurs at Many Sites in the Picornavirus Genome

The genetic mapping experiments of Cooper[10] and McCahon et al.[2] demonstrated the ability of independent *ts* mutants to recombine with each other in a great many different pair-wise combinations, consistent with the existence of a large number of potential recombination sites in the genome. To map these sites biochemically requires that the parental RNA sequences be distinguishable, so creating the problem that it is impossible to study homologous recombination without using heterologous sequences and thus interfering with the process you are trying to study. To determine whether recombination is a general or site-specific phenomenon, we therefore adopted two complementary strategies. First, we crossed two FMDV strains, O_1 and O_6, which belonged to the same serotype. Many different pair-wise combinations of *ts* mutants were crossed[20-22] with the aim of generating crossovers throughout the genome, which were then mapped approximately by RNAase T_1 fingerprinting. The second strategy, which has been adopted by several research groups,[27-32] was to analyze intertypic recombinants at the nucleotide sequence level (Sections II.A.5 through 7).

Figure 2 summarizes the results of the intratypic recombination experiments. We have analyzed 43 ts^+ progeny of nine crosses between different pairs of *ts* mutants. From these, a total of 17 distinguishable types of recombinant were identified. The main conclusions are as follows:

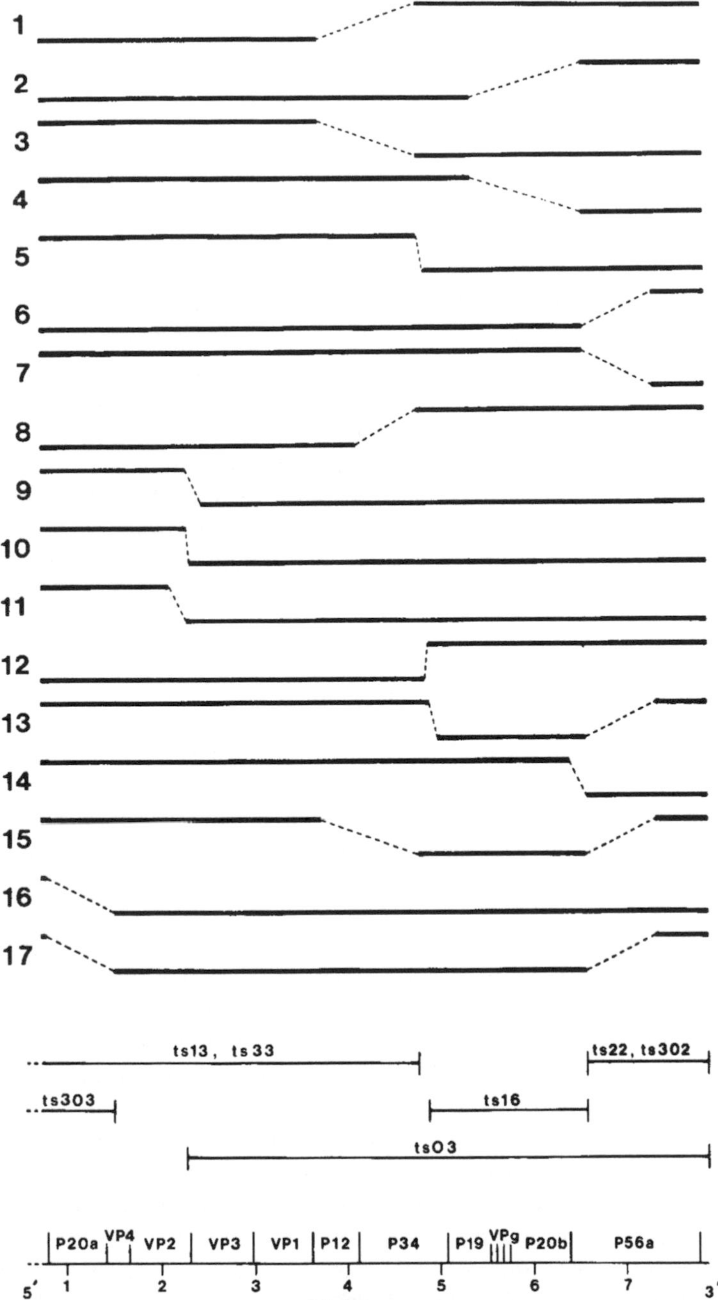

FIGURE 2. Multiple sites of recombination between FMDV subtype strains O_1 and O_6. The figure shows maps of 17 different kinds of recombinant, based on electrofocusing the virus-induced proteins and RNA fingerprinting. Regions of known parentage are represented by solid lines (O_1, upper; O_6 lower); dotted lines linking them indicate crossover regions. Loci of parental *ts* mutations, as deduced from these studies, are shown at the bottom. (From King, A. M. Q., McCahon, D., Saunders, K., Newman, J. W. I., and Slade, W. R., *Virus Res.*, 3, 343, 1985. With permission.)

1. Crossovers are located in many different regions scattered throughout the genome, with at least 12 different sites being needed to explain the data. Given the limited number of genetic markers on which the analysis is based, this result is consistent with a general, as opposed to a site-specific, mechanism of recombination.
2. Most recombinants were produced by a single genetic crossover, although three, Rec 13, 15, and 17, were formed by two crossovers each. Such a high proportion of double recombinants in single growth cycle experiments suggests either that we have grossly underestimated the recombination frequency, or, more probably, that the occurrence of one recombination event increases the chance of a second in the same genome.
3. The only region of the genome in which recombination is known *not* to have occurred is in the VP1 and adjoining VP3 genes. Since this is the most highly variable region of the picornavirus genome,[23,24] it is possible that recombination in this region was prevented by nucleotide sequence heterogeneity, or that recombination did occur, but gave rise to nonfunctional gene products.
4. All but five of the 43 ts^+ progeny examined proved to be recombinant. The few revertants that did arise were confined to two crosses in which there was no enhancement in the yield of ts^+ progeny over the controls. This is an important finding since it verifies that selection at the restrictive temperature does primarily detect recombination, and thus validates all the painstaking genetic mapping studies of Cooper, Lake, McCahon, and their associates.

5. *Genetic Exchanges Between Distantly Related Picornaviruses*

To study the nucleotide sequence requirements of recombination, and to assess the potential of recombination for changing viruses, it is important to know the extent to which distantly related strains can exchange genetic information. Several laboratories have recently reported the isolation of recombinants between different serotypes of the same genus. Those recombinants for which nucleotide sequences have been determined at the crossovers are summarized in Figure 3 and 4.

a. Recombination in Tissue Culture

Tolskaya et al.[3] isolated a series of recombinants possessing the antigenicity of type 3 poliovirus, together with the guanidine resistance of the type 1 genome. RNAase fingerprinting showed that each recombinant genome was comprised of a 5′ half inherited from the type 3 parent and a 3′ half from the type 1 parent.[33] When one of these recombinants was back-crossed with a type 1 virus, selecting this time with anti-type 3 serum, a double recombinant, entirely type 1 except for a short type 3 stretch in the middle of the genome, was formed.[34] Nucleotide sequences in the region of the crossover have been determined for two of these recombinants (Figure 3A). These confirm that each recombinant was formed by a single, homologous recombination event near the middle of the genome, between the loci of the selectable markers indicated on Figure 3A. These sequences define the crossover positions to within stretches of seven and eleven nucleotides, respectively, in which the parental sequences are identical. Figure 3A also shows the region in which crossovers occurred in a remarkable set of 13 type 1/type 2 poliovirus recombinants isolated by Kirkegaard and Baltimore.[17] These crossovers all lay within a short stretch of 190 nucleotides (shown in black on Figure 3A) separating the two selectable markers g and *ts*.

McCahon et al.,[35] in our laboratory, have described a variety of recombinants, made between different European serotypes of FMDV (O × A), and also between European and South African serotypes (O × SAT2, and A × SAT2; the terms "European" and "South African" refer to the two main evolutionary lines of FMDV[36]). In crosses between the most distantly related strains, the frequency of recombination was too low to be detected by the infectious center test, although recombinants did prove to be present among the ts^+ progeny.

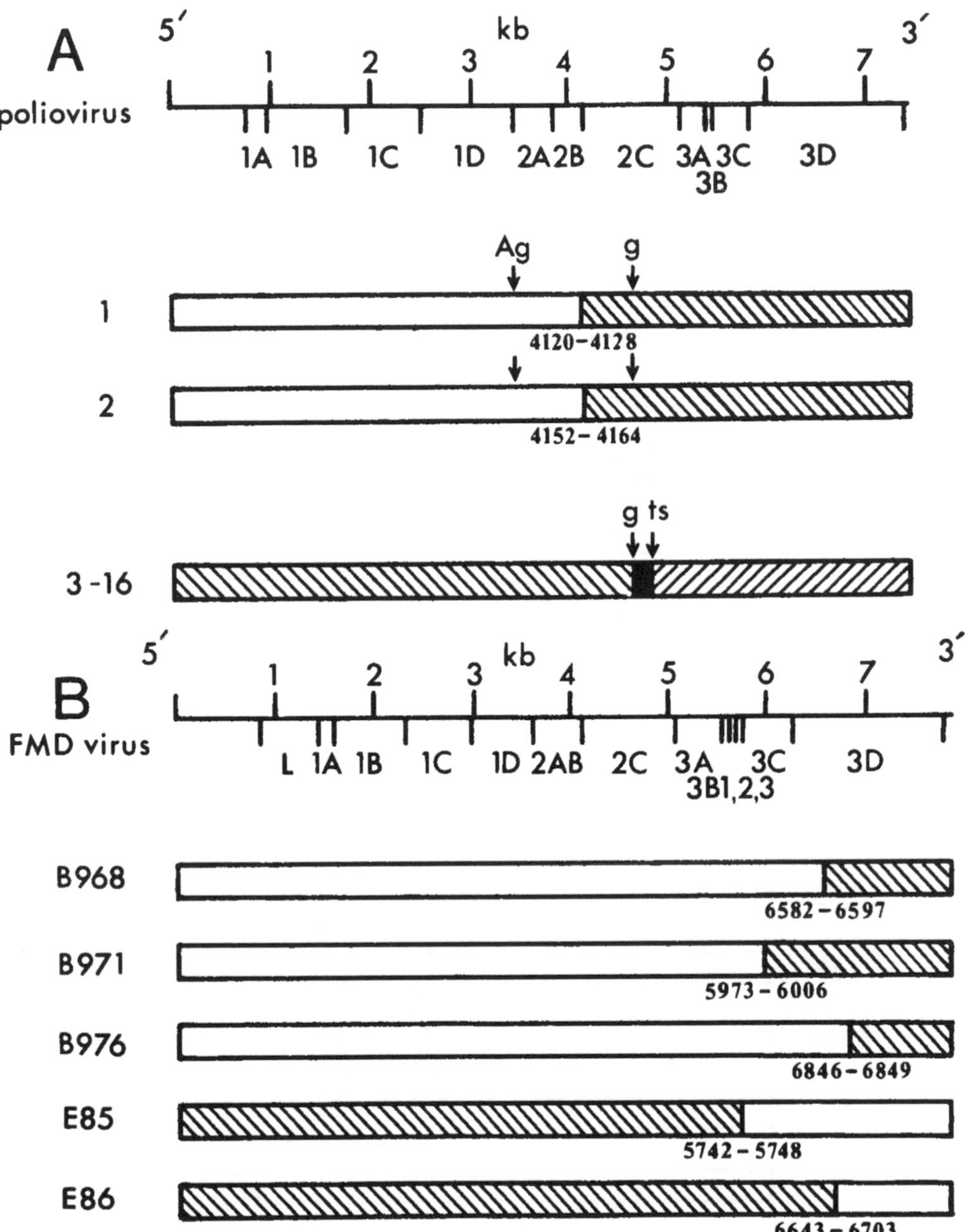

FIGURE 3. Intertypic recombinants of picornaviruses, isolated in tissue culture. Coding regions of viral proteins are indicated using the unified nomenclature.[25] Loci of selectable markers are indicated by arrows: Ag, antigen; g, guanidine-resistance.[26] Parental origins are denoted by (A) poliovirus: slant left lines, type 1; slant right lines, type 2; open field, type 3. (B) FMDV: open field, type O; slant left lines, type SAT2. Numbers beneath each recombinant give the sequence limits of the cross-over; positions in the poliovirus genome are numbered according to Reference 27; FMDV, Reference 23. Sources: recombinants 1 and 2, Reference 28; 3 through 16, Reference 17; B968 through E86, Reference 29.

The recombinant genomes were first mapped by T_1 RNAase fingerprinting, and locations of the crossovers further refined by sequencing appropriate restriction fragments of cDNA.[29] The positions of the crossovers for five type O/type SAT2 recombinants isolated from two crosses, are shown in Figure 3B. The genetic crossovers of the two sets of FMDV recombinants differ in orientation, but are all confined to the 3′-terminal one third of the genome.

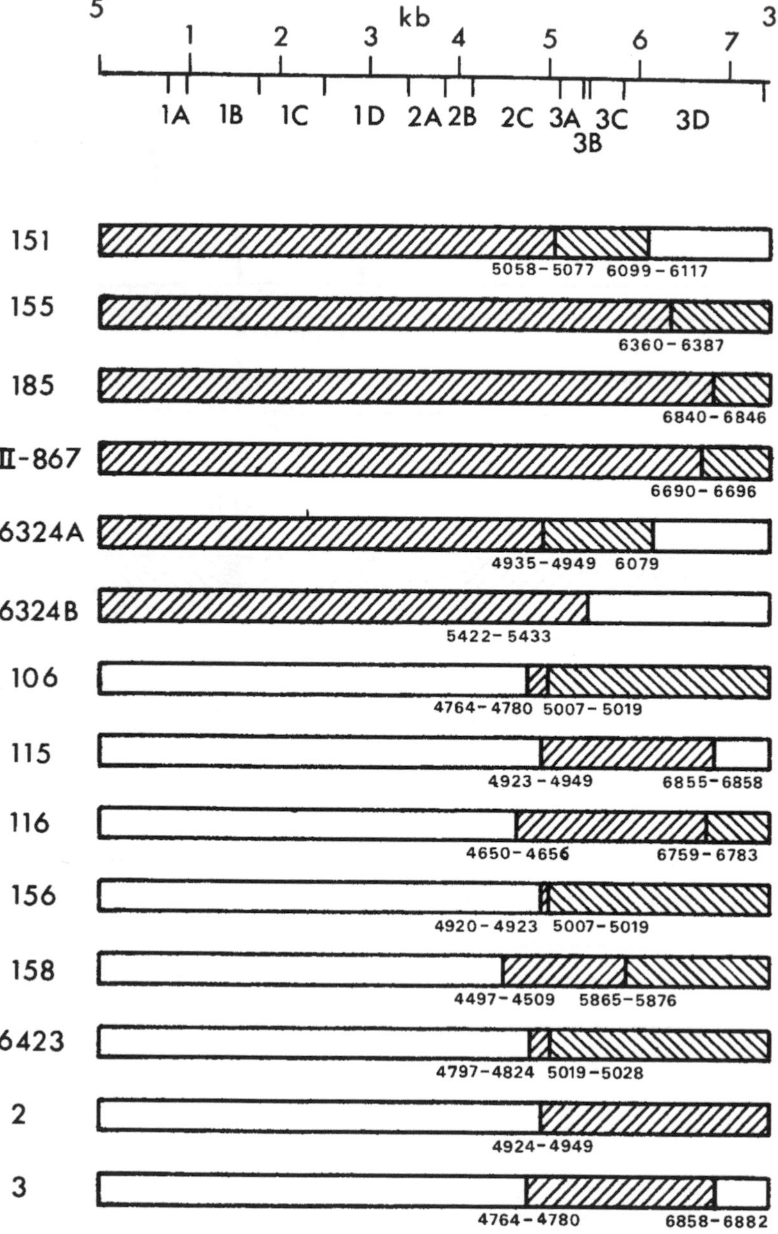

FIGURE 4. Intertypic recombinants of poliovirus vaccine strains. Parental origins are denoted by slant left lines, type 1; slant right lines, type 2; open field, type 3. Sources: recombinant 151, Reference 30; 155 through 6423, Reference 31; 2 and 3, Reference 32.

b. Recombination in Poliovirus Vaccinees

Poliovirus vaccine is widely given orally as a mixture of live attenuated strains of all three serotypes. In 1984, Kew and Nottay[30] reported the startling discovery of a recombinant between the type 2 and type 1 Sabin vaccine strains, isolated from a contact of a vaccinee. The implication is that genetic recombination between the strains of vaccine virus, multiplying within in the vaccinee, had contributed towards the reversion of the virus to neurovirulence. Since that first observation, vaccine-associated poliomyelitis cases have turned out to be a rich source of intertypic recombinants. Figure 4 summarizes all the vaccine-

```
B968        6570       6580       6590       6600       6610       6620
      CTCCAAgCAC AaGGGAGACA CaAAGATGTC TGAGGA                                01
                              AAGATGTC TGAGGAaGAC AAAGAGCTGT TCaGaCttTG  SAT2

B971        5969       5979       5989       5999       6009       6019
      TTGAGTTTGA GATcAAAGTA AAAGGACAGG ACATGCTCTC AGACGC                     01
                     AAAGTA AAAGGACAGG ACATGCTCTC AGACGCtGCG CTCATGGTGt  SAT2

B976        6819       6829       6839       6849       6859       6869
      ACCCGAggcT GAGGCtGCcC TGAAGCTCAT GGAGAAaAG                             01
                                             AGg GagTACAAgT TcacTTGcCA  SAT2

E85         5708       5718       5728       5738       5748       5758
                                                     GTCACt GAGAGTGGtG  01
      TGAAGAAaCC TGTCGCTTTG AAAGTGAAAG CaAAGAATtT GATaGTCAC                 SAT2

E86         6674       6684       6694       6704       6714       6724
                                       A TCAAGGGTGT tGAtGGaCTC GACGCCATGG  01
      cAAATGCCCC ACTGAGCATc TAtGAGGCtA TCAAGGGT--                            SAT2
```

FIGURE 5. Nucleotide sequences in the region of genetic crossovers of FMDV intertypic recombinants.[29] The figure shows the parental sequences that are present in the recombinants; the overlap in each case defines the limits of the crossover. Upper case letters denote bases common to both parents; lower case letters refer to sites at which the parental sequences differ. Maps of these recombinants are illustrated in Figure 3A.

derived recombinants that have been reported to date. Those reported by Kew and Nottay[30,31] are comprised of six antigenically type 2 recombinants (including isolate number 151, originally described,[30] and now known to have a second crossover) and six type 3 recombinants. This work is complemented by that of Minor et al.,[32] who did an in-depth study of the evolution of vaccine polioviruses in a normal, healthy child. Astonishingly, two entirely different recombinants appeared sequentially over a period of several weeks before the infection was finally cleared.

6. *Is a Minimum Length of Sequence Homology Required at the Crossover Site?*

From the data on 46 crossover sites summarized in Figures 3 and 4, we can discern two features common to picornavirus recombinants. First, all recombinants are produced by faithful and accurate homologous crossovers, i.e., there are no insertions, deletions or mismatches. Second, crossovers are entirely confined to the 3′ half of the genome. Since this is where picornavirus sequences are most conserved, most crossover sites could be located only to within a stretch of bases where the recombinant, and both parental sequences, were identical.

From the limits of each of these homology regions, given in Figures 3 and 4, it is clear that an extensive region of perfectly matching sequence is not necessary at the precise point of crossover. The sequences of the five FMDV recombinants given in Figure 5 exemplify the wide variation in the length of the homology regions (from 2 to 32 nucleotides) seen in all of the studies. The extensive study of Kirkegaard and Baltimore[17] reveals no bias either for or against tracts of matching sequence at crossover sites; the mean length of homology (approximately five nucleotides) is consistent with a random distribution of crossovers. In one case, there was no homology region at all, and this was therefore the only crossover that could be located precisely to a single internucleotide bond.

7. *Recombination is a General Phenomenon*

Although crossovers between different serotypes were restricted to conserved regions within the 3′ half of the genome, their sequences provide additional support for the idea that recombination is intrinsically a general process; that is, crossovers between isogenic parents occur anywhere in the genome. The reasons are twofold. First, the sequence data demonstrate many independent crossover sites (28 in poliovirus alone). If the FMDV data are included, recombination is seen to occur in every 3′ nonstructural protein gene except 2A. Most convincing of all is the work of Kirkegaard and Baltimore,[17] showing nine different crossover sites within a region of just 190 bases. This complements the intersubtype crosses[22] described in Section II.A.4, which showed recombination in many different regions of the genome, including 5′ proximal regions. Together, these two studies imply the existence of many hundreds of potential recombination sites throughout the genome.

Second, the corollary of a large number of recombining sites is the absence of any discernible sequence specificity. Crossovers do not occur near putative mRNA splicing signals, or viral replication signals, nor do crossovers necessarily occur near sequences predicted to fold into stable secondary structures. Kew and Nottay[31] noted an AA dinucleotide at or adjacent to all 20 of the crossovers that they have reported. However, this was probably a coincidence, since it was not seen in the other studies (e.g., Figure 5).

B. Homologous Recombination in Coronaviruses

For more than 20 years, picornaviruses appeared to be uniquely endowed with the ability to recombine in RNA. Recently however, Lai et al.[8] reported the isolation of a recombinant between *ts* mutants of two strains, A59 and JHM, of mouse hepatitis virus (MHV), a coronavirus. The fingerprint of the genomic RNA revealed that the first three kilobases at the 5′ end were inherited from strain JHM, the remainder being like A59. This places the genetic crossover somewhere in gene A which codes for the viral RNA polymerase. The small mRNAs derived from the 3′ end of the genome were, as expected, identical to those of the A59 parent, except for their leader sequences which, being derived from the 5′ terminus of the genome, resembled JHM.

Lai and co-workers have since characterized several more recombinants isolated under conditions that select against only one of the parents.[37] The results appear to indicate an extremely high recombination frequency in coronaviruses, although the method used to isolate MHV recombinants, a type of infectious center method, does not allow a quantitative estimate of recombination frequency to be made. It will be necessary to confirm this conclusion by analyzing the progeny of a single growth cycle using a standard yield test.

The recombinant RNA fingerprints that have been published to date are of seven different types. Each is consistent with a single, homologous genetic crossover, with the sites being scattered throughout the 5′ half of the coronavirus genome. This region encodes the putative polymerase, and is presumably a highly conserved region of the genome. Thus, on present evidence, coronaviruses appear to share with picornaviruses the ability to undergo high-frequency homologous recombination at many different sites within conserved regions of the genome.

C. Homologous Recombination Among Other RNA Viruses

1. Introduction

Outside the picornavirus and coronavirus families there is little definitive evidence that homologous recombination occurs among RNA viruses, and a large body of experience, particularly with negative strand viruses, that says it does not (Section III.C). Some of the rearrangements in brome mosaic virus (BMV) RNA, recently reported by Bujarski and Kaesberg,[38] fit our definition of homologous recombination, but since others do not, these will be dealt with under the heading of nonhomologous recombination (Section III.B.C). Evidence of homologous recombination at the level of RNA has been adduced for two other groups of viruses, described below. No doubt, many more examples remain to be discovered.

2. Tobraviruses

The region 200 to 1100 nucleotides from the 3′ end of the two genome segments of the bitartite virus, tobacco rattle virus, are identical.[39] It is difficult to see how that degree of conservation can be maintained by selection only, and the evidence strongly suggests that these terminal sequences are constantly being exchanged by recombination.

3. Reversiviruses

The life cycle of retroviruses (Volume II, Chapter 1), of other reversiviruses (caulimoviruses, Volume II, Chapter 2 and hepatitis B virus, Volume II, Chapter 3), and of retrotransposons (Volume II, Chapter 4) goes through an RNA intermediate, the replication by reverse transcriptase of which involves a template-switching step. Since this takes place on a positive strand genomic RNA template by a virus-encoded polymerase, the phenomenon is highly relevant to this review. Reverse transcription starts close to the 5′ end of the RNA template, releasing a short "strong-stop" DNA molecule which then acts as a primer for reinitiating synthesis at a complementary sequence situated at the 3′ end of the template. The resulting sequence rearrangement has been demonstrated elegantly for cauliflower mosaic virus by the work of Hohn's group in the form of both intra-[40] and intermolecular[41] recombinants (reviewed in Chapter 2 in Volume II). The significance of this site-specific template jumping is, first, that it could exemplify a general mechanism of homologous recombination in retroviruses occurring during RNA replication and, second, it introduces the concept that run-off transcripts are highly recombigenic (Section II.F).

D. Mechanism of Homologous Recombination

The outstanding characteristics of RNA recombination, as performed by picornaviruses, are that it is (1) *homologous* (there are never any insertions or deletions), (2) *efficient* (a large fraction of genomes undergo recombination each growth cycle), and (3) *general* (it occurs anywhere in the genome). To achieve this combination of properties requires some mechanism for aligning the parental RNA sequences. Since no cellular process remotely fitting this description has ever been described, and since different families of RNA viruses appear to vary in their ability to recombine (Section III.C), we must look to the virus itself for the source of this unusual activity.

These considerations favor a template-switching (or "copy-choice") mechanism, as originally proposed by Cooper et al.[42] Since the poliovirus replicative intermediate is single-stranded in vivo,[43] the growing RNA strand must unwind from its template as it is synthesized. This provides an opportunity for the growing RNA strand to dissociate from the template and prime synthesis of a recombinant RNA molecule on a second template. The correct site of reinitiation is presumably determined by base-pairing between the template and the 3′ end of the primer. Kirkegaard and Baltimore[17] recently studied the ability of polioviruses to recombine under conditions in which the replication of one parent was selectively blocked before infecting with the other. This confirmed that recombination is a replicative process, as expected for a copy-choice mechanism. However, this requirement for active replication applied to only one of the parents, with the results suggesting that the polymerase switches template preferentially during synthesis of minus-sense RNA.

There are plenty of precedents for RNA virus polymerases either switching templates, as in coronavirus mRNA synthesis[44] and the reverse transcriptase reaction (Section II.B) or, alternatively, initiating synthesis from heterologous primers, as in mRNA synthesis by orthomyxoviruses[45] and bunyaviruses.[46] The end product in all cases is a recombinant nucleic acid molecule. The fact that these are all normal and necessary viral functions implies that the polymerases involved are extremely adept at performing such recombinational processes. Indeed, Makino et al.[37] have gone even further in arguing that genetic recombination in coronaviruses and the leader-primed synthesis of mRNA are different manifestations of exactly the same process, and that this explains why the distribution of genetic crossovers in coronaviruses is skewed towards the 5′ end of the genome.

One difficulty with this copy-choice model is that regions in which the parental nucleotide sequences match perfectly, given in Figures 3 and 4, are far too short in many cases either to hold the primer and template together at the cross-over site or, more importantly, to specify the homologous site in preference to all the other potential sites. Presumably, the primer RNA finds its correct position on the template by forming longer stretches of base-pairs, that can tolerate some mismatching. In this connection, it will be recalled that recombination frequency is extremely sensitive to sequence differences between the parental RNAs (Section II.A.3). The 100-fold reduction in recombination frequency observed by Kirkegaard and Baltimore[17] in the intertypic cross, compared with the intratypic control, cannot be explained by a comparable reduction in the number of potential crossover sites (there were too many sites for that). It follows that much of the effect of sequence divergence on recombination frequency is due simply to the fact that intertypic crossovers are intrinsically less probable. This can be explained on the copy-choice model by competition between homologous and heterologous templates for the primer; the primer has a much higher probability of annealing to a complementary strand of its own parental type than it has of forming a heteroduplex with the other type.

E. Do RNA Viruses Recombine in Nature?

Studies on polioviruses vaccinees show that recombination can occur in the infected animal. Kew and Nottay[31] estimate that out of the approximately 10 cases of vaccine-associated poliomyelitis per year in the U.S., 15% involve intertypic recombinants. It is not only among poliomyelitis cases that recombinants are found; the isolation of two different recombinants from a single, healthy child suggests that recombination may be the norm rather than the exception. It may be argued that the production of recombinants by vaccinees is a rather artificial situation. However, the demonstration of recombination in nature has been brought a step nearer by dramatic new evidence of recombination between a vaccine strain and a wild strain.[47] Several field isolates from an outbreak of poliomyelitis in Honduras in 1984 were found to possess the capsid protein genes of the Sabin type 1 vaccine strain, whereas the (normally more conserved) terminal regions of the genome diverged in nucleotide sequence by more than 10%. Clearly these terminal regions must have been derived by recombination from wild strains of poliovirus. These wild parental strains have not been identified, although it likely that they also belong to type 1.

Although none of these examples of recombination in the host can be regarded as entirely "natural", it seems safe to conclude that whenever conditions for recombination arise in nature, then, sooner or later, it will happen. That suitable conditions do sometimes arise is shown by the recovery of up to three different serotypes of FMDV from a single animal in the field.[48] The answer to the question posed above is, therefore, "yes", but with one qualification: picornavirus serotypes are free to exchange genetic information anywhere within the 3′ half of the genome, but apparently not within the 5′ proximal sequences that determine antigenicity. What benefit such limited genetic exchanges might confer is considered in the next section.

F. Why do RNA Viruses Recombine?

Many RNA viruses exchange information by either genetic recombination or reassortment of genome segments, but the selective advantage of doing so has never been established. Some unsegmented viruses, like Sindbis, seem to get along quite well without undergoing homologous recombination. For those that do, the function may be an unselected side effect of the replication mechanism, just as reassortment is unavoidable between isogenic isolates of a segmented RNA virus.

The following paragraphs offer some suggestions as to the selective advantages of recombination.

1. Recombination Generates New Virus Variants

The difficulty with this theory is that it does not explain why recombinants should have a selective advantage; in general, the reverse will be true. An advantage would be conferred on the virus by genetic crossovers in the capsid genes that altered antigenicity, but crossovers within the coat protein genes are highly restricted, and none has ever been observed in the major immunogenic protein VP1, even between closely related strains (Section II.A.4). Thus, although recombinant variants might arise occasionally, any advantage would be exceptional. Whether a rapidly mutating virus would be able to maintain the capacity for recombination, solely to take advantage of such rare events, is doubtful.

2. Recombination is an RNA Repair Mechanism

According to this theory, recombination between overlapping fragments of viral RNA might generate complete, viable genomes. This could explain the rescue of UV-inactivated poliovirus[49] and FMDV[5] at high multiplicity of infection. Run-off replication of incomplete templates could provide a rich source of free primers for copy-choice recombination. A precedent for this is the switch that reverse transcriptase makes at the 5′ end of the template (Section II.C). However, this mechanism of repair must be regarded as speculative, since it has yet to be shown that subgenomic fragments can take part in recombination.

3. Recombination Eliminates Deleterious Mutations

Much has been written about the high mutation rate of RNA viruses. Reanny[50] has suggested that the infidelity of RNA replicases limits the size of RNA genomes, and that reassortment of genome segments (and hence recombination also) may serve to reduce this "noise". If there has to be a reason for viruses to recombine, then this one is the most plausible on several counts: (1) the process would normally involve recombination between isogenic viruses, which occurs at high frequency, throughout the genome; (2) missense mutations are constantly arising in RNA in virus populations and are much more likely to be deleterious than advantageous; (3) the selective advantage of recombination between defective mutants is experimentally demonstrable in tissue culture, and it is also a plausible explanation for recombination between attenuated strains of poliovirus in vaccinees; and (4) the theory helps to explain why not all unsegmented RNA viruses have the ability to recombine. Sindbis virus does not need to recombine because complementation enables mixtures of defective mutants to grow efficiently. By this means, the virus population can tolerate many genetic defects. However, complementation is inefficient in picornaviruses.[1] To the extent that their replicative genes function inefficiently in *trans*, deleterious mutations will be propagated poorly. In these circumstances, the only way in which picornaviruses can utilize their nondefective genes is by genetic recombination. (This argument also disposes of the objection that recombination merely shuffles mutations around, destroying as many wild-type genomes as it makes; in fact, recombination does confer an advantage on a genome population that has a nonrandom frequency distribution of genetic defects of the kind expected here.) In the case of coronaviruses, the advantage of recombination presumably derives from the fact that these viruses possess the largest of all known genomic RNA molecules.

In the face of a mutation frequency of approximately 10^{-4} per nucleotide,[51] the really hard thing to explain is how RNA viruses succeed in propagating themselves without changing. How is it, for example, that the lengthy derivation the type 3 Sabin poliovirus vaccine strain from its virulent progenitor resulted in only ten base substitutions in the entire genome?[52] This author proposes that picornaviruses have adopted a distinctive replicative strategy in which the ability to recombine and inability to complement are two elements ensuring high-fidelity reproduction during passage at high multiplicity.

III. NONHOMOLOGOUS RECOMBINATION

A. Sindbis Virus

Nonhomologous recombination has been reported in two positive strand RNA viruses, Sindbis and brome mosaic virus. In the course of determining the RNA sequence of several DI preparations of Sindbis virus, Monroe and Schlesinger[53] found that two of them had independently acquired part of the cellular RNA molecule, aspartyl tRNA, at their 5′ termini. There was no significant homology between the viral and tRNA sequences at either crossover site. Alphaviruses readily generate DI particles, and an intriguing aspect of Sindbis virus DI RNAs is that they frequently contain parts of this particular tRNA. This contrasts with the DI RNAs of negative strand viruses in which intermolecular exchanges appear to be rare, with there being but one reported instance, a mosaic DI RNA of influenza virus.[54]

B. Brome Mosaic Virus

At present, BMV provides the sole example of genetic recombination in a plant RNA virus. BMV is a tripartite virus in which the three species of genomic RNA have a similar, though not identical, sequence at the 3′ end which folds into a tRNA-like structure. Bujarski and Kaesberg[38] recently showed that a deletion mutant lacking a 20-nucleotide stem loop from RNA segment 3 was still infectious for plants. After prolonged propagation of this mutant RNA 3 in the presence of wild-type RNA segments 1 and 2, a variety of revertant forms of RNA 3 were detected. In all cases, the missing stem-loop sequence had been restored by substituting an homologous region from one of the other two genome segments. Analysis of their nucleotide sequences showed that each of the five recombinant species of RNA 3 examined must have been generated by an independent recombination event. Three recombinants had a sequence of normal length and were produced by equal (i.e., homologous) crossovers, whereas the other two contained duplications of 40 and 10 nucleotides, respectively, and must therefore have been generated by unequal crossovers.

C. Mechanism of Nonhomologous Recombination

DI RNAs, of which the two Sindbis virus/tRNA recombinants are examples, are generally assumed to be produced by a copy-choice mechanism in which the polymerase dissociates from the template and reinitiates elsewhere, using its growing RNA strand as the primer.[55] Pfefferkorn[56] has reviewed the compelling evidence that Sindbis virus rarely, if ever, undergoes homologous recombination. This apparent inability to recombine homologously contrasts with its facility for rearranging its RNA in a nonhomologous fashion. Indeed, none of the families of unsegmented RNA virus that are most noted for their ability to make DI RNAs (toga-, rhabdo-, and paramyxoviruses) have yet been shown to recombine homologously. It is also worth noting that the sequence rearrangements seen in DI RNAs have entirely different characteristics from those of homologous recombinants, in that the former have a highly nonrandom distribution along the genome and tend to occur near sequence motifs resembling transcriptional signals.[57,58] Thus, although both types of RNA recombination can be explained by polymerase jumping, there appears to be a fundamental division between those viruses that characteristically recombine homologously and those that recombine nonhomologously.

The biological basis for this difference was proposed in Section II.F. It also makes mechanistic sense that the two processes should tend to be mutually exclusive. Where the polymerase is free to indulge in nonhomologous template switching, the proportion of equal crossovers occurring by chance will be very low. The choice between homologous and nonhomologous rearrangements presumably reflects the way in which viral replication is normally initiated. For poliovirus RNA synthesis to be initiated, some kind of preformed primer is obligatory, and it is easy to imagine that the replicase has no way of reinitiating synthesis unless its growing RNA strand can find a complementary template sequence with which to hybridize.

BMV is the only virus in which both homologous and nonhomologous recombination has been described.[38] It is possible that BMV manifests two different mechanisms for genetic recombination. Alternatively, since the recombinant viruses were detected only after several days of growth, it may be that the frequency of homologous recombination is relatively low in BMV. This would explain why the conserved 3′ terminal noncoding sequences of its three genome segments are nonidentical; if BMV were to undergo high-frequency recombination as in picornaviruses, it would be expected that a single species of 3′ terminal sequence would quickly become fixed in the population of replicating genome segments, just as we see in tobraviruses. Since BMV also makes viable nonhomologous recombinants, a phenomenon never found in picornaviruses, it is the preference of this author to classify BMV provisionally among those viruses that undergo nonhomologous recombination, along with Sindbis virus, with whose replicase it shares amino acid sequence homology.[59]

D. Nonhomologous Recombination in Evolution

DI particles have been isolated from almost every animal virus in which they have been sought, including picornaviruses. Thus, even though the frequency of such events varies widely, it is reasonable to speculate that most, of not all, viruses occasionally undergo intermolecular rearrangements of a nonhomologous type. The paucity of reported examples may simply reflect the difficulty of devising experimental conditions under which the products of such processes will have a selective advantage. On an evolutionary time scale, however, the traffic of RNA genes from virus to virus, or from host cell to virus, by nonhomologous recombination, may be profoundly important. Evidence suggestive of past gene translocations can be seen in the sequence homology, cited earlier, between the replicase genes of such diverse RNA viruses as Sindbis, BMV, and tobacco mosaic viruses,[59] and in the tandem triplication of the VPg gene of FMDV.[60]

ACKNOWLEDGMENTS

I am grateful to Karla Kirkegaard, Vadim Agol, Olen Kew, Rebecca Rico-Hesse, and Stephen Ortlepp for the generous provision of prepublication material; to Brian Harrison for pointing out the evidence for recombination in tobraviruses; and to Gail Wertz, in whose laboratory this review came to be written, for her hospitality. Work by S.A.O. in our laboratory was supported by the Biomolecular Engineering Program of the commission of the European Communities, under research contract number GB1-2-010-UK to D. McCahon.

REFERENCES

1. **Cooper, P. D.,** Genetics of picornaviruses, in *Comprehensive Virology,* Vol. 9, Fraenkel-Conrat, H. and Wagner, R. R., Eds., Plenum Press, New York, 1977, 133.
2. **McCahon, D., Slade, W. R., Priston, R. A. J., and Lake, J. R.,** An extended genetic recombination map of foot-and-mouth disease virus, *J. Gen. Virol.,* 35, 555, 1977.
3. **Tolskaya, E. A., Romanova, L. I., Kolesnikova, M. S., and Agol, V. I.,** Intertypic recombination in poliovirus: Genetic and biochemical studies, *Virology,* 124, 121, 1983.
4. **Emini, E. A., Leibowitz, J., Diamond, D. C., Bonin, J., and Wimmer, E.,** Recombinants of Mahoney and Sabin strain poliovirus type 1: analysis of in vitro phenotypic markers and evidence that resistance to guanidine maps in the nonstructural proteins, *Virology,* 137, 74, 1984.
5. **Pringle, C. R.,** Evidence of genetic recombination in foot-and-mouth disease virus, *Virology,* 25, 48, 1965.

6. **Hirst, G. K.,** Genetic recombination with Newcastle disease virus, polioviruses, and influenza virus, *Cold Spring Harbor Symp. Quant. Biol.*, 27, 303, 1962.
7. **Ledinko, N.,** Genetic recombination with poliovirus type 1: studies of crosses between a normal horse serum-resistant mutant and several guanidine-resistant mutants of the same strain, *Virology*, 20, 107, 1963.
8. **Lai, M. M. C., Baric, R. S., Makino, S., Keck, J. G., Egbert, J., Leibowitz, J. L., and Stohlman, S. A.,** Recombination between nonsegmented RNA genomes of murine coronaviruses, *J. Virol.*, 56, 449, 1985.
9. **McCahon, D., and Slade, W. R.,** A sensitive method for the detection and isolation of recombinants of foot-and-mouth disease virus, *J. Gen. Virol.*, 53, 333, 1981.
10. **Cooper, P. D.,** A genetic map of poliovirus temperature sensitive mutants, *Virology*, 35, 584, 1968.
11. **Lake, J. R., Priston, R. A. J., and Slade, W. R.,** A genetic recombination map of foot-and-mouth disease virus, *J. Gen. Virol.*, 27, 355, 1975.
12. **King, A. M. Q., McCahon, D., Newman, J. W. I., Crowther, J., and Carpenter, W. C.,** Electrofocusing structural and induced proteins of aphthovirus, *Curr. Top. Microbiol.*, 104, 219, 1983.
13. **King, A. M. Q. and Newman, J. W. I.,** Temperature-sensitive mutants of foot-and-mouth disease virus with altered structural polypeptides. I. Detection by electrofocusing, *J. Virol.*, 34, 59, 1980.
14. **King, A. M. Q., Slade, W. R., Newman, J. W. I., and McCahon, D.,** Temperature-sensitive mutants of foot-and-mouth disease virus with altered structural polypeptides. II. Comparison of recombination and biochemical maps, *J. Virol.*, 34, 67, 1980.
15. **Saunders, K. and King, A. M. Q.,** Guanidine-resistant mutants of aphthovirus induce the synthesis of an altered nonstructural polypeptide, P34, *J. Virol.*, 42, 389, 1982.
16. **Lowe, P. A., King, A. M. Q., McCahon, D., Brown, F., and Newman, J. W. I.,** Temperature-sensitive RNA polymerase mutants of a picornavirus, *Proc. Natl. Acad. Sci. U.S.A.*, 78, 4448, 1981.
17. **Kirkegaard, K. and Baltimore, D.,** The mechanism of RNA recombination in poliovirus, *Cell*, 47, 433, 1986.
18. **Pringle, C. R. and Slade, W. R.,** The origin of hybrid variants derived from sub-type strains of foot-and-mouth disease virus, *J. Gen. Virol.*, 2, 319, 1968.
19. **Mackenzie, J. S. and Slade, W. R.,** Evidence for recombination between two different immunological types of foot-and-mouth disease virus, *Aust. J. Exp. Biol. Med. Sci.*, 53, 251, 1975.
20. **King, A. M. Q., McCahon, D., Slade, W. R., and Newman, J. W. I.,** Recombination in RNA, *Cell*, 29, 921, 1982.
21. **Saunders, K., King, A. M. Q., McCahon, D., Newman, J. W. I., Slade, W. R., and Forss, S.,** Recombination and oligonucleotide analysis of guanidine-resistant foot-and-mouth disease virus mutants, *J. Virol.*, 56, 921, 1986.
22. **King, A. M. Q., McCahon, D., Saunders, K., Newman, J. W. I., and Slade, W. R.,** Multiple sites of recombination within the RNA genome of foot-and-mouth disease virus, *Virus Res.*, 3, 373, 1985.
23. **Forss, S., Strebel, K., Beck, E., and Schaller, H.,** Nucleotide sequence and genome organization of foot-and-mouth disease virus, *Nucleic Acids Res.*, 12, 6587, 1984.
24. **Stanway, G., Cann, A. J., Hauptmann, R., Hughes, P., Clarke, L. D., Mountford, R. C., Minor, P. D., Schild, G. C., and Almond, J. W.,** The nucleotide sequence of poliovirus type 3 Leon 12 a1b: compaison with poliovirus type 1, *Nucleic Acids Res.*, 11, 5629, 1983.
25. **Reuckert, R. R. and Wimmer, E.,** Systematic nomenclature of picornaviral proteins, *J. Virol.*, 50, 957, 1984.
26. **Pincus, S. E., Diamond, D. C., Emini, E. A., and Wimmer, E.,** Guanidine-selected mutants of poliovirus: mapping of point mutations to polypeptide 2C, *J. Virol.*, 57, 638, 1986.
27. **Toyoda, H., Kohara, M., Katoaka, Y., Suganuma, T., Omata, T., Imura, N., and Nomoto, A.,** Complete nucleotide sequences of all three poliovirus serotype genomes. Implication for genetic relationship, gene function and antigenic determinants, *J. Mol. Biol.*, 174, 561, 1984.
28. **Romanova, L. I., Blinov, V. M., Tolskaya, E. A., Viktorova, E. G., Kolesnikova, M. S., Guseva, E. A., and Agol, V. I.,** *Virology*, 155, 202, 1986.
29. **Ortlepp, S. A., Newman, J. W. I., and King, A. M. Q.,** unpublished observations, 1986.
30. **Kew, O. M. and Nottay, B. K.,** Evolution of the oral polio vaccine strain in human occurs by both mutation and intermolecular recombination, in *Modern Approaches to Vaccines*, Chanock, R. and Lerner, R., Eds., Cold Spring Harbor Laboratory, Cold Spring Harbor, N.Y., 1984, 357.
31. **Kew, O. M. and Nottay, B. K.,** personal communication, 1986.
32. **Minor, P. D., John, A., Ferguson, M., and Icenogle, J. P.,** Antigenic and molecular evolution of the vaccine strain of type 3 poliovirus during the period of excretion by a primary vaccinee, *J. Gen. Virol.*, 67, 693, 1986.
33. **Agol, V. I., Grachev, V. P., Drozdov, S. G., Kolesnikova, M. S., Kozlov, V. G., Ralph, N. M., Ramonova, L. I., Tolskaya, E. A., Tyufanov, A. V., and Viktorova, E. G.,** Construction and properties of intertypic poliovirus recombinants: first approximation mapping of the major determinants of neurovirulence, *Virology*, 136, 41, 1984.

34. **Agol, V. I., Drozdov, S. G., Grachev, V. P., Kolesnikova, M. S., Kozlov, V. G., Ralph, N. M., Romanova, L. I., Tolskaya, E. A., Tyufanov, A. V., and Viktorova, E. G.,** Recombination between attenuated and virulent strains of poliovirus type 1: derivation and characterization of recombinants with centrally located crossover points, *Virology,* 143, 467, 1985.
35. **McCahon, D., King, A. M. Q., Roe, D. S., Slade, W. R., Newman, J. W. I., and Cleary, A. M.,** Isolation and biochemical characterization of intertypic recombinants of foot-and-mouth disease virus, *Virus Res.,* 3, 87, 1985.
36. **Robson, K. J. R., Harris, T. J. R., and Brown, F.,** An assessment by competition hybridization of the sequence homology between the RNAs of the seven serotypes of FMDV, *J. Gen. Virol.,* 37, 271, 1977.
37. **Makino, S., Keck, J. G., Stohlman, S. A., and Lai, M. M. C.,** High-frequency RNA recombination of murine coronaviruses, *J. Virol.,* 57, 729, 1986.
38. **Bujarski, J. J. and Kaesberg, P.,** Genetic recombination between RNA components of a multipartite plant virus, *Nature (London),* 321, 528, 1986.
39. **Bergh, S. T., Koziel, M. G., Huang, S., Thomas, R. A., Gilley, D. P., and Seigel, A.,** The nucleotide sequence of tobacco rattle virus RNA-2 (CAM strain), *Nucleic Acids Res.,* 13, 8507, 1985.
40. **Grimsley, N., Hohn, T., and Hohn, B.,** Recombination in a plant virus: template-switching in cauliflower mosaic virus, *EMBO J.,* 5, 641, 1986.
41. **Dixon, L., Nyffenegger, T., Delley, G., Martinez-Ixquierdo, J., and Hohn, T.,** Evidence for replicative recombination in cauliflower mosaic virus, *Virology,* 150, 463, 1986.
42. **Cooper, P. D., Steiner-Pryor, A., Scotti, P. D., and Delong, D.,** On the nature of poliovius recombinants, *J. Gen. Virol.,* 23, 41, 1974.
43. **Richards, O. C., Martin, S. C., Jense, H. G., and Ehrenfeld, E.,** The structure of poliovirus replicative intermediate RNA. Electron microscope analysis of RNA cross-linked in vivo with psoralen derivative, *J. Mol. Biol.,* 173, 325, 1984.
44. **Makino, S., Stohlman, S. A., and Lai, M. M. C.,** Leader sequences of murine coronavirus mRNAs can be freely reassorted: evidence for the role of free leader RNA in transcription, *Proc. Natl. Acad. Sci. U.S.A.,* 83, 4204, 1986.
45. **Plotch, S. J., Bouloy, M., Ulmanen, I., and Krug, R. M.,** A unique cap(m7GpppXm)-dependent influenza virion endonuclease cleaves capped RNAs to generate the primers that initiate viral RNA transcription, *Cell,* 23, 847, 1981.
46. **Bishop, D. H. L., Gay, M. E., and Matsuoko, Y.,** Nonviral heterogeous sequences are present at the 5′ ends of one species of snowshoe hare bunyavirus S complementary RNA, *Nucleic Acids Res.,* 11, 6409, 1983.
47. **Rico-Hesse, R. and Kew, O. M.,** personal communication, 1986.
48. **Hedger, R. S.,** The Maintenance of FMD in Africa, Ph.D. thesis, University of London, England, 1976.
49. **Drake, J. W.,** Interference and multiplicity reactivation in polioviruses, *Virology,* 6, 244, 1958.
50. **Reanny, D.,** The molecular evolution of viruses, in *The Microbe: Part I, RNA Viruses,* Mahy, B. W. J. and Pattison, J. R., Eds., Cambridge University Press, Cambridge, 1984, 1975.
51. **Steinhauer, D. A. and Holland, J. J.,** Direct method for quantitation of extreme polymerase error frequencies at selected single base sites in viral RNA, *J. Virol.,* 57, 219, 1986.
52. **Stanway, G., Hughes, P. J., Mountford, R. C., Reeve, P., Minor, P. D., Schild, G. C., and Almond, J. W.,** Comparison of the complete nucleotide sequences of the genomes of the neurovirulant poliovirus P3/Leon/37 and its attenuated Sabin vaccine derivative P3/Leon 12a1b, *Proc. Natl. Acad. Sci. U.S.A.,* 81, 1539, 1984.
53. **Monroe, S. S. and Schlesinger, S.,** Common and distinct regions of defective-interfering RNAs of Sindbis virus, *J. Virol.,* 49, 865, 1984.
54. **Jennings, P. A., Finch, J. T., Winter, G., and Robertson, J. S.,** Does the higher order structure of the influenza virus nucleoprotein guide sequence rearrangements in influenza viral RNA?, *Cell,* 34, 619, 1983.
55. **Lazzarini, R. A., Keene, J. D., and Schubert, M.,** The origins of defective interfering particles of the negative-strand RNA viruses, *Cell,* 26, 145, 1981.
56. **Pfefferkorn, E. R.,** Genetics of togaviruses, in *Comprehensive Virology,* Vol. 9, Fraenkel-Conrat, H. and Wagner, R. R., Eds., Plenum Press, New York, 1977, 209.
57. **Meier, E., Harmison, G. G., Keene, J. D., and Schubert, M.,** Sites of copy choice replication involved in generation of vesicular stomatitis virus defective-interfering particle RNAs, *J. Virol.,* 51, 515, 1984.
58. **Re, G. G., Morgan, E. M., and Kingsbury, D. W.,** Nucleotide sequences responsible for generation of internally deleted Sendai virus defective interfering particles, *Virology,* 146, 27, 1985.
59. **Ahlquist, P., Strauss, E. G., Rice, C. M., Strauss, J. H., Haseloff, J., and Zimmern, D.,** Sindbis virus proteins nsP1 and nsP2 contain homology from several RNA plant viruses, *J. Virol.,* 53, 563, 1985.
60. **Forss, S. and Schaller, H.,** A tandem repeat gene in a picornavirus, *Nucleic Acids Res.,* 10, 6441, 1982.

Chapter 8

THE GENERATION AND AMPLIFICATION OF DEFECTIVE INTERFERING RNAs

Sondra Schlesinger

TABLE OF CONTENTS

I. Introduction 168

II. DI RNAs Derived from Positive Strand RNA Viruses 168
- A. Picornaviruses 168
- B. Togaviruses 169
  - 1. Sequence Organization of DI RNAs of Sindbis and Semliki Forest Virus 170
  - 2. Expression of Cloned Alphavirus DI cDNAs and Deletion Mapping of the Sindbis DI Genome 171
    - a. Semliki Forest Virus 171
    - b. Sindbis Virus 172
  - 3. Insertion of Foreign Sequences into a Sindbis DI Genome 173
  - 4. Generation and Accumulation of Alphavirus Defective Interfering Particles in Different Host Cells 174
- C. Qβ Bacteriophage 174

III. DI RNAs Derived from Negative Strand RNA Viruses 175
- A. Vesicular Stomatitis Virus (VSV) 175
  - 1. The Generation of DI RNAs 175
  - 2. DI RNAs and Nondefective Virus From Persistent Infections 177
- B. Sendai Virus 177
- C. Influenza Virus 177

IV. Summary of the Structures of DI Genomes 179

V. Mechanisms of Interference 179

VI. Conclusions and Perspectives 180

Acknowledgments 181

References 181

I. INTRODUCTION

The appearance of defective interfering (DI) particles is a well-recognized consequence of passaging virus at high multiplicity of infection. Those properties that define DI particles and distinguish them from other interfering phenomena, first delineated by Huang and Baltimore,[1] include the following: (1) they contain only a part of the viral genome; (2) helper virus is required for replication and encapsidation of the defective genome; (3) they replicate at the expense of the nondefective virus and interfere with the intracellular replication of the latter; and (4) interference is specific; it is most severe for the homologous virus, can occur with related viruses, but is not observed with unrelated viruses. In addition, the defective particles are encapsidated by the same constellation of viral proteins as the helper virus. In some instances, the accumulation of DI particles decreases the cytopathic effects of the virus infection and may be essential in the establishment of persistent infections.

This chapter is devoted exclusively to DI particles derived from RNA viruses and, based on the convention for RNA viruses, is subdivided into DI particles obtained from positive and negative strand RNA genomes.[2] RNA isolated from positive strand RNA viruses is infectious. DI RNAs can also express their biological activity when transfected into cells in the presence of helper virus.[3,4] The positive strand RNA genome functions as an mRNA — therefore, DI RNAs also have this potential. Negative strand RNA viruses must transcribe the genome RNA into mRNAs. These viruses contain, within the particles, proteins required for transcription; once the viral ribonucleoprotein has entered the cell, the first step in the replicative cycle is transcription of the genome. This transcription is defined as *primary transcription,* in contrast to transcription occurring after the synthesis of viral proteins which is termed *secondary transcription.*

Several earlier reviews document the RNA viruses from which DI particles have been isolated.[5-7] The presence of DI particles in these virus populations depends on two factors: the generation of defective genomes and their amplification. The generation of defective molecules requires only that aberrant replicative events occur. These errors will be detected, however, only if the resultant RNAs can be amplified, and amplification requires that the molecules be recognized by viral replicase and presumably by those proteins involved in the encapsidation process. The ability to identify what properties of defective RNAs are required for amplification requires that the sequence of these genomes and that of the nondefective genome from which they are derived be determined. This information is now available for several DI genomes and these are the ones featured here almost exclusively. Defining the essential sequences in a DI RNA, however, represents only one step in understanding the biological properties of these DI particles and, included within the framework of generation and amplification of DI RNAs, is a brief overview of the mechanisms of interference.

II. DI RNAs DERIVED FROM POSITIVE STRAND RNA VIRUSES

A. Picornaviruses

One of the most extensive early investigations of DI RNAs was carried out by Cole et al.[8] and Cole and Baltimore.[9,10] They isolated DI particles by high multiplicity passaging of type 1 poliovirus in suspension cultures of HeLa cells. It required more than 16 high-multiplicity passages to detect DI particles, and the defective RNAs ranged in size from 80 to 90% of that of the virion RNA. Later studies by Nomoto et al.[11] showed that the deletions in polio DI RNAs were located approximately 1300 to 3100 nucleotides from the 5′ end of the genome in the region coding for capsid proteins. Based on detailed electron microscopic analysis, Lundquist et. al.[12] found that deletions in independently isolated DI RNAs were always in the 5′ region of the genome. In addition, the deletions were heterogeneous, and in some cases, two or more deletions were detected in the same molecule.

The picornaviruses have been distinctive among RNA viruses with respect to both the number of passages required to detect defective particles and the relatively small size of the deletions. The 16 to 18 passages required by Cole et al.[8] to detect DI particles in their strain of poliovirus were less than the number required to detect DI particles in other picornaviruses. McClure et al.[13] carried out a detailed study of the appearance of DI particles by passaging of poliovirus and mengovirus. They found that in some cases the deletions in the genome RNA were so small (4 to 6% of the genome size) that it was not possible to separate DI particles from virions. They were able to detect DI RNAs only by high-resolution gel analysis of denatured RNAs. They observed that the accumulation of DI RNAs was dependent on both the strain of virus and the cell type in which the virus was grown. Thus, DI RNAs of mengovirus could be generated in baby hamster kidney (BHK) cells, but not in HeLa cells, even though the latter cell line was capable of replicating the deleted RNAs.

Recently, Radloff and Young[14] found that DI particles could be isolated by passaging of encephalomyocarditis virus in three different cell lines — HeLa, L929, and BHK cells. It took 40 to 50 passages for the DI particles to accumulate, but the cell lines all behaved similarly with respect to their ability to produce DI particles. Detection of DI particles in the Sabin vaccine strain of poliovirus type 1 required more than 30 passages.[15]

It isn't clear why so many passages are required to detect DI particles in picornaviruses. The finding by McClure et al.[13] that DI RNAs could be detected in some cell lines, but not in others strongly suggests that the generation of DI RNAs is dependent on host cell factors, an observation which has been also noted for other viruses. Since the detection of DI particles requires extensive amplification of the deleted RNA, it may be that only a very limited size range of RNA molecules is able to be replicated and/or encapsidated by the picornavirus proteins. Large deletions in a picornavirus genome may produce molecules unable to meet the packaging constraints of the icosahedral structure of the picornavirus nucleocapsid.

The deletions in the DI RNAs have all been mapped to a narrow region between 1300 and 3100 nucleotides from the 5′ terminus. A sequence analysis of the RNA from DI particles derived from the Sabin strain of poliovirus revealed that the deletions in these DI RNAs all occurred without changing the coding frame of the nucleotides downstream from the deletion. This finding led Kuge et al.,[15a] to propose that one or more of the viral nonstructural proteins is *cis*-acting, and, therefore, any genome in which the correct translation frame is not maintained would not survive. In light of these recent results it will be particularly interesting to know what types of deletions will be biologically active. It should be possible to construct various deletions in the infectious cDNA clones that have been described.[16,17] If the small size of the deletions is a consequence only of a size limitation in packaging, then it may be possible to replace picornavirus sequences with nonviral stretches. This type of approach could be used to determine if there are specific *cis*-acting sequences required for encapsidation.

B. Togaviruses

Detailed studies of DI particles in this family have been carried out with two members of the alphavirus genus — Sindbis virus and Semliki Forest virus (SFV).[18] The initial isolation and characterization of these DI particles have been reviewed.[19,20] Depending on the cells and the conditions of passaging, the first DI RNAs to be detected after about three to five high-multiplicity passages are about one half the size of the virion RNA.[21,22] These molecules soon disappear on subsequent passaging and are replaced by molecules one fourth to one fifth the size of the original genome. It has been assumed that larger DI RNAs give rise to the smaller ones since all of the oligonucleotides found in the smaller molecules are present in the larger RNAs. Naturally occurring DI RNAs fall into a size range that is a fractional integer of the genome size (12 kilobases). This observation, as well as the finding that the nucleocapsids of DI particles are essentially the same size and density as those of the infectious virions, suggest that DI nucleocapsids contain multiple species of DI RNA to make up a mass equivalent to that of virion RNA.[21-24]

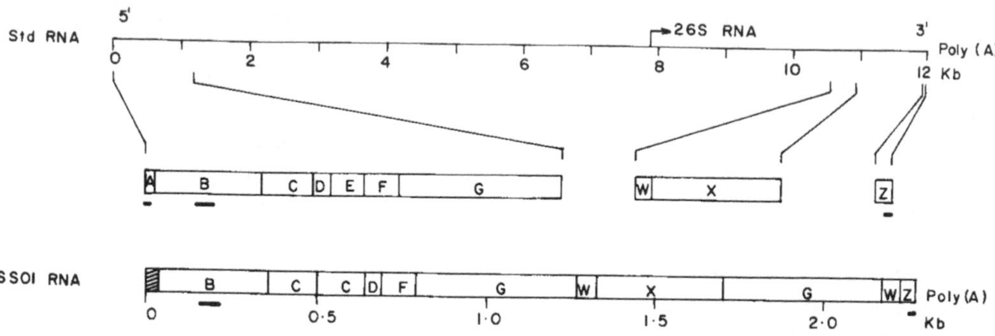

FIGURE 1. A comparison between the deduced sequence of the standard Sindbis HR virus RNA and the SSO1 DI RNA. The top line represents the total 12 kilobases (Kb) of standard virion RNA; the regions of homology with the DI RNA are expanded below. The bold underlines represent the conserved regions of the genome (see the text). Regions A to G are derived from the 5′ two thirds of the standard genome, which codes for the nonstructural proteins. Regions W to Z are derived from the 3′ one third of the standard genome, which codes for the structural proteins. The hatched area in SSO1 RNA is the $tRNA^{Asp}$. (Modified from Monroe, S. S. and Schlesinger, S., *J. Virol.*,, 49, 865, 1984. With permission from the American Society of Microbiology.)

1. Sequence Organization of DI RNAs of Sindbis and Semliki Forest Virus

The initial analysis of alphavirus DI RNAs based on nucleic acid hybridization and T1-ribonuclease resistant oligonucleotide mapping suggested that alphavirus DI RNAs contained one or more internal deletions of the virion RNA.[25,26] Sequence analyses, however, demonstrated that these RNAs are more complex than had originally been thought. Sequencing of two cloned complementary DNAs (cDNAs) derived from a population of DI RNAs of SFV showed that the DI genomes consist of repeated units derived from the 5′ two thirds of the viral genome and a region derived from the 3′ terminus.[27-29] A similar structure was found for a cDNA from a DI genome (SSO1) of Sindbis virus,[30] and a comparison of its sequence organization with that of the virion RNA is shown in Figure 1. About 75% of the sequences found in this DI genome come from the 5′ terminal 1200 nucleotides of the virion RNA. Although the DI genome is 20% the physical length of the virion RNA, because of the repeats, it has only 13% the sequence complexity of the larger RNA.

The sequence of the entire genome of Sindbis virus has been determined,[31] and there is sufficient sequence information for several related alphaviruses to identify regions of conservation. The four most highly conserved regions are (1) 19 nucleotides at the 3′ terminus,[32] (2) 21 nucleotides spanning the start of the subgenomic 26 S mRNA,[33] called the junction region, (3) 51 nucleotides near the 5′ terminus,[34] and (4) about 40 nucleotides at the 5′ terminus.[34] Sequence information from the defective genomes of Sindbis and Semliki Forest viruses reveals interesting correlates with respect to these four regions.[29,30] The 3′ termini of several different DI RNAs from both viruses are identical to those of the standard genomes, demonstrating that this region is also conserved in defective genomes. The junction region is missing from all alphavirus DI genomes examined to date. In contrast, the 51-nucleotide region is conserved in these genomes. The 5′ terminal sequences are the most problematical. They are the least conserved among the four conserved regions of alphavirus genomes and are not conserved in many of the defective genomes of Sindbis virus.

A most unexpected finding was that the 5′ end of the DI RNA diagrammed in Figure 1 is almost identical to the sequence of a rat $tRNA^{Asp}$.[35] This DI RNA and another independently generated one contain essentially all but the terminal nine 5′ nucleotides of the $tRNA^{Asp}$. The 3′ C–C–A terminus of the tRNA is covalently attached near the 5′ terminus of the DI RNA. In one DI RNA, the tRNA replaced the 5′ terminal 31 nucleotides of the virion RNA; in the other DI RNA, 23 nucleotides from the 49 S RNA were replaced. The incorporation of the $tRNA^{Asp}$ into DI RNA is almost certainly not a random event. Three of the four times that DI RNAs were generated in chicken embryo fibroblasts, the 5′ terminus was $tRNA^{Asp}$. The predominant species of $tRNA^{Asp}$ would be expected to constitute only about 5% of the total cellular tRNA, making it improbable that the same tRNA would be selected at random.

The presence of this tRNA in the DI genome raises the possibility that it plays some role in standard viral RNA replication. Although there is no evidence for such a role, there is precedent for an involvement of tRNA or tRNA-like structures in viral replication. A cellular tRNA serves as primer for the synthesis of retrovirus cDNA.[36] It is also well established that the 3′ termini of many plant viral RNAs have an amino acid accepting activity and have a tRNA-like structure.[37] The recent evidence of evolutionary relatedness between plant RNA viruses and Sindbis virus provides further grounds for thinking that a tRNA could be involved in Sindbis virus replication.[38]

The 5′ terminal sequences of six different Sindbis DI RNAs generated either in chicken embryo fibroblasts or in BHK cells have been identified.[39] Each independently generated DI RNA actually represents a population of molecules of DI RNA; however, in all cases but one, only a single 5′ terminus was found. The most common (three of six) and the most unusual 5′ terminus is the $tRNA^{Asp}$, but one of two DI RNAs (DI-1) generated in BHK cells has the 5′ terminus of 26 S RNA with a deletion from nucleotides 24 to 67 of the 26 S RNA. The other DI RNA population generated in BHK cells has the same 5′ end as the virion RNA. The DI RNA population generated in chicken cells that does not have a $tRNA^{Asp}$ 5′ terminus is the one in which several different 5′ termini were detected. They consisted of rearrangements of regions near the 5′ end. All six DI RNAs are replicated efficiently by the viral enzymes. Although it is possible that these 5′ terminal sequences all form some related stem-loop structure,[34] it appears that a variety of structures or sequences at the exact 5′ terminus are consonant with replication of the RNA.

Sequences lost from the DI RNAs must be those that are not required for either replication or packaging of these RNAs. It is also possible that the loss of a particular sequence can favor amplification of a defective genome and it may be significant that the region spanning the start of the subgenomic 26 S mRNA is deleted in the first DI RNAs to be detected during high-multiplicity passaging.[22,30] The loss of this region could provide a selective advantage for a DI RNA, the negative strand of which would serve as a template only for replication and would not be engaged in transcription of a subgenomic mRNA.

2. *Expression of Cloned Alphavirus DI cDNAs and Deletion Mapping of the Sindbis DI Genome*

The retention of viral sequences in a DI genome does not prove that they are essential for the viability of the DI RNA. The identification of essential sequences requires that they be deleted or modified and that the biological activity of the resulting molecules be determined. Procedures for specific modifications of DNA are well documented, and alterations of RNA can be carried out with their cDNA counterparts which then must be transcribed back into RNA. Two approaches have been taken to analyze alphavirus DI genomes.

a. Semliki Forest Virus

Jalanko and Söderland[40] inserted a cloned cDNA from a DI RNA of SFV into the late gene region of Simian virus 40 (SV40). The DI RNA was synthesized in monkey kidney cells that were coinfected with the SV40-Semliki Forest DI hybrid virus and helper SV40 virus. This clone lacks the 5′ terminus of the Semliki Forest DI RNA. When cells expressing this DI RNA were superinfected with SFV, the particle yield was reduced about twofold. There was no effect on the synthesis of virus-specific RNA, nor was there any evidence that the DI RNA was packaged. The data suggest that the DI RNA expressed from the SV40 vector does not contain recognition sites for replication by the SFV enzymes, but that it does retain packaging signals because it interferes with particle yield. Because the DI RNA itself was not packaged, essential information required for packaging may not have been retained. This type of study suggests that it would be possible to design an RNA that would effectively inhibit virion release but would not itself be packaged. Such an RNA could be a model for a new type of antiviral agent.

b. Sindbis Virus

To identify those sequences required for the replication and encapsidation of DI RNAs of Sindbis virus, Levis et al.[4] determined the effects of a series of deletions on the biological activity of the DI genome. To do this, they cloned the DI cDNA directly downstream from the promoter for the SP6 bacteriophage DNA-dependent RNA polymerase. The DI cDNA could then be transcribed into RNA in vitro and transfected into chicken embryo fibroblasts that are simultaneously infected with helper virus. Only a small fraction of the cells are transfected with DI RNA, but all the cells are infected with infectious virus so that every cell receiving a DI RNA will have the enzymes required to replicate that molecule. The transfected DI RNA becomes the predominant viral RNA species after two to three passages. The pattern of viral RNAs detected in infected cells during the formation of the third passage is shown in Figure 2. The concentration of RNA used in these transfection assays was varied over a 100-fold range. DI-25 is the cloned DI genome, the sequence of which is essentially identical to that of the naturally occurring DI RNA shown in Figure 1. By the third passage, even cells transfected with the lowest concentration of this DI RNA show substantial amplification of the DI RNA and interference with the synthesis of the viral 49 S and 26 S RNAs. DI-554 is a DI RNA obtained from a deleted cDNA and is described below.

The deletions constructed and their effect on the biological activity of the transcribed RNA are summarized in Figure 3. More than 90% of the sequences in the DI RNA have no essential role in the replication or packaging of the molecule. Only sequences in the 162 nucleotides at the 5′ terminus and sequences in the 19 nucleotides at the 3′ terminus are specifically required for replication and packaging of the genome. The 19 nucleotides at the 3′ terminus are those defined by Ou et al.[32] to be highly conserved among alphaviruses. Thus, these nucleotides appear to be necessary and may be sufficient to define the recognition site for initiation of negative strand synthesis.

The two regions of interest at the 5′ terminus are the extreme 5′ end and the 51 conserved nucleotides located from nucleotides 155 to 205 in the virion RNA.[34] The observations that naturally occurring DI RNAs vary at the extreme 5′ terminus and that this region is the least conserved among alphaviruses led to the proposal that the 51-nucleotide region plays a crucial role in one or both functions of replicase binding and of encapsidation.[34] Deletion of this region, however, does not destroy the biological activity of the DI RNA,[4] and, therefore, the function of these sequences remains in question. There is, however, a quantitative difference in the ability of the RNA to be amplified. The original DI cDNA clone (DI-25) is amplified to high levels after three high multiplicity of infection passagings, even when only 5 ng of RNA is used for transfection (Figure 2). In contrast, the clone with the 51 nucleotides deleted (DI-554) shows barely detectable levels of DI RNA at passage 3 when the transfecting RNA concentration is 5 or 50 ng. Detection of the DI RNA requires both replication and encapsidation; it is not known whether one or both of these activities is affected in the deleted DI-554 RNA.

Sequences upstream from the 51-nucleotide conserved sequence are essential for the biological activity of DI RNA. A cDNA with a deletion of the first seven nucleotides and one with a deletion from nucleotides 22 to 162 produce RNA transcripts that are not biologically active. The 5′ terminal seven nucleotides are part of the $tRNA^{Asp}$ sequence which is not present either in several other DI RNAs or in Sindbis virion RNA and must be only one of several sequences able to be recognized at the 5′ terminus.

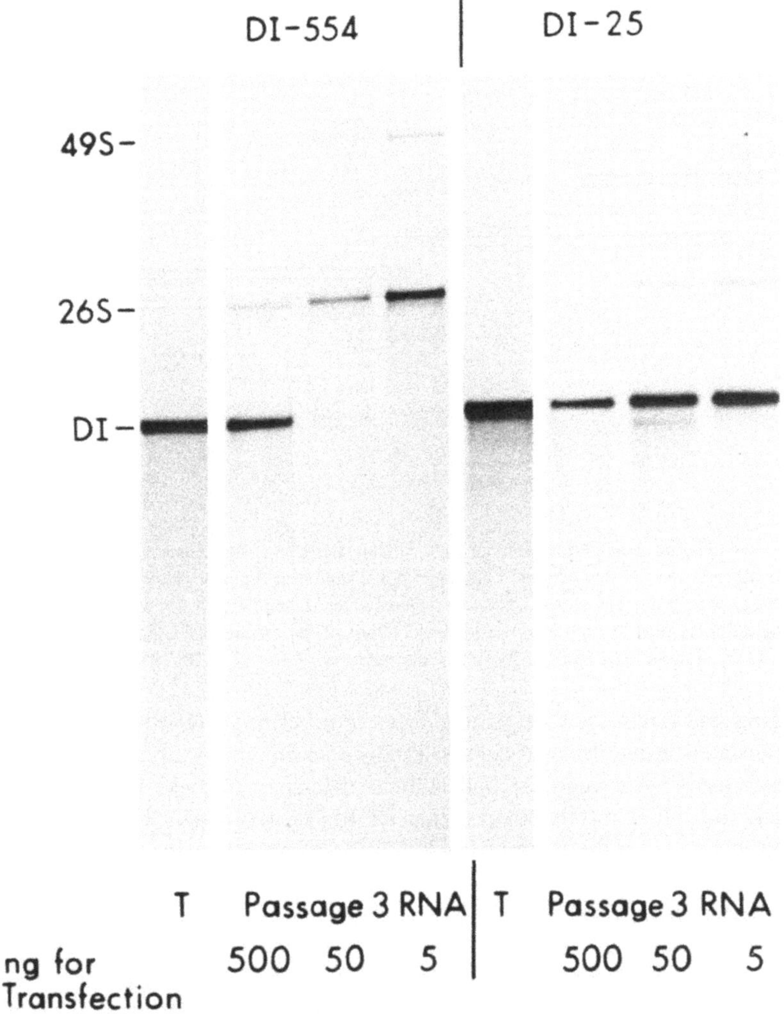

FIGURE 2. The amplification of DI RNAs of Sindbis virus in chicken embryo fibroblasts after transfection with varying concentrations of transcribed RNA. The intracellular RNA was analyzed during the formation of passage 3. T refers to the transcript used for transfection. DI-25 is derived from intact DI cDNA obtained from SSO1 RNA (Figure 1). DI-554 lacks the conserved 51-nucleotide sequence.

3. Insertion of Foreign Sequences into a Sindbis DI Genome

The conclusion that only a small fraction of Sindbis sequences are essential for propagation of the DI genomes suggested that the defective genomes might serve as vectors for the expression of foreign genes. To determine the feasibility of this idea, two foreign genes — the bacterial chloramphenicol acetyltransferase (CAT) gene and the vesicular stomatitis virus (VSV) glycoprotein gene, the G gene — were inserted into the DI genome in place of Sindbis sequences.[41] In these constructions, the viability of the DI genomes depended on the inclusion of the 51-nucleotide conserved region — a result that is not surprising given that the presence of these sequences does provide a quantitative advantage to the DI RNA (Figure 2).

The DI RNAs containing either the CAT gene or the VSV G protein gene are amplified

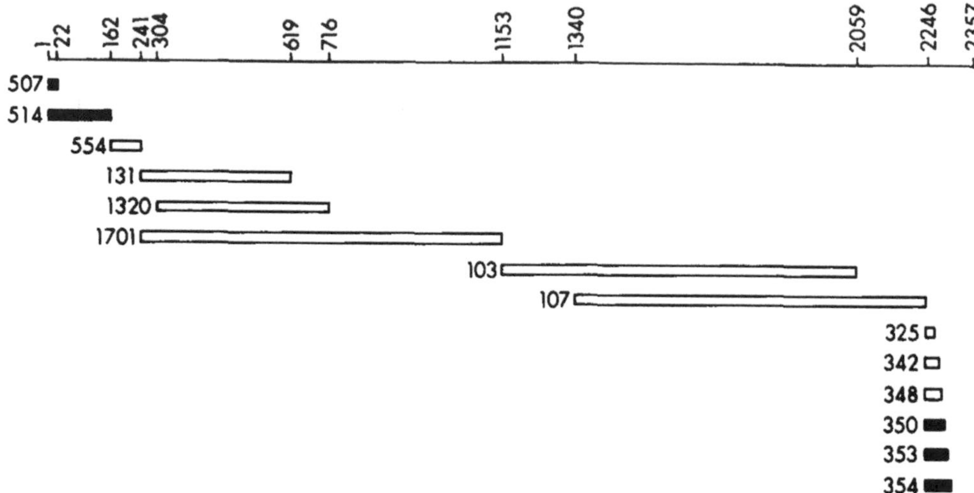

FIGURE 3. A deletion map of the genome of a Sindbis DI RNA. The deletions, which cover the entire DI genome, are all indicated on this diagram. The open boxes represent regions which were deleted without causing a loss of biological activity. The closed boxes are regions which, when deleted, led to a complete loss in biological activity. The numbers next to each box are the designations of the different deleted cDNA clones. (From Levis, R., Weiss, B. G., Tsiang, M., Huang, H., and Schlesinger, S., *Cell*, 44, 137, 1986. With permission.)

on passaging and retain the foreign sequences. Furthermore, DI RNA containing the CAT gene is translated in the infected cells to produce an enzymatically active chloramphenicol acetyltransferase.[41a] Although the translational efficiency of these RNAs has not yet been determined, the initial results suggest that DI RNAs may prove to be useful vectors for introducing genes — in the form of RNA — into eucaryotic cells.

4. Generation and Accumulation of Alphavirus Defective Interfering Particles in Different Host Cells

One of the first reports of the generation of DI particles of SFV described a much greater level of interference in mouse cells than in chicken embryo cells.[42] This observation was extended by Stark and Kennedy,[22] who compared the generation and amplification of DI RNAs in seven different cultured cell types. They found differences both in the time of first appearance and in the rate of accumulation of DI particles. In one line of HeLa cells, DI particles were not detected even after 200 passages.

The differences among cells could depend on the generation of DI RNAs, on the amplification of the DI RNA, or on the packaging efficiency of the RNA. Stark and Kennedy[22] showed that if DI particles were added to the one nonproducing HeLa cell line, the DI RNA could be replicated. They concluded that the absence of DI particles was due to their inability to be generated.

There have been several reports showing that DI particles obtained from vertebrate cells are unable to replicate and interfere in *A. albopictus* cells. Steacie and Eaton[44] showed that a stock of DI particles generated in chicken embryo fibroblasts replicates, but does not interfere, in mosquito cells. These authors also demonstrated quite clearly that DI particles generated in mosquito cells are heterogeneous with respect to their ability to replicate in vertebrate cells. On the basis of the studies of Tooker and Kennedy,[43] the results with mosquito cells may also be complicated by heterogeneity in the host cell population.

C. Qβ Bacteriophage

Qβ bacteriophage is one of a group of small single strand RNA phages that replicate in *Escherichia coli*.[45] Although it is not usually included among viruses that generate DI particles, the Qβ replicase and the variant RNAs replicated by the enzyme represent an important model for any study of RNA template recognition. The Qβ replicase is composed of one phage-coded polypeptide and three host-coded polypeptides.[46,47] The latter are the 30 S ribosomal protein S1 and the protein synthesis elongation factors EF-Tu and EF-Ts. The identification of host cell proteins as components of this replicase complex still represents the best example of a requirement for host proteins in the functioning of a viral RNA polymerase.

Qβ replicase binds to internal regions of Qβ RNA and initiates transcription at the 3′ terminus of the positive strand.[46,47] Based on studies of the binding of Qβ replicase to fragments of Qβ RNA, Meyer et al. conclude that productive interactions require that the enzyme bind to RNA and that the 3′ terminus be properly oriented so that it lies in the initiation site.[48] They have proposed that the enzyme recognizes nucleotide sequences and also a three-dimensional structure of the RNA.

Qβ replicase normally shows a high degree of template specificity for Qβ RNA. It also produces and replicates variant RNAs generated in vitro under conditions in which the only constraint on the RNA is to replicate fast.[49,50] In many respects this condition parallels the amplification of DI RNAs in vivo even though the size of a DI RNA often may not be the deciding factor in its selection. Qβ replicase also efficiently replicates in vitro a naturally occurring 6 S RNA and variant RNAs produced in the absence of added template.[47] Two of these RNA species, one of 218 nucleotides (midivariant RNA) and the other of 114 nucleotides (microvariant RNA), were sequenced by Mills et al.[51,52] A third RNA of 91 nucleotides (nanovariant RNA) was sequenced and analyzed by Schaffner et al.[53] The latter authors compared the sequences of the three variants and found almost no common sequence homologies among them except for the 3′ terminal $C–C–C(A)_{OH}$. The lack of sequence homology supports the concept that structure is more important for recognition than a particular linear sequence. This proposal was also made for the recognition of the 5′ terminus of alphaviruses[34] and the DI RNAs of Sindbis virus.[39] Nishihara et al., however, report stretches of sequence identities between the midivariant positive strand RNA and the Qβ RNA negative strand, and suggest that these regions of homology contribute to the recognition of midivariant RNA by the enzyme.[54]

Secondary structure for the variant RNAs recognized by Qβ replicase have been drawn based on computer predictions,[53,54] and in some cases, the structures have been confirmed experimentally.[55] It seems likely that the extensive information already available about these RNAs will continue to make them important tools in resolving the roles of sequence specificity and structure in enzyme recognition. The mechanism of replication of Qβ RNA and of variant RNAs is reviewed in Chapter 1 of Volume I of this series.

III. DI RNAs DERIVED FROM NEGATIVE STRAND RNA VIRUSES

A. Vesicular Stomatitis Virus (VSV)

1. The Generation of DI RNAs

DI RNAs derived from VSV have been the subject of extensive investigations in several laboratories and have also been the focus of several excellent reviews.[6,7,56] Nucleic acid hybridization analyses and sequencing of the termini of these RNAs demonstrated that there are several different classes. The most prevelent are illustrated in Figure 4. The majority of DI RNAs are derived from the 5′ end of the VSV genome and possess complementary termini (stems or panhandles). The extent of complementarity varies from about 46 to 200 nucleotides.[57] A second class of DI RNA molecules, termed snap-back,[58,59] are isolated as double-stranded molecules, in which one half of the molecule is essentially the complement of the other.

The discovery of these structures, particularly the variable stem or panhandle molecules,

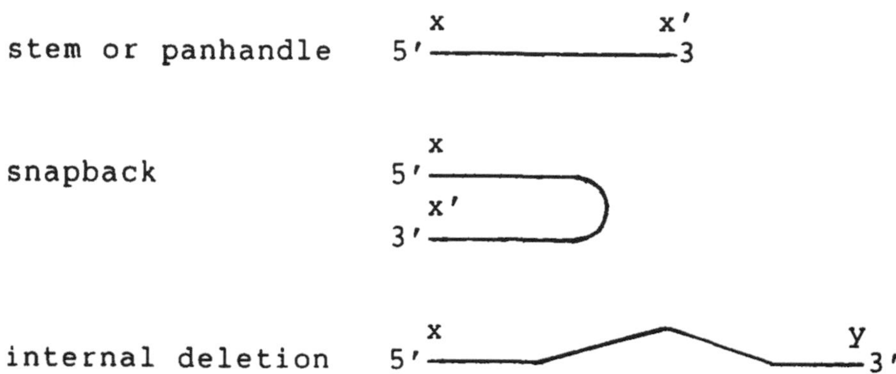

FIGURE 4. Sequence organization of DI RNAs derived from the nondefective genome of VSV. X and Y represent the 5′ and 3′ termini of the nondefective genome, respectively. X′ is the complement of X. Sequences deleted are indicated by /\. These structures have been described in more detail by Perrault[7] and by Lazzarini et al.[56]

in both VSV and Sendai virus DI RNA molecules led to a model to explain the generation of these copyback structures.[60,61] This model proposes that during transcription of the antigenome or plus strand, the polymerase falls off the template carrying the nascent genomic strand with it. Synthesis is reinitiated such that the enzyme "copies back" using the nascent strand as the template. The site at which reinitiation takes place determines the length of the stem. The finding that the length of the stems is variable, extending from 46 base-pairs to essentially the entire length of the RNA, in the example of the snap-back molecules, suggests that there are no specific sites at which reinitiation preferentially occurs. Based on their sequence studies, however, Meier et al. propose that specific sequences may enhance the generation of particular types of DI RNA.[62]

Several other classes of VSV DI RNAs have also been described. One class contains a single internal deletion and retains both the 5′ and 3′ ends of the genome.[7,56,63] Only one member of this class had been characterized until recently when Kang et al.[64] described the appearance of internally deleted molecules derived from a heat-resistant strain of VSV. DI RNAs that conserve the 3′ terminus of the virion genome will be transcribed by the viral polymerase to yield leader and messenger RNAs. Studies with the originally internally deleted DI RNA indicated that it was distinct from the other DI RNAs in being able to interfere with a heterologous strain of VSV as well as with the homologous strain.[7] (The differences in the biological properties of these two classes of DI RNAs are described in Section III.B on Sendai virus).

The other classes of VSV DI RNAs are more complicated and represent combinations of either stem and snap-back or stems and internal deletions.[7,65] These latter two classes are probably derived from DI RNAs rather than directly from nondefective virion RNA.[65] The variety of DI RNAs detected led both Lazzarini et al.[56] and Perrault[7] to propose a general model to account for the variety of DI RNA structures found. The crux of their models is that the polymerase is promiscuous and is able to switch templates without releasing the nascent strand. This model allows almost any structure to be generated; the ability of a structure to survive and be amplified depends on whether or not it can be replicated and encapsidated.

2. DI RNAs and Nondefective Virus from Persistent Infections

Holland and co-workers established persistent infections of cultured BHK cells by infecting cells with VSV stocks heavily enriched in DI particles. They have analyzed both the DI particle population and the infectious virions over more than 10 years.[6,66-69] One of the most interesting aspects of their study has been the extensive changes in both the virus and DI particle population that continue to occur. Perhaps the most significant change is the appearance of virus mutants resistant to the DI particles present in the original inoculum.[66] This type of mutant virus has also been isolated from cultures persistently infected with Sindbis,[70] rabies,[71] or lymphocytic choriomeningitis[72] viruses. The selection of these mutants is probably crucial for the survival of the nondefective virus which otherwise would be so suppressed by the DI RNAs that it would be lost from the culture.

A likely mechanism for the acquisition of resistance would be one in which changes in the replicase and in the replicase recognition sites occur such that the replicase binds to the nondefective RNA, but no longer recognizes the defective RNA. As a test of this mechanism for VSV, O'Hara et al. sequenced the 5′ and 3′ termini of 11 different viruses isolated from persistently infected cultures and of 13 viruses from undiluted passaging.[68,69] All of these viruses were not subject to interference by the original DI particles in the inoculum. Stepwise mutations were detected particularly in the first 54 nucleotides of the 5′ terminus and less frequently in the 3′ terminal nucleotides, consistent with the idea of terminal recognition sequences. In several instances, however, no change in either terminus was found. Therefore, resistance to DI RNA interference can also occur by changes in recognition sites or structures that are not at the extreme termini.

To obtain more information about the DI particles that appear at different stages during a persistent infection as well as those that accumulate by undiluted passaging, O'Hara et al. sequenced sixteen different DI RNAs.[69] In most cases, the RNA obtained from DI particles at a particular time in the persistent infection had the same terminal nucleotide sequences as the infectious virion isolated at the same time. Differences were detected, however, between DI RNA and nondefective RNA obtained from undiluted passaging of virus. These DI RNAs all contained 5′ complementary termini, and changes in one terminus were accompanied by a compensatory change in the other.

These DI RNAs show rearrangements and repeats as well as deletions. O'Hara et al.[69] point out that the variety of internal sequences suggests that the terminal stems should be the only regions of importance for replication and interference. Apparently, however, two DI particles that do not interfere well with their homologous virus have the same terminal RNA sequences as some DI particles that interfere well. These results taken together with their earlier finding that nondefective genomes can acquire resistance to interference without changes in terminal nucleotides suggest that internal sequences may be of some importance in the survival of DI particles. Their importance could be attributed to specific nucleotide sequences or to their contribution to an overall tertiary structure which can certainly be influenced by these sequences.

B. Sendai Virus

The observation that DI RNAs of Sendai virus were visualized as circles in the electron microscope suggested that the termini of these molecules are complementary.[73] Nucleic acid hybridization studies of the terminal stems of these molecules provided further evidence for complementarity.[61] They also showed that these RNAs contained sequences homologous to the infectious virion RNA, but that the ends of the nondefective genome are not complementary. This analysis by Kolakofsky and co-workers led them to propose the copyback model[61] for the generation of DI RNAs described above.

Further studies of the DI RNAs isolated from Sendai virus showed, however, that many of these RNAs are internally deleted molecules.[74-76] The existence of well-characterized DI RNAs of different structures made it possible for Re and Kingsbury to compare the efficiencies of the different DI RNAs.[77] They first compared a copyback DI RNA with one that was internally deleted and found that the copyback RNA completely outgrew the internally deleted species by the seventh passage. The size of the RNAs does not seem to be a factor, since a larger copyback RNA competes effectively with smaller internally deleted RNAs.

Re et al. suggest that the internally deleted RNAs do not compete effectively either because the 3′ terminus is not a strong promoter of RNA synthesis or because molecules that have a 3′ terminus identical to that of the nondefective RNA will, like the nondefective RNA, also be involved in transcription and therefore will be less effective in replication.[77] One of the internally deleted RNAs they analyzed turned out to be relatively resistant to competition by the copyback DI RNA. This particular internally deleted RNA is transcriptionally inactive due to a mutation in the 3′ terminal region. It contains two mutations near the 3′ terminus of the negative strand. A residues were inserted into positions 47 and 51 from the 3′ end in place of G and U residues, respectively. These changes provide some clues to the sequence requirements at this terminus.

The authors conclude that the 3′ terminal promoter is not as effective as the 5′ promoter based on their finding that internally deleted molecules are outcompeted by the copyback molecules. Transcriptional inactivity, however, also appears to confer some replicative advantage. In spite of the selective disadvantage, many of the DI RNAs they characterized are internally deleted and transcriptionally active. Perhaps these DI RNAs are generated more frequently and once they are amplified, suppress the replication of both the infectious genome and any newly generated DI RNA molecules. The competition described by Re et al. was with DI RNA molecules at essentially the same concentration. It would be interesting to know if copyback RNAs would compete effectively if they were introduced at a low level, compared to the internally deleted molecules.

C. Influenza Virus

The first description of defective particles in animal viruses was attributed to von Magnus[78] who reported the presence of "incomplete" virus in populations of influenza virus in embryonated chicken eggs. In spite of the early discovery of DI particles in this virus family, only recently has a detailed characterization of the genomes of these particles been accomplished. The genome of influenza virus is segmented and some of the difficulties in analyzing defective RNAs were due to the difficulties in distinguishing them from the genome segments and from the mRNAs transcribed from those segments. Only after the eight segments that make up the complete genome of the virus were identified was it possible to identify and analyze the DI RNAs. Two excellent and comprehensive reviews[79,80] have recently been published and only some of the highlights of the work with these DI RNAs are presented here.

Forty-one DI RNAs of influenza virus have been sequenced.[79-81] All of these RNAs arose by internal deletions and almost all (90%) contain just a single deletion. Three RNAs have two deletions, one has a rearrangement and one is a mosaic of two segments.[82] Most (35) of the defective RNAs are derived from the polymerase genes. There are three polymerase genes, PB1, PB2, and PA, but the largest number of deleted RNAs (26) come from the PB2 gene. There is no explanation for this selectivity. Nayak et al.[79,80] point out that although the polymerase genes are the largest RNA segments and therefore could be subject to more replicase errors, they also are replicated at a lower rate than other genes. If errors generating DI RNAs are random, then both the size of the gene and the number of molecules replicated should determine the distribution of deletions in the RNA segments. The bias in the distribution of DI RNAs detected must be due either to nonrandom errors made by the replicase or to selectivity in amplification. There did not appear to be any unique sequences in the polymerase genes to support the idea that deletions in these genes would be favored.

An examination of the sequences of the DI RNAs of influenza virus did not reveal any obvious sequence stretch or secondary structure to explain why particular deletions would be favored. Nayak et al.[79,80] propose that for influenza virus, as well as for other viruses in which DI RNAs are generated by internal deletions, the "promiscuous" polymerase may not be so promiscuous; it may not dissociate from its template and reattach at random, but instead, just skip or roll over looped-out regions of the template. In addition, it has been pointed out that influenza RNAs are replicated not as free RNA, but as ribonucleoprotein. Therefore, it may be the tertiary structure of the ribonucleoprotein that is crucial in guiding how the replicase generates deletions.[81]

IV. SUMMARY OF THE STRUCTURES OF DI GENOMES

Based on the available sequence information, DI RNAs can be placed into the following three classes:

1. Internal deletions. These genomes have also been termed 5′-3′[79] because both the 5′ and 3′ terminal sequences are retained. This type of deletion is found in DI RNAs derived from both positive and negative strand RNA genomes. These DI molecules may also have additional modifications such as repeats, rearrangements, or simple nucleotide changes.
2. 5′ Conserved molecules. These molecules retain the 5′, but not the 3′ terminus of the virion.[6,7,58] In all cases so far examined, the 3′ sequences are the inverted complement of the 5′ terminus. The two types of molecules are those which form a base-paired stem or panhandle and the snap-back molecules in which the 3′ half is the complement of the 5′ half. These types have only been described for DI RNAs derived from VSV and Sendai virus.
3. 3′ Conserved molecules. These molecules retain the 3′, but not the 5′ terminus of the virion genome. The only naturally occurring DI genomes that fall into this class at present are the DI genomes of Sindbis virus.[18] This designation emphasizes a similarity between these DI genomes rather than their unusual and different 5′ termini. The importance of the 3′ terminus of positive strand RNA viruses has been suggested by in vitro replication experiments. Thus, a naturally occurring variant of Qβ bacteriophage RNA still binds viral replicase, even when nucleotides at the 5′ terminus are missing, whereas fragments of this RNA lacking 3′ nucleotides are inactive.[54] A similar result was obtained in studies with the brome mosaic viral replicase which is able to use as template a 134-nucleotide fragment generated from the 3′ terminus of the viral RNA.[83]

V. MECHANISMS OF INTERFERENCE

The three classes of DI genomes not only are recognized by viral proteins, but they also are usually replicated at the expense of the nondefective viral genome. The mechanisms by which DI RNAs interfere with the replication of the nondefective genome are not understood, but considering the state of our knowledge of RNA replication, this is not surprising. The steps affected by DI RNA may depend on the strategy of replication of the nondefective virus and could differ for different classes of DI RNAs.

The studies by Cole and Baltimore[9,10] with poliovirus showed that the DI and nondefective standard viral genomes compete almost equally for viral specific products. Only standard RNA is able to direct the formation of capsid protein so there is a decrease in the synthesis of this product and a reduced yield of total particles. The interference in total yield is coupled to a 5 to 8% enrichment of the DI particles in a single cycle. This level of enrichment may be due to an ability of DI genomes to replicate faster because they are smaller, but Cole and Baltimore found that the DI and standard RNAs are synthesized in ratios that are essentially proportional to the ratio of DI and standard particles in the inoculum.

Both interference of the nondefective virus and enrichment of DI particles are more extensive in the other systems reviewed here. The small size of the DI genomes does not seem to be a deciding factor in enrichment because, at least for VSV, Sendai, and the alphaviruses, the most effective DI RNAs are not always the smallest. Several models to explain interference by the 5′ conserved DI RNAs of VSV and Sendai virus have been proposed.[6,7,56] They are based on the different recognition that must exist for the 5′ and 3′ termini of the virion genome. The 3′ terminus normally is involved in transcription of mRNAs as well as in the synthesis of the positive strand template. The 3′ terminus of the negative strand of the 5′ conserved DI RNAs is identical to the 3′ of the positive strand (stem class, Figure 4) and could be recognized by the replicase involved in negative strand synthesis. The DI RNA will be engaged only in the synthesis of positive and negative strands, whereas the standard genome is also occupied with mRNA transcription. It has also been proposed that modulation between replication and transcription is related to the intracellular concentration of the nucleoprotein.[84]

The mechanisms of interference by DI RNAs of alphavirus are also not clear. Stark and Kennedy[22] suggest that DI genomes of alphaviruses have a replicative advantage because they lack a region of the virion RNA — perhaps the junction region required for subgenomic RNA production. The negative strand of alphavirus RNAs is the template for both genomic and subgenomic RNAs. DI genomes, missing the junction region, would participate only in replication, not in the transcription of subgenomic mRNAs. It is also possible that the unusual 5′ termini of DI RNAs — the $tRNA^{Asp}$ and the 26 S RNA sequences — are (at the 3′ terminus of the negative strand) a better substrate for the replicase than the 5′ terminus of virion RNA.

Nayak et al. have described possible mechanisms of interference for influenza virus.[79,80] They suggest that the size of the DI RNA may actually be an important factor in interference. They also consider an alternative model in which it is the structure of the RNA that determines its selectivity. The DI RNA could either bind polymerase molecules more effectively or lack internal regulatory signals that attenuate the synthesis of the nondefective RNA. These two proposals are not mutually exclusive, but the structure model does offer some rationale for why the PB2 gene segment of the influenza genome is the one most often deleted.[79,80]

It is not clear to what extent the ability to be translated affects the interfering properties of any of the DI RNAs. Poliovirus DI RNAs are translated, but they also do not have very strong interfering abilities. Naturally occurring Sindbis DI RNAs can be translated,[85] although some of these RNAs have mutations in which stop codons have been produced.[29,30] The Sindbis DI RNAs that translate a foreign protein are not good interfering molecules, but the foreign sequences per se may be contributing to the poor interference. Some DI genomes of influenza virus can be translated, but how their interfering ability compares with that of nontranslating DI RNAs is not known.

VI. CONCLUSIONS AND PERSPECTIVES

The ubiquity of DI particles among RNA viruses suggests that they are an integral part of viral replication and may be an important element in controlling the yield of infectious virus particles. Their generation can be explained by errors made during replication. Based on the sequence of DI RNAs, replicases appear able to copy back on their template,[7,56] fall off the template and reinitiate on another template,[7,56] and roll over looped-out sequences in the template.[79,80] That a replicase can also carry out recombination with a host tRNA is a more unusual activity, but whether this is an error or a clue to a role for this tRNA in Sindbis RNA replication remains to be determined.[18]

The question of what sequences are essential for amplification of DI RNAs has been examined directly only for the DI RNAs of Sindbis virus.[4] Deletion mapping and the insertion of foreign sequences demonstrated that terminal sequences are the only ones that are absolutely required for amplification. Quantitative differences in the activity of the RNAs exist, however, and some of the altered RNAs are not as effective as the RNA identical in sequence to a naturally occurring DI RNA. Therefore, internal sequences or structure may be able to affect the amplification of these DI RNAs. The transfection and amplification assay requires that the RNA be both replicated and encapsidated. What effect deletions and insertions have on each of these steps can only be resolved when they can be analyzed independently.

The approach described for Sindbis DI RNAs should also be applicable to DI RNAs derived from other positive strand viruses. For negative strand viruses, however, the analysis may be more complex. These RNAs — and the virion RNAs from which they are generated — are transcribed and replicated as ribonucleoproteins, not as free RNA. Determining the biological activity of engineered negative strand DI RNAs may require in vitro encapsidation before they can be tested for their ability to be replicated.

Future directions in the investigations of DI RNAs include not only establishing the sequences and structures essential for the viability of these RNAs, but also determining the mechanisms of interference. These studies depend to a great extent on advances in our understanding of viral RNA replicases. A case in point is the implication that host cell factors can affect the generation of defective RNAs. That host cell protins are an important component of the Qβ bacteriophage replicase is well established[46,47] (Volume I, Chapter 1), and host cell proteins have also been isolated as components of the poliovirus replicase[86-89] (Volume I, Chapter 2). Evidence for their role in other RNA virus replicases is more indirect. How host cell factors can cause differential effects on DI RNA generation or amplification clearly requires more detailed knowledge of these replicase complexes. DI RNAs should provide useful probes for studying these complexes. The flexibility of the DI RNA structures, as well as their smaller size and perhaps higher affinity for replicase, are properties that may be valuable for analyzing RNA transcription and replication both in vivo and in vitro.

ACKNOWLEDGMENTS

I thank my colleagues who sent me their reprints and material prior to publication and I especially acknowledge helpful discussions with Donald Mills about Qβ bacteriophage.

REFERENCES

1. **Huang, A. S. and Baltimore, D.,** Defective viral particles and viral disease processes, *Nature (London)*, 226, 325, 1970.
2. **Baltimore, D.,** Expression of animal virus genomes, *Bacteriol. Rev.*, 35, 235, 1971.
3. **Bruton, C. J. and Kennedy, S. I. T.,** Defective-interfering particles of Semliki Forest virus: structural differences between standard virus and defective-interfering particles, *J. Gen. Virol.*, 31, 383, 1976.
4. **Levis, R., Weiss, B. G., Tsiang, M., Huang, H., and Schlesinger, S.,** Deletion mapping of Sindbis virus DI RNAs derived from cDNAs defines the sequences essential for replication and packaging, *Cell*, 44, 137, 1986.

5. **Huang, A. S. and Baltimore, D.,** Defective interfering animal viruses, in *Comprehensive Virology*, Vol. 10, Fraenkel-Conrat, H. and Wagner, R. R., Eds., Plenum Press, New York, 1977, chap. 2.
6. **Holland, J. J., Kennedy, S. I. T., Semler, B. L., Jones, C. L., Roux, L., and Grabau, E. A.,** Defective interfering RNA viruses and host-cell response, in *Comprehensive Virology*, Vol. 16, Fraenkel-Conrat, H. and Wagner, R. R., Eds., Plenum Press, New York, 1980, 137.
7. **Perrault, J.,** Origin and replication of defective interfering particles, *Curr. Top. Microbiol. Immunol.*, 93, 151, 1981.
8. **Cole, C. N., Smoler, D., Wimmer, E., and Baltimore, D.,** Defective interfering particles of poliovirus. I. Isolation and physical properties, *J. Virol.*, 7, 478, 1971.
9. **Cole, C. N. and Baltimore, D.,** Defective interfering particles of poliovirus. II. Nature of the defect, *J. Mol. Biol.*, 76, 325, 1973.
10. **Cole, C. N. and Baltimore, D.,** Defective interfering particles of poliovirus. III. Interference and enrichment, *J. Mol. Biol.*, 76, 345, 1973.
11. **Nomoto, A., Jacobson, A., Lee, Y. F., Dunn, J., and Wimmer, E.,** Defective interfering particles of poliovirus: mapping of the deletion and evidence that the deletions in the genomes of DI(1), (2) and (3) are located in the same region, *J. Mol. Biol.*, 128, 179, 1979.
12. **Lundquist, R. E., Sullivan, M., and Maizel, J. V., Jr.,** Characterization of a new isolate of poliovirus defective interfering particles, *Cell*, 18, 759, 1979.
13. **McClure, M. A., Holland, J. J., and Perrault, J.,** Generation of defective interfering particles in picornaviruses, *Virology*, 100, 408, 1980.
14. **Radloff, R. J. and Young, S. A.,** Defective interfering particles of encephalomyocarditis virus, *J. Gen. Virol.*, 64, 1637, 1983.
15. **Kajigaya, S., Arakawa, H., Kuge, S., Koi, T., Imura, N., and Nomoto, A.,** Isolation and characterization of defective-interfering particles of polio-virus Sabin 1 strain, *Virology*, 142, 307, 1985.
15a. **Kuge, S., Saito, I., and Nomoto, A.,** Primary structures of poliovirus defective interfering particle genomes and possible generation mechanisms of the particles, *J. Mol. Biol.*, 192, 473, 1986.
16. **Racaniello, V. R. and Baltimore, D.,** Cloned poliovirus complementary DNA is infectious in mammalian cells, *Science*, 214, 916, 1981.
17. **Semler, B. L., Dorner, A. J., and Wimmer, E.,** Production of infectious poliovirus from cloned cDNA is dramatically increased by SV40 transcription and replication signals, *Nucleic Acids Res.*, 12, 5123, 1984.
18. **Schlesinger, S. and Weiss, B. G.,** Defective RNAs of alphaviruses, in *The Togaviridae and Flaviviridae*, Schlesinger, S. and Schlesinger, M. J., Eds., Plenum Press, New York, 1986, 149.
19. **Stollar, V.,** Defective-interfering particles of togaviruses, *Curr. Top. Microbiol. Immunol.*, 86, 35, 1979.
20. **Stollar V.,** Defective-interfering alphaviruses, in *The Togaviruses: Biology, Structure, Replication*, Schlesinger, R. W., Ed., Plenum Press, New York, 1980, 427.
21. **Guild, G. M. and Stollar, V.,** Defective interfering particles of Sindbis virus. III. Intracellular viral RNA species in chick embryo cell cultures, *Virology*, 67, 24, 1975.
22. **Stark, C. and Kennedy, S. I. T.,** The generation and propagation of defective-interfering particles of Semliki Forest virus in different cell types, *Virology*, 125, 285, 1978.
23. **Guild, G. M. and Stollar, V.,** Defective interfering particles of Sindbis virus. V. Sequence relationships between SVstd 42S RNA and intracellular defective viral RNA, *Virology*, p. 175, 1977.
24. **Kennedy, S. I. T., Bruton, C. J., Weiss, B., and Schlesinger, S.,** Defective-interfering passages of Sindbis virus: nature of the defective virion RNA, *J. Virol.*, 19, 1034, 1976.
25. **Kennedy, S. I. T.,** Sequence relationships between the genome and the intracellular RNA species of standard and defective interfering Semliki Forest virus, *J. Mol. Biol.*, 108, 491, 1976.
26. **Dohner, D., Monroe, S., Weiss, B., and Schlesinger, S.,** Oligonucleotide mapping studies of standard and defective Sindbis virus RNA, *J. Virol.*, 29, 794, 1979.
27. **Lehtovaara, P., Söderlund, H., Keränen, S., Pettersson, R. F., and Kääriäinen, L.,** 18S defective-interfering RNA of Semliki Forest virus contains a triplicated linear repeat, *Proc. Natl. Acad. Sci. U.S.A.*, 78, 5353, 1981.
28. **Söderlund, H., Keränen, S., Lehtovaara, P., Petersson, R. F., and Kääriäinen, L.,** Structural complexity of defective-interfering RNAs of Semliki Forest virus as revealed by analysis of complementary DNA, *Nucleic Acids Res.*, 9, 3403, 1981.
29. **Lehtovaara, P., Söderlund, H., Keränen, S., Petterson, R. F., and Kääriäinen, L.,** Extreme ends of the genome are conserved and rearranged in the defective-interfering RNAs of Semliki Forest virus, *J. Mol. Biol.*, 156, 731, 1982.
30. **Monroe, S. S. and Schlesinger, S.,** Common and distinct regions of defective-interfering RNAs of Sindbis virus, *J. Virol.*, 49, 865, 1984.
31. **Strauss, E. G., Rice, C. M., and Strauss, J. H.,** Complete nucleotide sequence of the genomic RNA of Sindbis virus, *Virology*, 133, 92, 1984.
32. **Ou, J.-H., Strauss, E. G., and Strauss, J. H.,** Comparative studies of the 3′-terminal sequences of several alphavirus RNAs, *Virology*, 109, 281, 1981.

33. **Ou, J.-H., Rice, C. M., Dalgarno, L., Strauss, E. G., and Strauss, J. H.,** Sequence studies of several alphavirus genomic RNAs in the region containing the start of the subgenomic RNA, *Proc. Natl. Sci. U.S.A.*, 79, 5235, 1982.
34. **Ou, J.-H., Strauss, E. G., and Strauss J. H.,** The 5′-terminal sequences of the genomic RNAs of several alphaviruses, *J. Mol. Biol.*, 168, 1, 1983.
35. **Monroe, S. S. and Schlesinger, S.,** RNAs from two independently isolated defective-interfering particles of Sindbis virus contain a cellular tRNA sequence at their 5′ ends, *Proc. Natl. Acad. Sci. U.S.A.*, 80, 3279, 1983.
36. **Varmus, H. and Swanstrom, R.,** Replication of retroviruses, in *RNA Tumor Viruses*, Weiss, R., Teich, N., Varmus, H., and Coffin, J., Eds., Cold Spring Harbor Laboratory, Cold Spring Harbor, N.Y., 1982, 369.
37. **Haenni, A.-L., Joshi, S., and Chapeville, F.,** tRNA-like structures in the genomes of RNA viruses, *Prog. Nucleic Acid Res.*, 27, 85, 1982.
38. **Ahlquist, P., Strauss, E. G., Rice, C. M., Strauss, J. H., Haseloff, J., and Zimmern, D.,** Sindbis virus proteins nsP1 and nsP2 contain homology to nonstructural proteins from several RNA plant viruses, *J. Virol.*, 53, 536, 1985.
39. **Tsiang, M., Monroe, S. S., and Schlesinger, S.,** Studies of defective-interfering RNAs of Sindbis Virus with and without $tRNA^{Asp}$ sequences at their 5′ termini, *J. Virol.*, 54, 38, 1985.
40. **Jalanko, A. and Söderlund, H.,** The repeated regions of Semliki Forest virus defective-interfering RNA interferes with the encapsidation process of the standard virus, *Virology*, 141, 257, 1985.
41. **Schlesinger, S., Levis, R., Weiss, B. G., Tsaing, M., and Huang, H.,** Replication and packaging sequences in defective interfering RNAs of Sindbis virus, in *Positive Strand RNA Viruses*, Brinton, M. and Rueckert, R., Eds., Alan R. Liss, New York, 1987, 241.
41a. **Levis, R., Huang, H., and Schlesinger, S.,** Engineered defective interfering RNAs of Sindbis virus express bacterial chloramphenicol acetyltransferase in avian cells, *Proc. Natl. Acad. Sci. U.S.A.*, 84, 4811, 1987.
42. **Levin, J. G., Ramseur, J. M., and Grimley, P. M.,** Host effect on arbovirus replication: appearance of defective-interfering particles in murine cells, *J. Virol.*, 12, 1401, 1973.
43. **Tooker, P. and Kennedy, S. I. T.,** Semliki Forest virus multiplication in clones of Aedes albopictus cells, *J. Virol.*, 37, 589, 1981.
44. **Steacie, A. D. and Eaton, B. T.,** Properties of defective-interfering particles of Sindbis virus generated in vertebrate and mosquito cells, *J. Gen. Virol.*, 65, 333, 1984.
45. **Zinder, N. D.,** *RNA Phages*, Cold Spring Harbor Laboratory, Cold Spring Harbor, N.Y., 1975.
46. **Kamen, R. I.,** Structure and function of the Qβ RNA replicase, in *RNA Phages*, Zinder, N. D., Ed., Cold Spring Harbor Laboratory, Cold Spring Harbor, N.Y., 1975, chap. 2.
47. **Blumenthal, T.,** RNA replication: function and structure of Qβ-replicase, *Annu. Rev. Biochem.*, 48, 525, 1979.
48. **Meyer, F., Weber, H., and Weissmann, C.,** Interactions of Qβ replicase with Qβ RNA, *J. Mol. Biol.*, 153, 631, 1981.
49. **Mills, D. R., Peterson, R. L., and Spiegelman, S.,** An extracellular Darwinian experiment with a self-duplicating nucleic acid molecule, *Proc. Natl. Acad. Sci. U.S.A.*, 58, 217, 1967.
50. **Kacian, D. L., Mills, D. R., Kramer, F. R., and Spiegelman, S.,** A replicating RNA molecule suitable for a detailed analysis of extracellular evolution and replication, *Proc. Natl. Acad. Sci. U.S.A.*, 69, 3038, 1972.
51. **Mills, D. R., Kramer, F. R., and Spiegelman, S.,** Complete nucleotide sequence of a replicating RNA molecule, *Science*, 180, 916, 1973.
52. **Mills, D. R., Kramer, F. R., Dobkin, C., Nishihara, T., and Spiegelman, S.,** Nucleotide sequence of microvariant RNA: another small replicating molecule, *Proc. Natl. Acad. Sci. U.S.A.*, 72, 4252, 1975.
53. **Schaffner, W., Ruegg, K. J., and Weissmann, C.,** Nanovariant RNAs: nucleotide sequence and interaction with bacteriophage Qβ replicase, *J. Mol. Biol.*, 117, 877, 1977.
54. **Nishihara, T., Mills, D. R., and Kramer, F. R.,** Localization of the Qβ replicase recognition site in MDV-1 RNA, *J. Biochem.*, 93, 669, 1983.
55. **Mills, D. R., Priano, C., and Kramer, F. R.,** Requirement for secondary structure formation during coliphage RNA replication, in *Positive Strand RNA Viruses*, Brinton, M. A. and Rueckert, R., Eds., Alan R. Liss, New York, 1987, 35.
56. **Lazzarini, R. A., Keene, J. D., and Schubert, M.,** The origins of defective interfering particles of the negative-strand RNA viruses, *Cell*, 26, 145, 1981.
57. **Kolakofsky, D.,** Isolation of vesicular stomatitis virus defective interfering genomes with different amounts of 5′-terminal complementarity, *J. Virol.*, 41, 566, 1982.
58. **Perrault, J. and Leavitt, R. W.,** Characterization of snap-back RNAs in vesicular stomatitis defective interfering virus particles, *J. Gen. Virol.*, 37, 661, 1977.

59. **Schubert, M. and Lazzarini, R. A.,** Structure and origin of a snapback defective interfering particle RNA of vesicular stomatitis virus, *J. Virol.*, 37, 661, 1981.
60. **Huang, A. S.,** Viral pathogenesis and molecular biology, *Bacteriol. Rev.*, 41, 811, 1977.
61. **Leppert, M., Kort, L., and Kolakofsky, D.,** Further characterization of Sendai virus DI-RNAs: a model for their generation, *Cell,* 12, 539, 1977.
62. **Meier, E., Harmison, G. G., Keene, J. D., and Schubert, M.,** Sites of copy choice replication involved in generation of vesicular stomatitis virus defective-interfering particle RNAs, *J. Virol.*, 51, 515, 1984.
63. **Yang, F. and Lazzarini, R. A.,** Analysis of the recombination event generating a vesicular stomatitis virus deletion defective interfering particle, *J. Virol.*, 45, 766, 1983.
64. **Kang, C. Y., Schubert, M., and Lazzarini, R. A.,** Frequent generation of new 3′-defective interfering particles of vesicular stomatitis virus, *Virology,* 143, 630, 1985.
65. **Nichol, S. T., O'Hara, P. J., Holland, J. J., and Perrault, J.,** Structure and origin of a novel class of defective interfering particle of vesicular stomatitis virus, *Nucleic Acids Res.*, 12, 2775, 1984.
66. **Horodyski, F. M. and Holland, J. J.,** Viruses isolated from cells persistently infected with vesicular stomatitis virus show altered interaction with defective-interfering particles, *J. Virol.*, 36, 627, 1980.
67. **Horodyski, F. M., Nichol, S. T., Spindler, K. R., and Holland, J. J.,** Properties of DI particle resistant mutants of vesicular stomatitis virus isolated from persistent infections and from undiluted passages, *Cell,* 33, 801, 1983.
68. **O'Hara, P. J., Horodyski, F. M., Nichol, S. T., and Holland, J. J.,** Vesicular stomatitis virus mutants resistant to defective-interfering particles accumulate stable 5′-terminal and fewer 3′-terminal mutations in a stepwise manner, *J. Virol.*, 49, 793, 1984.
69. **O'Hara, P. J., Nichol, S. T., Horodyski, F. M., and Holland, J. J.,** Vesicular stomatitis virus defective interfering particles can contain extensive genomic sequence rearrangements and base substitutions, *Cell,* 36, 915, 1984.
70. **Weiss, B., Levis, R., and Schlesinger, S.,** Evolution of virus and defective-interfering RNAs in BHK cells persistently infected with Sindbis virus, *J. Virol.*, 48, 676, 1983.
71. **Kawai, A. and Matsumoto, S.,** Interfering and non-interfering defective particles generated by a rabies plaque variant virus, *Virology,* 76, 60, 1977.
72. **Jacobson, S., Dutko, F. J., and Pfau, C. J.,** Determinants of spontaneous recovery and persistence in MDCK cells infected with lymphocytic choriomeningitis virus, *J. Gen. Virol.*, 44, 113, 1979.
73. **Kolakofsky, D.,** Isolation and characterization of Sendai DI-RNAs, *Cell,* 8, 547, 1976.
74. **Re, G. G., Gupta, K. C.,, and Kingsbury, D. W.,** Genomic and copy-back 3′ termini in Sendai virus defective interfering RNA species, *J. Virol.*, 45, 659, 1983.
75. **Re, G. G., Morgan, E. M., and Kingsbury, D. W.,** Nucleotide sequences responsible for generation of internally deleted Sendai virus defective interfering genomes, *Virology,* 146, 27, 1985.
76. **Hsu, C.-H., Re, G. G., Gupta, K. C., Portner, A., and Kingsbury, D. W.,** Expression of Sendai virus defective-interfering genomes with internal deletions, *Virology,* 146, 38, 1985.
77. **Re, G. G. and Kingsbury, D. W.,** Nucleotide sequences that affect replicative and transcriptional efficiencies of Sendai virus deletion mutants, *J. Virol.*, 58, 578, 1986.
78. **von Magnus, P.,** Incomplete forms of influenza virus, *Adv. Virus Res.*, 2, 59, 1954.
79. **Nayak, D. P., Chambers, T. M., and Akkina, R. K.,** Defective-interfering (DI) RNAs of influenza viruses: origin, structure, expression, and interference, *Curr. Top. Microbiol. Immunol.*, 114, 103, 1985.
80. **Nayak, D. P., Chambers, T. M., and Akkina, R. K.,** Structure of defective-interfering (DI) RNAs of influenza viruses and their role in interference, in *Influenza Viruses,* Krug, R., Ed., Plenum Press, New York, in press.
81. **Jennings, P. A., Finch, J. T., Winter, G., and Robertson, J. S.,** Does the higher order structure of the influenza virus ribonucleoprotein guide sequence rearrangements in influenza viral RNA?, *Cell,* 34, 619, 1983.
82. **Fields, S. and Winter, G.,** Nucleotide-sequence heterogeneity and sequence rearrangements in influenza virus cDNA, *Gene,* 15, 207, 1981.
83. **Miller, W. A., Bujarski, J. J., Dreher, T. W., and Hall, T. C.,** Minus-strand initiation by brome mosaic virus replicase within the 3′ tRNA-like structure of native and modified RNA templates, *J. Mol. Biol.*, 187, 537, 1986.
84. **Blumberg, B. M. and Kolakofsky, D.,** An analytical review of defective infections of vesicular stomatitis virus, *J. Gen. Virol.*, 64, 9, 1839, 1983.
85. **Migliaccio, G., Castagnola, P., Leone, A., Cerasuolo, A., and Bonatti, S.,** mRNA activity of a Sindbis virus defective-interfering RNA, *J. Virol.*, 55, 877, 1985.
86. **Dasgupta, A., Zabel, P., and Baltimore, D.,** Dependence of the activity of the poliovirus replicase on a host cell protein, *Cell,* 19, 423, 1980.
87. **Baron, M. H. and Baltimore, D.,** Purification and properties of a host cell protein required for poliovirus replication in vitro, *J. Biol. Chem.*, 257, 12359, 1982.

88. **Dasgupta, A.,** Purification of host factor required for in vitro transcription of poliovirus RNA, *Virology*, 128, 245, 1983.
89. **Young, D. C., Tuschall, D. M., and Flanegan, J. B.,** Poliovirus RNA-dependent RNA polymerase and host cell protein synthesize product RNA twice the size of poliovirion RNA in vitro, *J. Virol.*, 54, 256, 1985.

Chapter 9

DELETION MUTANTS OF DOUBLE-STRANDED RNA GENETIC ELEMENTS FOUND IN PLANTS AND FUNGI

Donald L. Nuss

TABLE OF CONTENTS

I. Introduction 188

II. Deletion Mutants of Wound Tumor Virus 188
- A. Discovery and Characterization of Wound Tumor Virus Deletion Mutants 189
- B. Characterization of Wound Tumor Virus Remnant dsRNAs 192
- C. Sequence Organization of the Wound Tumor Virus Genome 192

III. Deletion Mutants of "Killer" dsRNAs of Yeast 196
- A. Characterization of "Killer" Virus Remnant dsRNAs 197

IV. Double-Stranded RNAs Associated with Hypovirulence Conversion in *Endothia parasitica* 200
- A. Structural Properties of Hypovirulence-Associated dsRNAs 200

V. Mechanism Responsible for Generation of Remnant dsRNAs 201

VI. Remnant dsRNAs as Experimental Tools 203

VII. Concluding Remarks 204

VIII. Addendum 205

Acknowledgment 206

References 206

I. INTRODUCTION

Genetic elements composed of double-stranded (ds) RNA are found in plants and fungi in a variety of forms. Plant-infecting reoviruses contain 10 or 12 discrete segments of dsRNA packaged as a set within one virus particle.[1-3] The genome segments of a second group of plant dsRNA viruses, the cryptic or temperate viruses,[4,5] vary with respect to number (two to five) and size ($\sim$1.6 to 0.8 $\times$ 10^6 daltons).[4] It is unclear whether the dsRNAs of these viruses are encapsidated in the same or in different particles.[5] The presence of dsRNAs in plants is often associated with infection by RNA viruses that possess single-stranded RNA genomes. These molecules, which represent replicative intermediate forms of viral genomes, are convenient diagnostic markers for such infections.[6,7] High-molecular-weight (greater than 5 kilobase-pairs) dsRNAs of ill-defined origin and function have been detected in a number of apparently healthy plants, e.g., the $\sim$14-kilobase-pair dsRNAs associated with *Phaseolus vulgaris* cv. Black Turtle soup.[8,9] As early as 1974, viruses or virus-like particles (VLPs) had been detected in isolates of over 60 species of fungi.[10] A common feature of these mycoviruses is a dsRNA genome[10,11] (whose replication is reviewed by Breunn in Volume I). In certain fungal species, the associated dsRNA genetic elements resemble plasmids more than they resemble viral genomes and can influence the virulence of their fungal hosts.[11,12]

Mutant molecules derived from dsRNA genetic elements by deletion mutation events have been described in detail for a plant-infecting reovirus[13-15] and the viruses associated with "killer" strains of yeast.[16-24] A recent analysis[25] of the dsRNAs associated with a hypovirulent strain of *Endothia parasitica,* the cause of chestnut blight, suggests that the complex banding patterns observed for hypovirulence-associated dsRNAs are also related to deletion mutation events. In this regard, close inspection of dsRNAs prepared from a number of plant and fungal sources reveals a heterogenous population of molecules of diverse size present in low concentration (e.g., References 7, 14, 23, and 25 to 29, this chapter, Figures 1 and 9). It is likely that these heterogenous subpopulations of molecules represent remnants of the major dsRNA species and that deletion mutation events are not uncommon during the replication of these elements.

Remnants of dsRNA genetic elements were initially viewed as a curiosity. However, recent studies which successfully employed analogous molecules to investigate genome structure-function relationships in other viral systems (References 30 to 32 and this volume) have generated new interest in deleted dsRNA molecules. Remnant dsRNAs retain structural domains required for transcription, replication, and, in the appropriate context, packaging.[13,21-25] As a result, synthetic and natural remnant molecules are ideal reagents for studies aimed at identifying and characterizing *cis*-regulatory structural domains of dsRNA genetic elements.

This chapter will summarize the limited information available regarding deletion mutants of dsRNA genetic elements found in plants and fungi. Discussion will focus on the properties of the remnant molecules, possible mechanisms responsible for their generation, and their potential as experimental tools.

II. DELETION MUTANTS OF WOUND TUMOR VIRUS (WTV)

Plant-infecting reoviruses are characterized by a double-stranded RNA genome composed of 10 or 12 discrete segments, a multiplication cycle which involves replication both in plant hosts and insect vectors, and the ability to induce tumors in plants hosts (for recent reviews see References 1 to 3 and 33). Two subgroups, or genera, have been created to deal with differences in genome segment number, vector specificity, and morphology exhibited by different members of the plant reovirus group.[34] The phytoreoviruses, consisting of three members, contain 12 dsRNA segments with a combined molecular weight of approximately 16×10^6 daltons and are transmitted by leafhoppers (*Cicadellidae, Jassoidea*). The Fiji viruses, with nine members, contain ten dsRNA segments and are transmitted by planthoppers (*Delphacidae, Fulguroidea*). Detailed discussions of morphological differences between members of the two genera can be found in several review articles.[2,3] All plant-infecting reoviruses, with the exception of rice dwarf virus (a phytoreovirus), induce tumors in plant hosts.[2]

In terms of molecular biology, the best characterized of the plant-infecting reoviruses is WTV.[1] The appeal of WTV as the model system for studying the molecular biology of this virus group is twofold. First, there exists a considerable body of basic biological information generated during the past 4 decades concerning the physical and morphological characteristics of the virus particle and the interaction of the virus with its insect vectors and plant hosts.[1,33,35-39] Second, a continuous cell culture system derived from the insect vector capable of supporting viral replication is available providing a powerful tool for the study and manipulation of WTV outside of the plant host.[36,38,40-41] Progress in defining the molecular biology of WTV during the past 6 years include the identification and partial characterization of the WTV primary transcription and translation products,[42-44] the cloning of full-length cDNA copies of 9 of the 12 WTV genome segments,[45] and the complete nucleotide sequence analysis of several of the segments (Reference 45 and unpublished data). In addition, considerable information has been obtained regarding the regulation of viral gene expression in infected insect vector cells[46,47] and the nature of remnant dsRNAs associated with transmission defective mutants generated in systemically infected plants.[13]

A. Discovery and Characterization of Wound Tumor Virus Deletion Mutants

Deletion mutants of WTV were discovered by Black and co-workers during the course of characterizing WTV isolates that had lost the ability to be transmitted by the leafhopper vector.[14,48,49] Transmission of WTV is a complex process that involves acquisition of the virus by the leafhopper vector during feeding on an infected plant, sequential multiplication in the organs and hemolymph of the insect, and passage of the virus to uninfected plants via the salivary fluid during subsequent feeding.[50-58] Infection of the leafhopper by WTV results in no obvious pathology,[36,54,55] and, once infected, the insect retains the ability to transmit the virus for the remainder of an apparently normal life span.[54,56,57] Efficient transmission is limited, however, to only two species of leafhoppers: *Agallia constricta* van Duzee and *Agalliopsis novella* (say).[58]

It became apparent as early as 1958 that WTV isolates purified from plants that had been maintained by vegetative propagation for several years after inoculation differed in the efficiency with which they were transmitted by the leafhopper vector.[14,36,49,59,60] That is, maintenance of the virus in infected plants without the opportunity of passage through the insect vector led, within several years, to a reduction or loss in the transmissibility of the virus population. Reduction in transmissibility was accompanied by a corresponding reduction in the specific infectivity of the virus isolates on cultured cells derived from the vector.[14,49]

Electrophoretic analysis of the genome segments of WTV isolates that had partially (deficient) or completely (defective) lost the ability to be transmitted by the vector revealed patterns that differed from the wild-type transmissible virus pattern and from each other in several respects.[14] Wild-type WTV contains equimolar amounts of 12 genome segments that range in size from 851 base-pairs to approximately 4 kilobase-pairs (Figure 1). In some transmission-deficient isolates, there was a reduction in the concentration of one or several wild-type segments and the appearance of lower-molecular-weight remnants of wild-type segments. In some isolates, a wild-type segment was reduced in concentration unaccompanied by the appearance of a remnant. Occasionally a remnant appeared to completely replace a wild-type segment. By observing the change in genome pattern of an isolate with

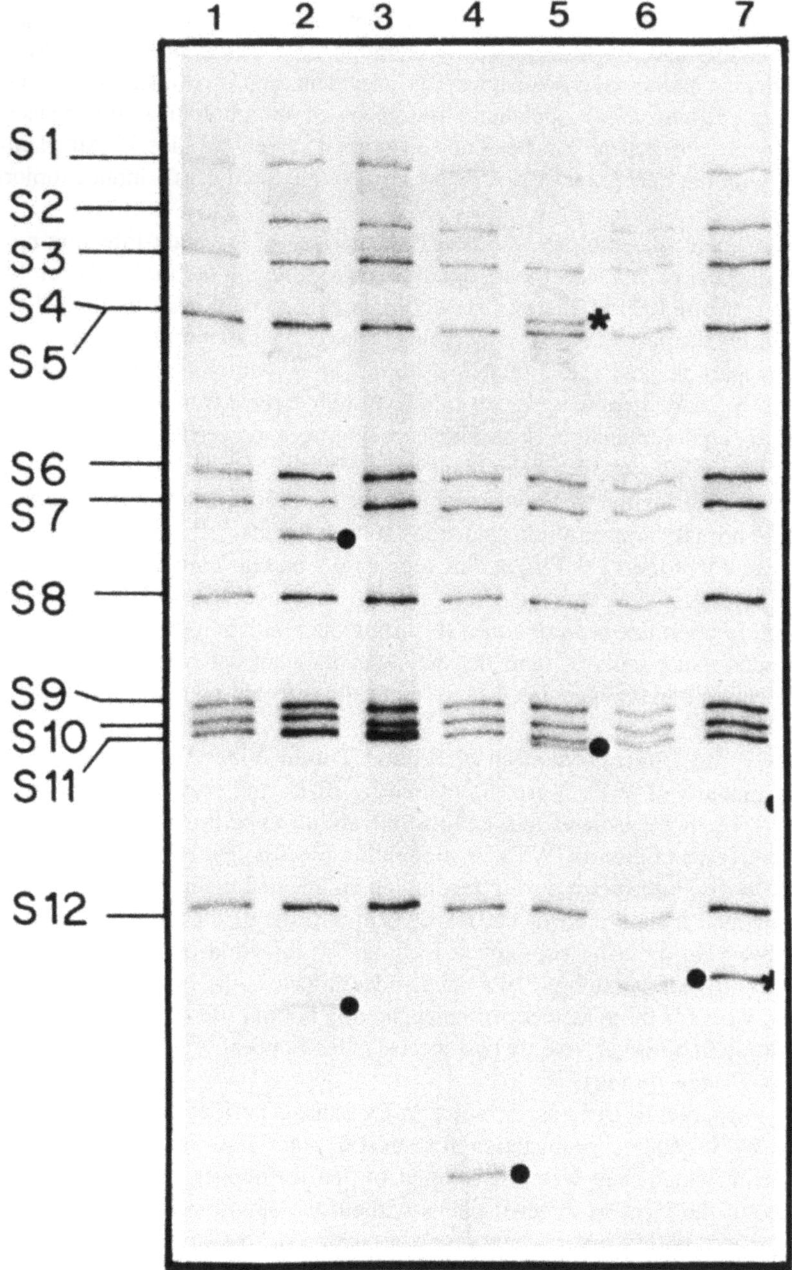

FIGURE 1. Autoradiograph of [^{32}P]-end-labeled genome segments of standard and selected transmission-defective isolates of WTV analyzed by polyacrylamide gel electrophoresis. Genome segments are designated S1 to S12[108] based on relative electrophoretic mobilities. S12 is 851 base-pairs,[45] while S1 has an estimated size of approximately 4 kilobase-pairs.[108] The dots to the right of individual lanes indicate the positions of remnant RNAs observed among the segments of individual transmission-defective isolates. The asterisks indicate remnants that have been characterized (see Figures 2 and 3). Lanes: 1, transmissible WTV; 2, MS7(57); 3, -S5(64); 4, -S5(60); 5, -S2(70); 6, 10% S1(60); 7, 10% S1(49). The nomenclature for the transmission defective isolates is taken from Reddy and Black.[15] (From Nuss, D. L., in *Double-Stranded RNA Viruses*, Bishop, D. H. L. and Compans, R. W., Eds., Elsevier, Amsterdam, 1983, 415. With permission. Copyright 1983 Elsevier Science Publishing Co., Inc.)

time of maintenance, it was possible to correlate the reduction in concentration or loss of a wild-type segment with the increases in concentration of a remnant, thus allowing a tentative determination of the segment from which a particular remnant was dreived. An examination of 28 isolates revealed deletion events in 4 of the 12 WTV genome segments: segment S1, S2, S5, and S7, with approximately 70% of the mutation events occurring in segments S1 and S5. In several cases, remnants of identical electrophoretic mobility were generated from the same genome segment in different infected plants.

At this point, it is useful to define the term "isolate" as used by Black and co-workers and retained in this chapter to describe the transmission-deficient and -defective WTV mutants. As indicated by these investigators,[14] a transmission-deficient isolate is a mixed population of transmissible and nontransmissible virus particles. While a transmission-defective isolate is also a mixed population of virus particles, as indicated by the complex genome patterns,[14,15] the ability of the particles of these populations to replicate in the vector appears to have been completely lost. The term "isolate" was introduced to describe these mixed populations because each population was physically isolated from each other within plants that were maintained by vegetative propation. That is, after introduction into a plant host by an infected leafhopper, the resulting virus population evolved in the isolated environment of the infected plant without the introduction of exogenous virus or the opportunity to replicate in the insect vector.

Starting with plants containing transmission defective isolates, Reddy and Black[15] next selected cuttings which contained virus populations that exhibited a progressive loss of genome segments previously shown[14] to be present in less than wild-type concentrations. This technique was viewed as a form of geographic segregation of virus particles within the plant. Apparently, the proportion of virus particles exhibiting mutant or wild-type genome patterns varies throughout the plant. Consequently, cuttings taken from one part of the plant may contain a virus population quite different from that found in a cutting taken from another part of the same plant. Using this selection method, virus isolates were obtained that appeared to lack segments S2, S5, or their remnants. Other selected isolates contained 10% of the normal complement of segment S1 or a full complement of a remnant of segment S7 in place of wild-type segment S7. As would be expected, the same selection technique can be used to select a virus population containing a high proportion of transmissible virions from transmission-deficient, but not from transmission-defective, isolates.

Following the 1977 publication by Reddy and Black[15] outlined above, the selected isolates were maintained in vegetatively propagated plants without further selection or analysis for several years. In 1981, cuttings of sweet clover plants infected with the selected isolates obtained from Black were analyzed in the author's laboratory in order to determine whether the isolates had remained stable or had incurred further deletion mutations. The analysis, which involved polyacrylamide gel electrophoresis of [^{32}P]pCp-labeled genome segments, revealed that continued passage of the selected isolates in the absence of negative selective pressure, resulted in the reestablishment of a genome pattern similar to that exhibited by the transmission-defective population from which the isolates were selected[13] (Figure 1). That is, many of the remnants of genome segments and, in one case, an intact segment thought to be absent in the selected population, were again detectable. The analysis technique used by Reddy and Black[15] was sensitive enough to detect the presence of a particular genome segment or remnant in 1000 virion paticles. Apparently, the remnants shown in Figure 1 were present in the selected isolates at a concentration of less than one per 1000 virions and increased during the 5-year period of unselected passage to a concentration as high as one in every virus particle (e.g., the largest remnant in lane 5, shown to be a remnant of segment S2).[13] This increase in concentration in the absence of negative selective pressure suggests that the remnants are preferentially replicated and packaged relative to wild-type segments, i.e., a remnant dsRNA displaces its cognate wild-type dsRNA segment. It is important to note that the isolate reported to lack genome segment S2 was still clearly devoid of that segment,[13] and that all six selected isolates tested contained undetectable or low levels of genome segment S5.[13]

B. Characterization of WTV Genome Remnant dsRNAs

Purified WTV particles contain a transcriptase that catalyzes, in vitro, the synthesis of message-sense transcripts from each of the 12 genome segments.[42,43,61,62] Analysis of the transcription products synthesized by the six selected transmission-defective WTV isolates characterized in Figure 1 confirmed expectations that the packaged remnant molecules are actively transcribed by the virion-associated transcriptase.[44,63] Since the genome remnants are transcribed, replicated, and packaged into virus particles in systemically infected plants, they must retain the nucleotide sequence domains required for the efficient execution of these events. The genome remnants can, therefore, be viewed as naturally occurring reagents useful for defining, to a first approximation, the location of these sequence domains.

One of the first questions to be addressed concerning the WTV genome remnants involved the nature of the deletion event, i.e., whether the remnants were generated by internal deletion or early termination events. To answer this question, the two remnants, designated in Figure 1 by asterisks, were characterized in detail.[13] Molecular hybridization analysis revealed that the designated remnant in lane 7 of Figure 1 was derived from genome segment S5 while the designated remnant in lane 5 is related to segment S2. To determine whether the genesis of these remnant RNAs involved a terminal- or internal-deletion event, the 3′ termini of the wild-type and remnant RNAs were labeled with [^{32}p]pCp, denatured, and digested partially or completely with ribonuclease T1. Since this nuclease cleaves specifically at the 3′ side of guanosine residues, partial digestion of the denatured dsRNAs yielded a series of 3′-end-labeled oligonucleotides for each RNA strand. By polyacrylamide gel analysis of the digests, the guanosine positions relative to the 3′ terminal end of each strand of the wild-type and remnant RNAs were compared simultaneously. Separate analysis of the 12 individual WTV genome segments by this method had revealed a distinctive pattern for each segment.[13] The partial and complete ribonuclease T1 digestion patterns for segment S2 and its remnant were indistinguishable for at least 40 nucleotides from the 3′ end of each RNA strand (Figure 2). Similar results were obtained for genome segment S5 and its remnant (Figure 3). The finding that the ends of the wild-type genome segment were retained to form a viable remnant RNA suggests a critical role for the terminal nucleotide sequences in genome function. Further discussion of this point requires a review of what is currently known about the sequence organization of the WTV genome.

C. Sequence Organization of the Wound Tumor Virus Genome

The dsRNA genome of WTV has been converted to ds cDNA and cloned into plasmid pBR322.[45] Apparent full-length clones of 9 of the 12 genome segments have been identified. Sequence analysis of cloned cDNA copies of several genome segments and direct analysis of the 3′ termini of the dsRNAs revealed that each WTV genome segment possesses the sequence ...UGAU–OH at the terminus corresponding to the 3′ end of the plus strand and ...AAUACC–OH at the terminus corresponding to the 3′ end of the negative strand (Figures 4 and 5). The fact that each of the WTV genome segments have termini consisting of identical nucleotide sequences adds emphasis to the conclusion drawn from analysis of the remnant RNAs regarding the possible role of the termini in genome function.

The complete nucleotide sequence of two of the WTV genome segments has been determined. Genome segment S12 consists of 851 nucleotides[45] and possesses one long open reading frame (ORF) that starts with the first AUG triplet (residues 35 to 37) and extends 534 nucleotides, as shown in schematic form in Figure 6. A second ORF, in phase with the first and consisting of 40 codons, is present, beginning at residues 674 to 676. The long

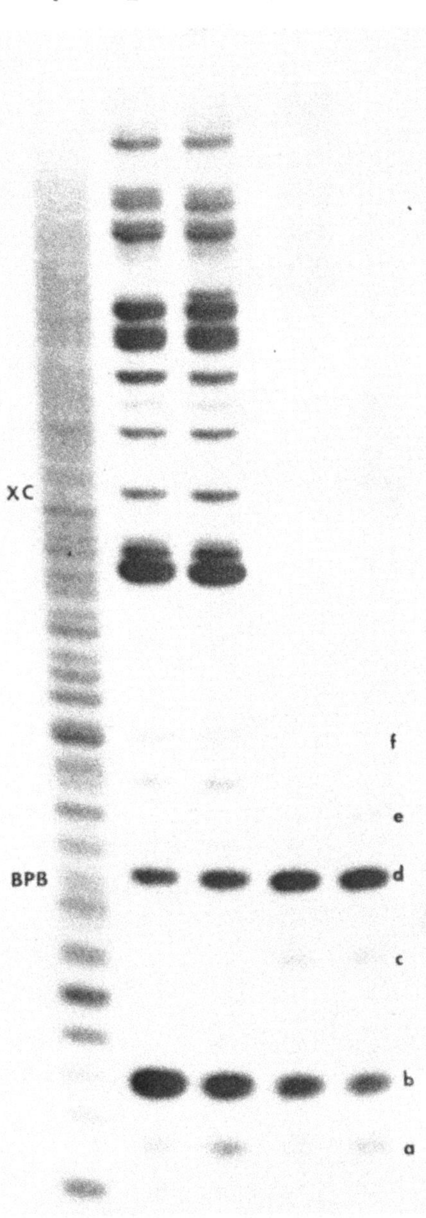

FIGURE 2. Polyacrylamide gel analysis of partial and complete ribonuclease T1 digests of [^{32}P]pCp-end-labeled genome segment S2 and its remnant, indicated by the asterisk in lane 5 of Figure 1. Lane 1, oligonucleotide ladder generated by partial alkaline digestion of end-lableed genome RNA. Lane 2, partial digest of genome segment S2. Lane 3, partial digest of remnant of S2. Lanes 4 and 5 are complete digests of the two RNAs, respectively. The migration positions of bromophenol blue (BPB) and xylene cyanol (XC) markers are shown in the left margin. Oligonucleotides of interest are indicated in the right margins. The predominant 3′ terminal RNase T1 resistant oligonucleotide of S2 and its remnant are oligonucleotide b and d. For complete details see Reference 13. (From Nuss, D. L. and Summers, D., *Virology*, 133, 276, 1984. With permission. Copyright 1984 Academic Press, Inc.)

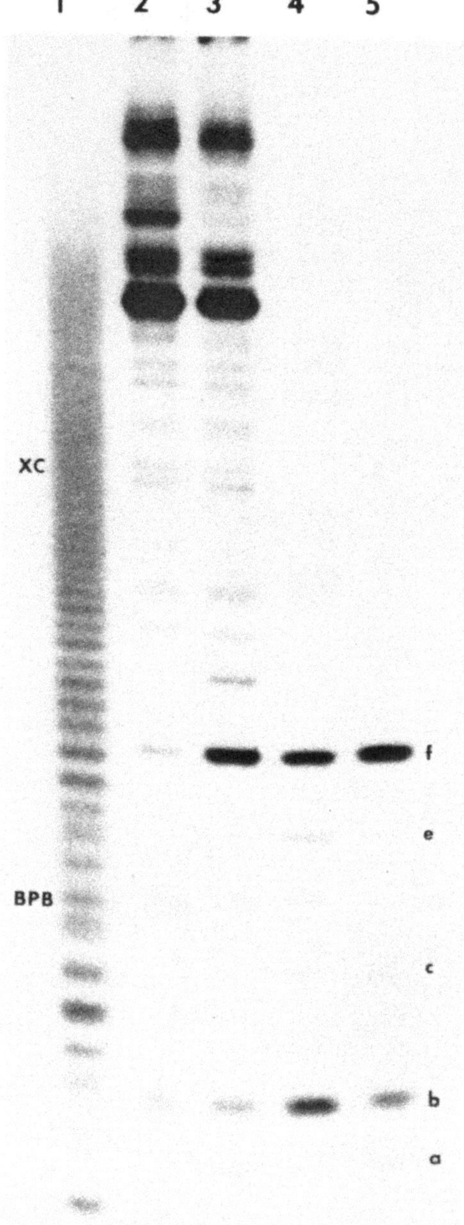

FIGURE 3. Polyacrylamide gel electrophoretic analysis of partial and complete ribonuclease T1 digests of [^{32}P]pCp-end-labeled genome segment S5 and its remnant, indicated by asterisk in lane 7 of Figure 1. Lane 1, oligonucleotide ladder generated by partial alkaline digestion of end-labeled genome RNA, Lanes 2 and 3, partial digests of genome segment S5 and its remnant, respectively. Lanes 4 and 5, complete digests of the two RNAs. Oligonucleotides b and f are the predominant 3′ terminal ribonuclease T1-resistant oligonucleotides of S5 and its remnant. (From Nuss, D. L. and Summers, D., *Virology*, 133, 276, 1984. With permission. Copyright 1984 Academic Press, Inc.)

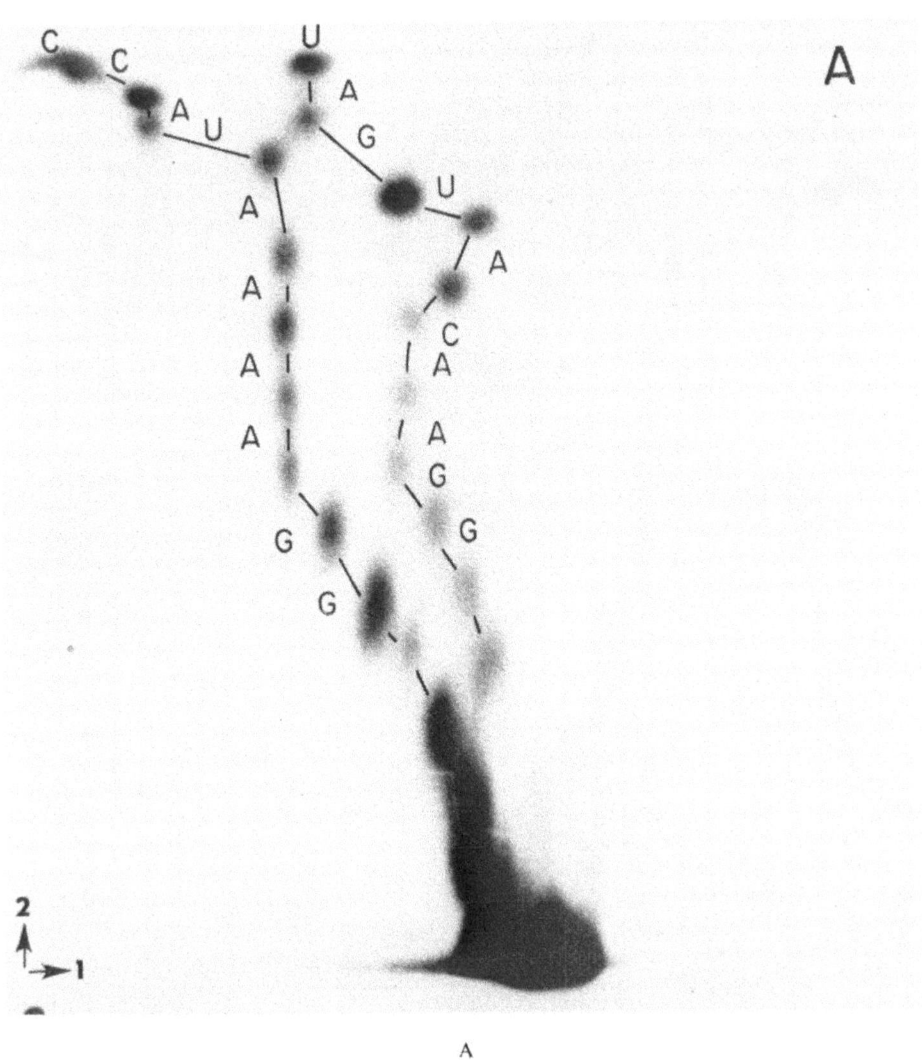

A

FIGURE 4. Two-dimensional oligonucleotide fingerprint analysis of partial alkaline digestion products of [^{32}P]pCp-labeled WTV genome segment S3 (A) and segment S9 (B). In each panel, the left tract represents the 3′ terminal sequence of the (−) strand, while the right tract represents the 3′ terminal sequence of the (+) strand. (From Asamizu, T., Summers, D., Motika, M. B., Anzola, J. V., and Nuss, D. L., *Virology*, 144, 398, 1985. With permission. Copyright 1984 Academic Press, Inc.)

ORF of segment S12 has the coding potential for a polypeptide of 178 amino-acid residues which would have a molecular weight of 19,171 daltons, a value consistent with the apparent molecular weight estimated (19,000 daltons) for the smallest of the previously identified WTV primary gene products, Pns12.[42] There currently exists no information concerning the possible function of the smaller ORF. The molecular anatomy of genome segment S5, thought to encode the 76-kdalton outer coat component P5,[37] is more typical of a reovirus genome segment in that it contains one large ORF which comprises 92.3% of the segment (Figure 7).

Completion of the sequence analysis of segment S5 has permitted a detailed examination of the sequence information retained in a viable genome remnant and has provided an opportunity to compare sequence information of several segments in an attempt to identify domains possibly involved in gene function. A series of nuclease protection experiments using labeled SP6 and T7 RNA polymerase-generated transcripts corresponding to segment

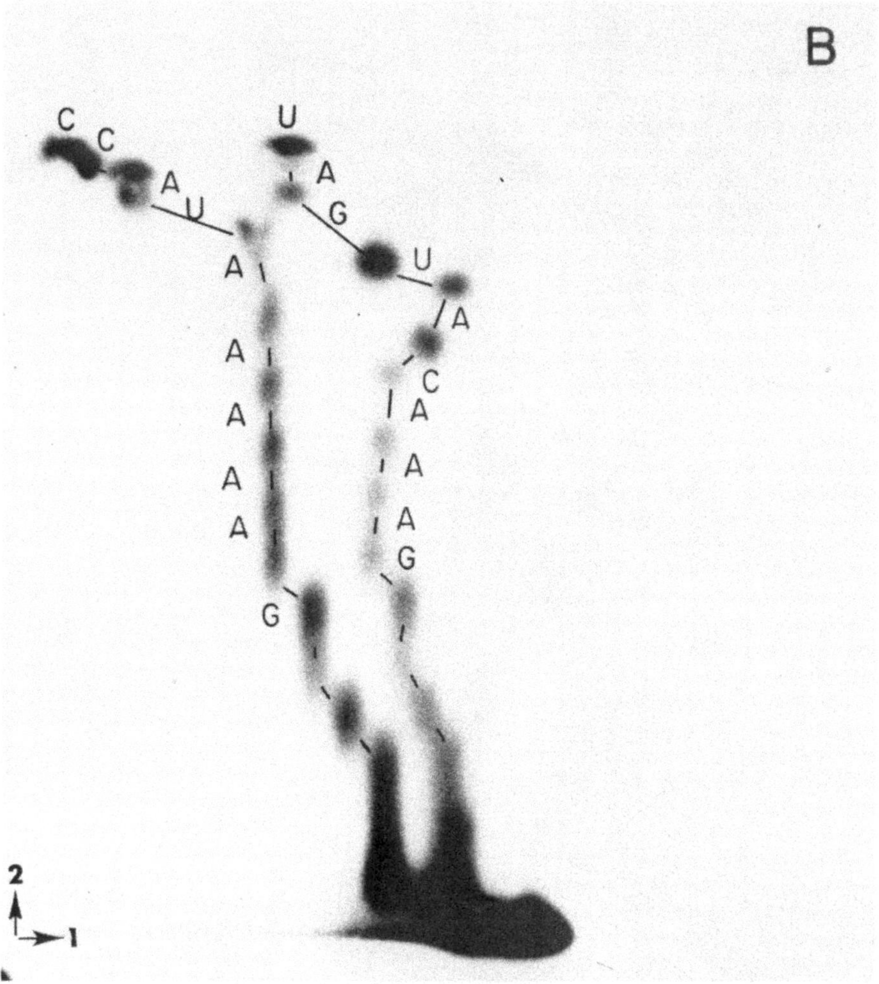

FIGURE 4B

S5 and remnant dsRNA mapped the deletion boundary to ~300 nucleotides from the 5′ end of the plus strand and 440 nucleotides from the 3′ end of the plus strand (Anzola, J., Asamizu, T., and Nuss, D., unpublished). As shown in Figure 8, when the nucleotide sequence comprising the 5′ and 3′ termini of the plus strand of segment S5 (sequences within the conserved region of the viable remnant of this segment) are placed in close approximation, a region of potential base-pairing becomes apparent. A similar region of potential base-pairing exists for the 3′ and 5′ termini of genome segment S12. Since this writing, several remnant RNAs derived from segment S5 were completely sequenced and the terminal nucleotide sequences of all 12 WTV genome segments were determined. This additional information is discussed in the Addendum (Section VIII).

III. DELETION MUTANTS OF "KILLER" dsRNAs OF YEAST

Some strains of yeast produce a toxin to which they are immune, but which is lethal to other strains. The cytoplasmic determinants responsible for this phenomenon in "killer" strains of *Saccharomyces cerevisiae* are dsRNA molecules (reviewed in References 64 to 67). The toxin and immunity component are encoded by a 1.8- to 1.9-kilobase-pair dsRNA

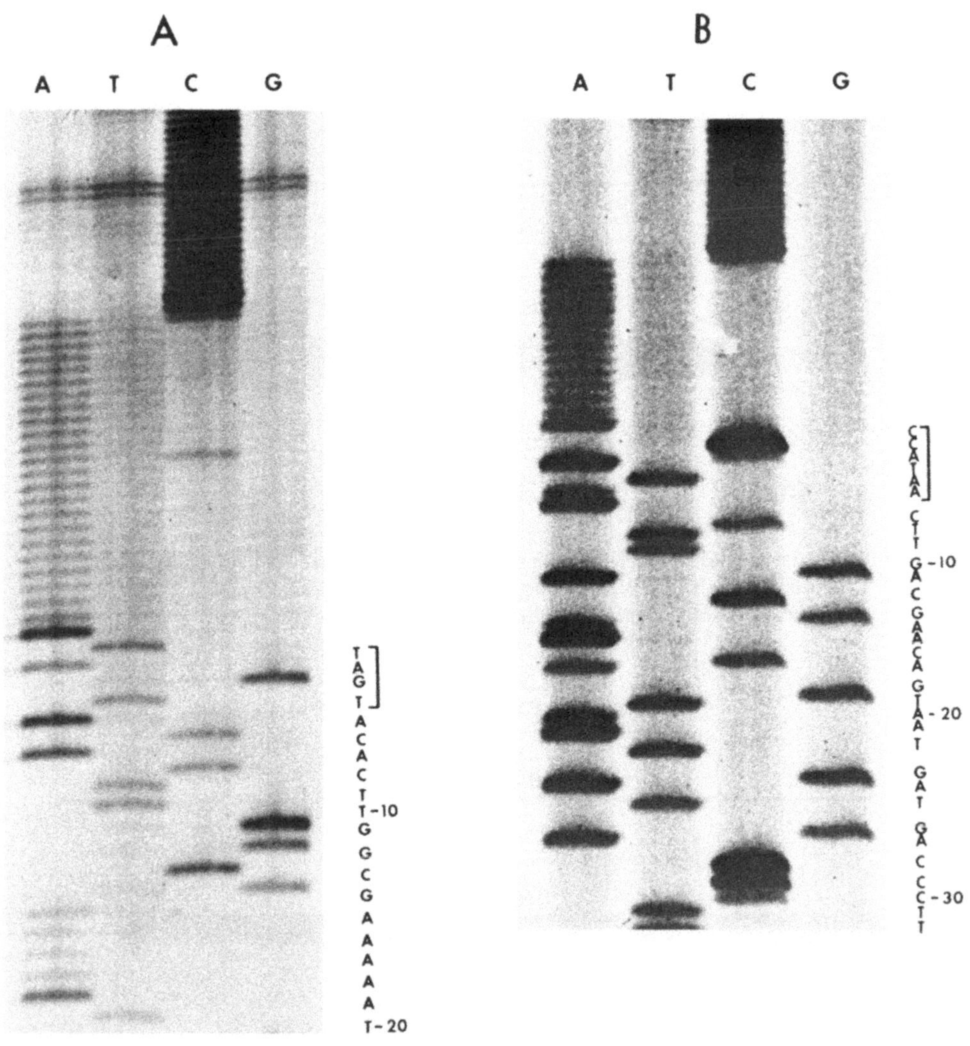

FIGURE 5. Dideoxy chain-terminator sequencing analysis of the 3′ terminal end of the positive (A) and negative (B) strands of WTV segment S12 subcloned into bacteriophage M13. The terminal sequences common to all WTV genome segments are indicated by brackets. The oligo(dA) and oligo(dC) tracts were added to the genome segments during cDNA synthesis and cloning. (From Asamizu, T., Summers, D., Motika, M. B., Anzola, J. V., and Nuss, D. L., *Virology*, 144, 398, 1985. With permission. Copyright 1984 Academic Press, Inc.)

species, designated M.[68,69] Replication of M-dsRNA is dependent upon a 4.5- to 4.9-kilobase-pair dsRNA species, designated L.[70] L-dsRNAs are also found in most nonkiller strains of yeast.[71] Both L- and M-dsRNAs are encapsidated in a VLP composed of one polypeptide which is encoded by L-dsRNA.[70,72-74] While *S. cerevisiae* killer viruses are not infectious, they can be transmitted by mating or cytoplasmic mixing.[17,75] Interestingly, replication and expression of the killer virus dsRNAs is dependent upon a number of chromosomal genes.[64-67]

A. Characterization of "Killer" Virus Remnant dsRNAs

Deletion mutants of *S. cerevisiae* killer viruses were first described by Somers[16] in 1973. Unlike strains carrying the wild-type dsRNA genome, the strains harboring the deletion mutants produced neither toxin nor immunity.[16,18,19] Moreover, nonkiller mutants lacked the

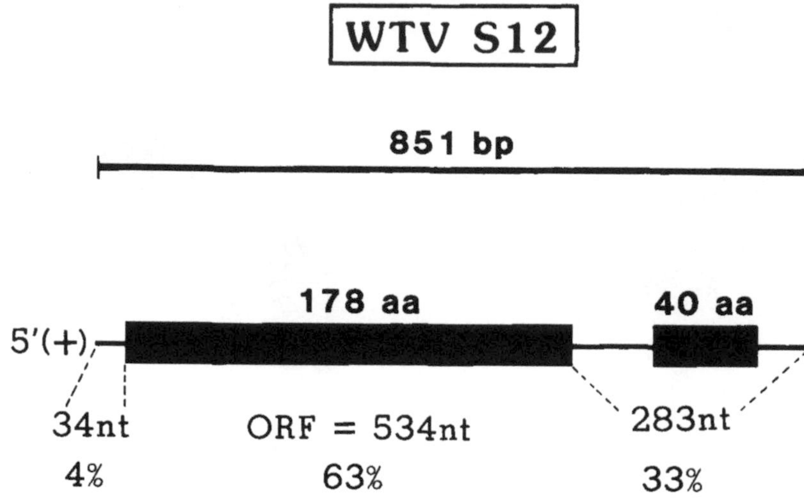

FIGURE 6. Schematic representation of the sequence organization of WTV segment S12. Thin bars indicate 5′ and 3′ noncoding regions, while the thick bars represent long open reading frames. Actual sequence information is presented in Reference 45.

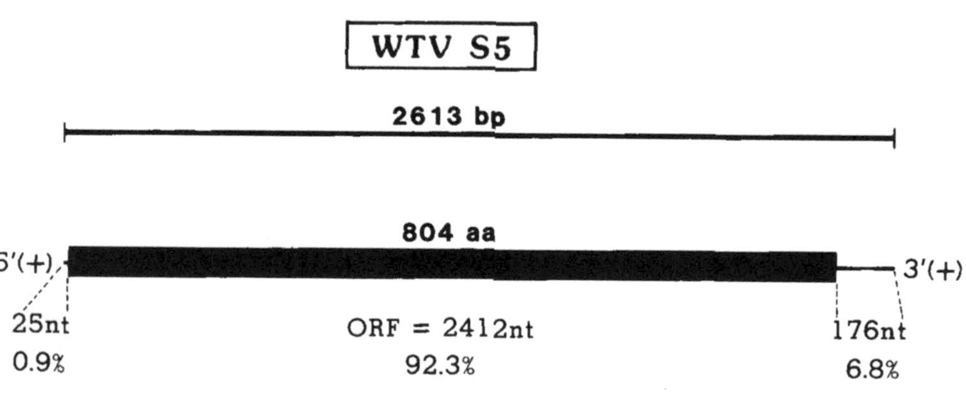

FIGURE 7. Schematic representation of the sequence organization of WTV segment S5. Actual sequence information to appear in Anzola et al., *Proc. Natl. Acad. Sci. U.S.A.*, December 1987.

M-dsRNA, but contained smaller dsRNAs denoted as S-dsRNAs. Such mutants were classified as "suppressive" because the S-dsRNA prevented or "suppressed" the replication of the wild-type M-dsRNA when killer and mutant strains were mated.[16,18-20] That is, the S-dsRNA displaced the M-dsRNA in diploids formed by mating killer and suppressive strains. Results obtained by measuring the rate at which S-dsRNA displaces M-dsRNA after mating suggests that the two molecules are competing for a factor(s) involved in their replication and that S-dsRNA competes more effectively.[20] The ability to displace the wild-type dsRNA is one property shared by S-dsRNA and WTV dsRNA remnants. A second similarity is that in both cases, the remnants are generated by internal deletion events.[21-24]

Two of the yeast S-dsRNAs (S3 and S14) have been characterized by heteroduplex mapping[22-24] and by sequence analysis either at the RNA or cDNA level.[21,22,24] While the entire sequence of S14 has been reported,[24] the entire sequence of the wild-type M-dsRNA from which it was derived has not yet been reported. The deletion boundary for the yeast viral S3-dsRNA is located 232 bases from the 5′ terminus of the M-dsRNA plus strand and resumes 550 bases from the 3′ terminus (Table 1). This represents a deletion of greater than

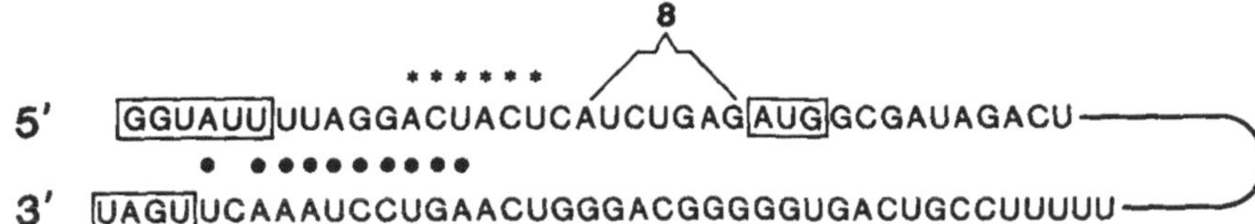

WTV T12

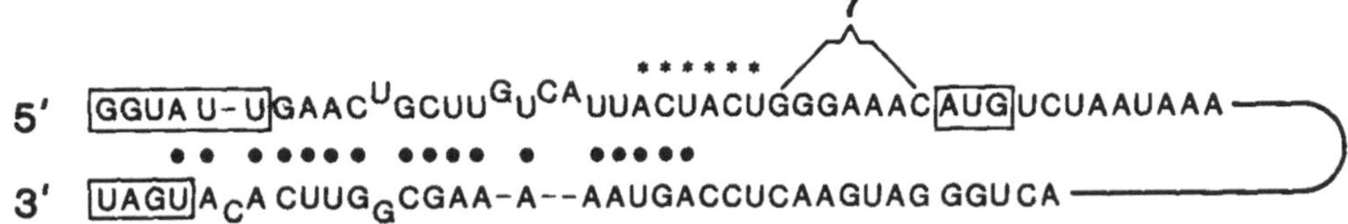

FIGURE 8. The 5′ and 3′ nucleotide sequences of transcripts corresponding to WTV segments S5 and S12, presented to show regions of potential base pairing. The conserved terminal sequences found in all WTV transcripts are boxed. The sequence ACUACU found in both T5 and T12 either eight or seven nucleotides (indicated by bracket) from the initiator AUG is indicated by stars. The potential interaction between 5′ and 3′ ends of the WTV transcripts is purely speculative. The estimated free energies for the base-paired regions of WTV T5 and WTV T12 were calculated as −18.2 and −17.7, respectively, using the thermodynamic criteria compiled by Salser.[109]

Table 1

Remnant RNA	Deletion breakpoint		
	230 Nucleotides from 5′ end of M-dsRNA		550 nucleotides from 3′ end of M-dsRNA
Yeast "killer" S3	5′~~GUCGAUCACC*	UGGGGUUCAU ---- ----	*CACAAGCACACUC --
	250 nucleotides from 5′ end of M-dsRNA		540 nucleotides from 3′ end of M-dsRNA
Yeast "killer" S14	5′~~CGUAGCGAGC*	GAUGCAGGUG ---- ----	*CUCACCUUGAG ----
	110 nucleotides from 5′ end		760 nucleotides from 5′ end
CPV "dwarf"	5′~~CAAUACAACUAU**	AACAACAGCU GAUAUUAACA	**AUCAUAGCUAUGGA--

Note: Sequence elements for yeast "killer" dsRNA remnants were taken from Thiele *et al.*[23] and Lee *et al.*[24] The plus strand sequence is shown. The majority of sequence information including the position numbers were taken from Lee *et al.*, while the M-dsRNA sequence following the first break point of S14 was taken from Thiele *et al.* The sequence elements for the CPV polyhedrin gene and the "dwarf" gene were taken from Reference 76. Asterisks are included to highlight the nucleotides bracketing the deletion breakpoint. Sequence context for deletion breakpoints of WTV segment S5-related remnant RNAs will appear in the article by Anzola et al., referred to in the Addendum (Section VIII).

1 kilobase-pair of the internal portion of M-dsRNA. The deletion boundary for S14-dsRNA is in the same region of the M-dsRNA interrupting the wild-type sequence at position 253 from the 5′ end of the plus strand, and resuming at position 540 from the 3′ terminus. Other than the fact that a cytidine residue is found on either side of the deletion break point, little additional insight into the mechanism of deletion mutant formation can be gained from the existing partial sequence context of the boundary region.

The complete sequence for a wild-type dsRNA segment and its corresponding remnant has been reported in the case of the polyhedrin gene of cytoplasmic polyhedrosis virus, an insect-infecting reovirus.[76] The remnant dsRNA, 315 base-pairs in length, was apparently generated by a simple deletion of a 635-base-pair middle portion of the 950-base-pair polyhedrin gene with the deletion break point occurring at position 121 from the 5′ end of the plus strand, and position 191 from the 3′ end of the plus strand of the parental segment (Table 1). In this case, the same dinucleotide sequence, AU, is found on either side of the break point analogous to the occurrence of the cytidine residues bracketing the deletion break point of S3- and S14-dsRNAs (Table 1). Determination of the complete sequence context at the deletion breakpoint for additional mutant dsRNAs may allow the identification of sequences that mimic virus transcription or replication control signals, as has been reported for Sendai virus, a nonsegmented negative strand RNA virus.[30,31]

IV. DOUBLE-STRANDED RNAs ASSOCIATED WITH TRANSMISSIBLE HYPOVIRULENCE IN *ENDOTHIA PARASITICA*

The fungal pathogen *E. parasitica,* is responsible for the virtual elimination of the American Chestnut tree (*Castanea dentata* [Morsh] Bork). However, some chestnut trees in Europe and a few localized areas of North American have survived the disease due to the presence of hypovirulent strains of *E. parasitica* that act as natural biological control agents (for reviews see References 12 and 77 to 79). Hypovirulent strains are debilitated in their ability to kill susceptible chestnut trees, display abnormal culture morphology, harbor cytoplasmic dsRNA molecules, and can transmit this dominant phenotype to normal virulent strains by hyphal anastomosis.[12,26,77-79] Correlative evidence strongly suggests that dsRNA is the cytoplasmically transmissible determinant responsible for the hypovirulence phenotype.[12,26,77]

Double-stranded RNAs associated with different hypovirulent strains vary with respect to size, concentration, and sequence homology.[26,28,29,80] Examples of dsRNAs isolated from a European and a North American hypovirulent strain are shown in Figure 9. Note the family of minor bands migrating faster than the large dsRNAs in both preparations. Minor dsRNA species have been reported previously to be associated with hypovirulent strains[29] and are visible upon close inspection of gel photographs presented in several publications.[26,28,29,80] In addition, dsRNA banding patterns change with time of culturing, often with the appearance of smaller dsRNA bands.[26,81] Since the dsRNAs are persistently maintained in *E. parasitica*, functional products of deletion mutation events, i.e., remnants, might be expected to accumulate with time if such events occur with any frequency. The minor dsRNA species referred to above may represent such deletion products. A recent analysis of the dsRNAs associated with a North American hypovirulent strain provides support for this possibility.[25]

A. Structural Properties of Hypovirulence-Associated dsRNAs

Analysis of dsRNAs, purified from the North American hypovirulent strain, Grand Haven 2 (GH2), revealed three major dsRNA species,[81,82] (Figure 9) of approximately 9.0 (L-RNA), 3.5 (M-RNA) and 1.0 kilobase-pairs (S-RNA), respectively. Based on differential sensitivities of appropriately end-labeled dsRNAs to specific ribonucleases and electrophoretic patterns of ribonucleases T1 digestion products, the following general picture of the structural properties of the GH2 dsRNAs has emerged (Figure 10). The 5′ terminus of the (+) strand (polarity arbitrarily assigned) of each dsRNA species is rich in guanosine residues (~15 nucleotides in length) and is terminated at the 5′ end with either a guanosine or an adenosine residue. The 3′ terminus of the (−) strands were found to be uniformly terminated with a cytidine residue. The 3′ termini of the (+) strands contain a stretch of 30 to 60 adenosine residues while the 5′ termini of the (−) strands appear to contain a complementary stretch of polyuridylic acid.

Molecular hybridization analysis showed that L- and M-RNA share extensive sequence

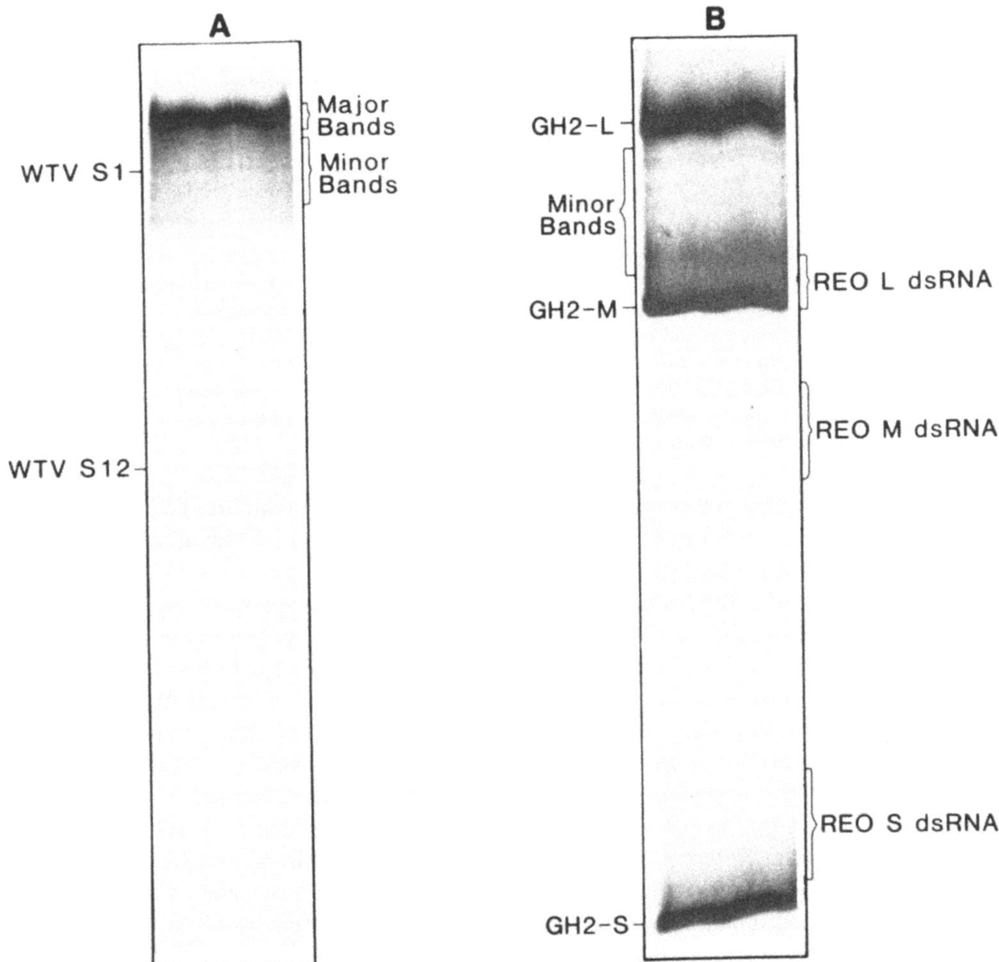

FIGURE 9. Polyacrylamide gel electrophoretic analysis of dsRNAs associated with hypovirulent strains of *Endothia parasitica*. Lane A, dsRNAs isoalted from the European hypovirulent strain 713.[80,110] Lane B, dsRNA isolated from the North American strain GH2.[81] Genome dsRNAs of WTV (positions shown at left of Lane A) or human reovirus,[111] (positions shown at right of Lane B) were electrophoresed as references. Based on their electrophoretic migration relative to WTV and reovirus dsRNAs, the estimated sizes of the dsRNAs of strain 713 are 6 to 7 kilobase-pairs, while the estimated sizes of the L-, M-, and S-dsRNAs of strain GH2 are 9, 3.5, and 1 kilobase-pairs, respectively. Note minor dsRNA band in each preparation.

homology, i.e., hybridize under stringent conditions. Moreover, electrophoretic migration patterns of partial and complete RNase T1 digests of end-labeled L- and M-RNAs were indistinguishable whether the two RNAs were labeled at the 3′ or 5′ ends. The combined results are consistent with the possibility that M-RNA was derived from L-RNA by an internal deletion event. In this regard, shorter versions of M-dsRNA have been observed, suggesting the occurrence of subsequent deletion events.[81] S-RNA is unrelated to L- and M-RNAs by the criteria listed above. Further analyses of dsRNAs found in other hypovirulent *E. parasitica* strains should reveal whether the complex banding patterns observed for these dsRNAs is related to the frequent occurrence of internal deletion events.

V. MECHANISMS RESPONSIBLE FOR GENERATION OF REMNANT dsRNAs

Available information concerning the replication of dsRNA genetic elements in eukaryotic

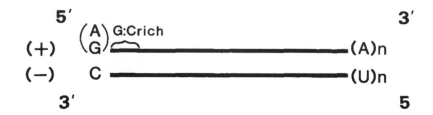

FIGURE 10. Schematic representation of general structural features of the dsRNAs associated with the GH2 hypovirulent strain of *E. parasitica*. The RNA strand containing the 3′ terminal polyadenylic acid sequences is arbitrarily designated as the positive strand (+) while the complementary strand with the 5′ terminal polyuridylic acid sequence is designated as the minus strand (−). Also indicated is the heterogeneity detected at the 5′ end of the (+) strand for each dsRNA. The 15- to 20-nucleotide-long G:C-rich region (not drawn to scale) found at the terminus corresponding to the 5′ end of the (+) strand of each dsRNA is also shown. Information from analysis contained in Reference 25.

systems has been obtained primarily from studies of human reovirus replication in Mouse L cells (see References 83 to 86 for reviews). Reovirus particles contain all of the enzymes necessary for synthesis and modification of functional virual transcripts.[85] Once activated by partial uncoating in the lysosome, infecting viral particles synthesize capped, full-length copies of the plus strand of each of the ten reovirus dsRNA genome segments.[83-86] The resulting transcripts act both as messenger RNA and, once associated with viral polypeptides in nascent subviral particles, as template for the synthesis of minus-strand RNA (references listed in Reference 86). Only one minus strand is synthesized per plus strand template. It is important to stress that available evidence indicates that it is viral single strand RNA and not viral dsRNA that is packaged into nascent subviral particles. Neither minus strand nor dsRNA are found free in the cytoplasm, but are found associated with subviral particles (references listed in Reference 86). Synthesis of transcripts from dsRNA is completely asymmetric (only transcripts of positive polarity are synthesized) and conservative (the original plus strand is not displaced) (references listed in References 83 to 86). A similar model exists for replication of the yeast "killer" virus dsRNA (references listed in References 67, 87, and 88), while essentially no information exists concerning the transcription and replication of dsRNAs associated with *E. parasitica.*

Remnants of dsRNA genetic elements could be generated via any of several pathways. Internal deletion events could occur during transcription of the dsRNA genome yielding deleted transcripts that, once packaged, would serve as template for the synthesis of a remnant dsRNA. Alternatively, the deletion event could occur during replication with a wild-type transcript serving as template. This mechanism would result in the production of a heteroduplex molecule within nascent virus particles, the plus strand of which would contain an internal single-strand loop. Since the deleted minus strand would serve as template during subsequent transcription, the resulting transcripts would be an internally deleted form of the parental wild-type segment. The deletion mutation event would presumably by perpetuated by packaging and replication of the remnant transcripts. Whether the deletion event occurs during transcription or replication, it is most likely that it proceeds by a copy-choice mechanism involving the intra- or intermolecular "jumping" of the viral transcriptase/replicase. There is no evidence of splicing events for dsRNA genetic elements or their transcripts.

Given the current state of knowledge concerning the mechanisms of dsRNA transcription and replication, it is difficult to predict the mechanism(s) responsible for the generation of remnant dsRNAs. Additional sequence analysis of remnant RNA deletion boundaries may provide some insight, as it has in the case of Sendai virus.[30,31] Additional information regarding precise details of normal dsRNA transcription and replication and the structural organization of the virus particle would certainly by useful. Conversely, continued study of the deletion mutation events associated with dsRNA genetic elements may provide insight into the normal mechanism of dsRNA transcription and replication.

VI. REMNANT dsRNAs AS EXPERIMENTAL TOOLS

It is now possible to introduce an artificial DNA phase into the life cycle of RNA viruses. Full-length cDNA copies of several viral genomes were shown to be infectious when introduced into the appropriate cell.[89-93] For several viruses, RNA transcribed from cloned viral cDNA, rather than the cloned cDNA molecule, was infectious.[94-97] This approach is being applied to WTV in an effort to identify regulatory sequences involved in genome RNA function using the remnant of segment S5 as a model. A series of transcription vectors have been constructed that yield transcripts which contain all or a portion of the sequence elements retained in the remnant RNA.[112] The resulting transcripts are being tested for their ability to be replicated and packaged when introduced into cultured insect vector cells infected with helper wild-type virus. A similar approach has been successfully applied to identify regulatory sequences found in defective interfering genomes of Sindbis virus.[32]

Since each of the WTV segments have identical terminal nucleotide sequences, and these sequences are conserved in remnant RNAs, the possibility was considered that only RNA molecules with 3′ and 5′ termini identical to authentic WTV RNAs would be functional. Consequently, advantage was taken of a recently developed transcription vector that allows precise control of the 5′ terminal sequence of resulting transcripts. Vector pPM1[98] consists of a modified bacteriophage lambda P_R promoter fused to a unique *SmaI* site. In *in vitro* run-off transcription reactions catalyzed by *E. coli* DNA-dependent RNA polymerase, transcription initiates at the first nucleotide of DNA fragments inserted at the *SmaI* site (Figure 11), yielding transcripts with the desired 5′ terminal sequence. The sequence at the 3′ terminus of a transcript can be controlled by tailoring the run-off cleavage site. For example, by adding the trinucleotide 5′–ATC–3′ to the terminus corresponding to the 3′ end of any WTV genome segment, an *EcoRV* site is generated. Digestion of the vector with *EcoRV* yields a blunt-end run-off site with the terminal sequence 5′–TGAT–3′ (the conserved WTV sequence). Thus, a transcription run-off reaction programed with an *EcoRV* linearized vector similar to that shown in Figure 11 yields a transcript with 5′ and 3′ terminal sequences identical to authentic WTV transcripts.

WTV sequence elements containing the tailored *EcoRV* site have also been placed under the control of phage RNA polymerase promoter sequences in the pGem™-3 vector (Promega Biotec®) (Figure 12). These vectors offer several advantages in terms of yield and specificity (i.e., transcription proceeds only from the vector promoter sequence and not from promoter sequences on contaminating *E. coli* DNA[99-101]). Transcription vectors have also been constructed that yield transcripts which have the potential of conferring an easily assayable activity when introduced into cells infected with helper virus. For example, the gene encoding resistance to Kanamycin[102] has been modified by adding flanking sequences corresponding to the 5′ and 3′ termini of WTV segment S5 and inserted into the two transcription vectors shown in Figures 11 and 12 (Anzola, J. and Nuss, D., unpublished). Since this writing, we have devised a strategy by which existing cDNA clones can be conveniently manipulated to produce precisely tailored synthetic transcripts.[112] The ability to produce transcripts that contain all or a portion of the structural elements involved in the regulation of WTV genome functions should lead to the eventual determination of the minimal sequence information required for efficient transcription, replication, and packaging of these dsRNA genetic elements. Naturally occurring functional remnants of dsRNA segments provide the starting point for the construction of appropriate transcription vectors.

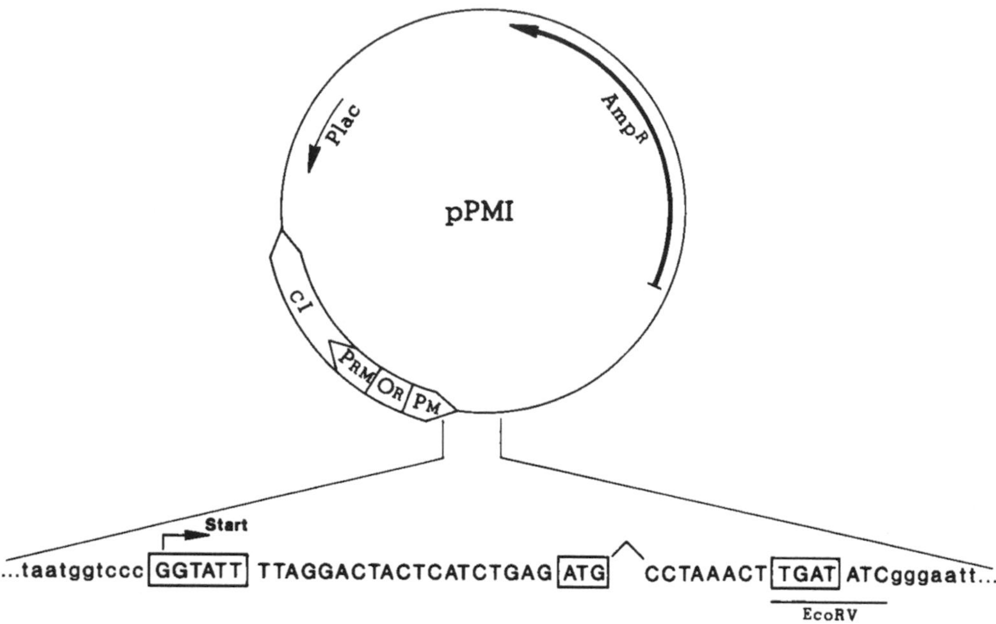

FIGURE 11. Schematic diagram indicating important features of transcription vector pPM1 construct containing structural elements of WTV segment S5. Lower case characters indicate pPM1 sequences around the modified P_R transcription initiation site, while large characters indicate nucleotide seqeunces of the resulting transcript. Boxed areas include the conserved WTV terminal sequences and the initiation codon for the transcript of WTV S5. Note the *EcoRV* site which allows control of the transcription run-off site. Vectors containing varying lengths of the 5′ and 3′ WTV S5 seqeunces have been constructed, providing large quantities of transcripts that contain all or a portion of the S5 sequences retained in the natural remnant of S5 including 5′ and 3′ terminal sequences identical to the authentic transcripts.[112] Transcripts synthesized in the presence of m^{7}GpppA will contain the natural capped 5′ terminus of WTV transcripts.[44] For a complete description of vector pPM1, see Reference 98.

VII. CONCLUDING REMARKS

Multiplication of WTV in systemically infected plants and of yeast killer virus and hypovirulence-associated dsRNA in their respective fungal hosts can be considered persistent infections. RNA virus genomes are known to evolve at a rapid rate under such conditions.[103] A common event in persistently infected cells is the generation of defective interfering (DI) viral genomes (reviewed in Reference 104). The remnant dsRNAs described in this chapter fit several of the criteria established for DI genomes: (1) remnants are shortened forms of the viral genome that retain, in this case, both termini; (2) in the case of WTV and the yeast killer virus, remnant RNAs interfere with (displace) the replication of the parental genome RNAs; and (3) interference is highly specific, e.g., remnants of WTV S5 displace only S5, and the S-RNAs of yeast killer viruses displace M-RNA, but not L-RNA. It is possible, therefore, that remnant dsRNAs could function as modulators of parental genome replication and contribute to the evolution of dsRNA genetic elements.

The origin and evolution of dsRNA genetic elements is an interesting topic for speculation. Given the large size of some dsRNAs, e.g., the >11-kilobase-pair dsRNAs found in some apparently healthy plants[8,9] and the 9-kilobase-pair L-RNA found in strain GH2 of *E. parasitica*,[25] it is likely that they encode several gene products, i.e., are polycistronic. The dsRNA genomes of the reoviruses are, in general, composed of monocistronic segments. The remnant dsRNAs may provide a clue as to the evolutionary relatedness of different dsRNA genetic elements. For example, one could envision the evolution of monocistronic segmented viral dsRNA genomes from large polycistronic dsRNAs by a combination of

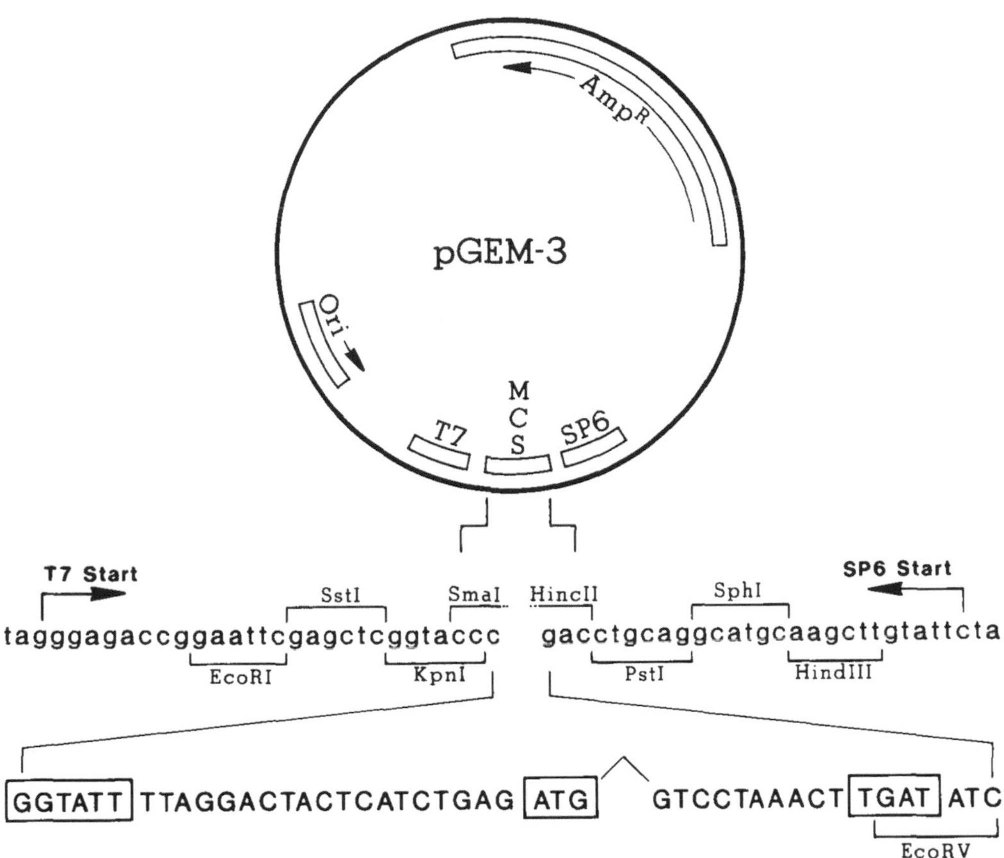

FIGURE 12. Schematic diagram of transcription vector pGem[Tm]-3 constructs containing structural elements of WTV segment S5. In this case, transcripts contain 28 nucleotides of nonviral sequence at the 5′ terminus. The presence of the tailored *EcoRV* site allows control of the sequence at the 3′ terminus of the transcript. Constructions have utilized the T7 promoter rather than the SP6 promoter because the sequence at the *Sph*I site introduces an AUG in the transcript upstream of the viral initiation codon. (pGen[Tm]-3 is a product of Promega Biotec®, Madison, Wisc.)

successive internal deletion and point mutation events. Conversely, in light of recent reports of fused dsRNA genome segments of rotaviruses,[105-107] the generation of large polycistronic dsRNA genetic elements from monocistronic dsRNA segments is also a possibility. Elucidation of the genome organization of the large, potentially polycistronic dsRNA genetic elements and their remnant RNAs is likely to provide insight into the reality of such processes.

Studies utilizing DI genome RNAs of positive and negative strand RNA viruses have recently provided precise information regarding regulatory elements involved in genome function.[30-32] As indicated in the previous section, remnant dsRNAs provide similar opportunities for defining structure-function relationships for dsRNA genetic elements. What were initially viewed as curiosities may find important applications as experimental reagents.

VIII. ADDENDUM

The following information will appear in a December 1987 issue of *Proc. Natl. Acad. Sci. U.S.A.* in an article entitled "Segment-Specific Inverted Repeats Found Adjacent to Conserved Terminal Sequences in Wound Tumor Virus Genome and Defective Interfering RNAs" by J. V. Anzola, et al.

Three DI RNAs (587-776 base-pairs) derived from WTV genome segment S5 (2613 base-pairs) were cloned and sequenced. Each DI RNA was generated by a single, simple internal deletion event. Although the DI RNAs were in continuous passage in systemically infected plants for more than 20 years, remarkably, their nucleotide sequences were identical to that of corresponding portions of segment S5 present in infrequently passaged, standard, transmission-competent virus. The positions of the deletion breakpoints within the DI RNAs indicate that the minimal sequence information required for replication and packaging of segment S5 resides within 319 base-pairs from the 5′ end of the (+)-strand and 205 base-pairs from the 3′ end. Sequence analysis of the termini of each WTV genome segment revealed the presence of a 6- to 9-nucleotide, segment-specific inverted repeat located immediately adjacent to the conserved terminal sequences shared by all 12 WTV genome segments.

The combined information currently available on WTV DI RNAs provides an emerging view of principles that govern the replication and packaging of a segmented double-stranded RNA genome. (1) Sequence information required for replication and packaging of a genome segment is located within the terminal domains. (2) Packaging of one pair of terminal structures excludes the packaging of a second copy of the same pair of terminal structures. The apparent exclusion of multiple copies of the same pair of terminal structures implies that each segment must contain at least two operational recognition sequence domains (sorting signals): one that specifies that it is a viral and not a cellular RNA (perhaps the conserved terminal sequences), and a second that specifies that it is a particular RNA segment (perhaps the segment-specific inverted repeats). It also implies that packaging of the 12 individual WTV segments must involve 12 different and specific protein-RNA and/or RNA-RNA interactions.

ACKNOWLEDGMENT

The skillful assistance of Ada Hernandez in the preparation of this chapter is gratefully acknowledged.

REFERENCES

1. **Nuss, D. L.,** Molecular biology of wound tumor virus, *Adv. Virus Res.*, 29, 57, 1984.
2. **Francki, R. I. B. and Boccardo, G.,** The plant reoviridae, in *The Reoviridae*, Joklik, W. F., Ed., Plenum Press, New York, 1983, 505.
3. **Shikata, E.,** Reoviruses, in *Handbook of Plant Virus Infectious and Comparative Diagnosis*, E. Kurstak, Ed., Elsevier/North-Holland, Amsterdam, 1981, 423.
4. **Boccardo, G., Lisa, V., and Milne, R. G.,** Cryptic viruses of plants, in *Double-Stranded RNA Viruses*, Compans, R. W. and Bishop, D. H. L., Eds., Elsevier, New York, 1983, 425.
5. **Boccardo, G., Milne, R. G., Luisoni, E., Lisa, V., and Accotto, G. P.,** Three seedborne cryptic viruses containing double-stranded RNA isolated from white clover, *Virology*, 147, 29, 1985.
6. **Gould, A. R. and Francki, R. I. B.,** Immunochemical detection of dsRNA in healthy and virus-infected plants and specific detection of viral dsRNA by hybridization to labeled complementary DNA, *J. Virol. Methods*, 2, 277, 1981.
7. **Dodds, J. A., Morris, J. J., and Jordan, R. L.,** Plant viral double-stranded RNA, *Annu. Rev. Phytopathol.*, 22, 151, 1984.
8. **Grill, L. K. and Garger, S. J.,** Identification and characterization of double-stranded RNA associated with cytoplasmic male sterility in *Vicia faba*, *Proc. Natl. Acad. Sci. U.S.A.*, 78, 7043, 1981.
9. **Wakarchuk, D. A. and Hamilton, R.,** Cellular double-stranded RNA in *Paeolus vulgaris*, *Plant Mol. Biol.*, 5, 55, 1985.
10. **Lemke, P. A. and Nash, C. H.,** Fungal viruses, *Biol. Rev.*, 38, 29, 1974.

11. **Lemke, P. A.,** Fungal viruses in agriculture, in *Beltsville Symposium in Agricultural Research. I. Virology and Agriculture,* Romberger, A. J., Ed., Allanheld, Montclair, N.J., 1977, 159.
12. **Day, P. R. and Dodds, A. J.,** Viruses of pathogenic fungi, in *Viruses and Plasmids in Fungi,* Lemke, P. A., Ed., Marcel Dekker, New York, 1979, 201.
13. **Nuss, D. L. and Summers, D.,** Variant dsRNAs associated with transmission-defective isolates of wound tumor virus represent terminally conserved remnants of genome segments, *Virology,* 133, 276, 1984.
14. **Reddy, D. V. R. and Black, L. M.,** Deletion mutations of the genome segments of wound tumor virus, *Virology,* 61, 458, 1974.
15. **Reddy, D. V. R. and Black, L. M.,** Isolation and replication of mutant population of wound tumor virions lacking certain genome segments, *Virology,* 80, 336, 1977.
16. **Somers, J. M.,** Isolation of suppressive mutants from killer and neutral strains of *Saccharomyces cerevisiae, Genetics,* 74, 571, 1973.
17. **Cande, J. and Fink, G. R.,** A mutant of *Saccharomyces cerevisiae* defective for nuclear fusion, *Proc. Natl. Acad. Sci. U.S.A.,* 73, 3651, 1976.
18. **Vodkin, M., Katterma, F., and Fink, G. R.,** Yeast killer mutants with altered double-stranded ribonucleic acid, *J. Bacteriol.,* 117, 681, 1974.
19. **Sweeney, T. K., Tate, A., and Fink, G. R.,** A study of the transmission and structure of double-stranded RNAs associated with the killer phenomenon in *Saccharomyces cerevisiae, Genetics,* 84, 27, 1976.
20. **Ridley, S. P. and Wickner, R. B.,** Defective interference in killer system of *Saccharomyces cerevisiae, J. Virol.,* 45, 800, 1983.
21. **Bruenn, J. and Kane, W.,** Relatedness of the double-stranded RNAs present in yeast virus-like particles, *J. Virol.,* 26, 762, 1978.
22. **Fried, H. M. and Fink, G. R.,** Electron microscopic heteroduplex analysis of "killer" double-stranded RNA species from yeast, *Proc. Natl. Acad. Sci.,* 75, 4224, 1978.
23. **Thiele, D. J., Hanning, E. M., and Leibowitz, M. J.,** Genome structure and expression of a defective interfering mutant of the killer virus of yeast, *Virology,* 137, 20, 1984.
24. **Lee, M., Pietras, D. F., Nemeroff, M. E., Corstanje, B. J., Field, L. J., and Bruenn, J. A.,** Conserved regions in defective interfering viral double-stranded RNAs from a yeast virus, *J. Virol.,* 58, 402, 1986.
25. **Tartaglia, J., Paul, C. P., Fulbright, D. W., and Nuss, D. L.,** Structural properties of double-stranded RNAs associated with a North American hypovirulent strain of *Endothia (cryphonectria) parasitica, Proc. Natl. Acad. Sci. U.S.A.,* 83, 9109, 1986.
26. **Anagnostakis, S. L. and Day, P. R.,** Hypovirulence conversion in *Endothia parasitica, Phytopathology,* 69, 1226, 1979.
27. **Morris, T. J. and Dodds, J. A.,** Isolation and analysis of double-stranded RNA from virus-infected plant and fungal tissue, *Phytopathology,* 69, 854, 1979.
28. **Dodds, J. A.,** Revised estimates of the molecular weights of dsRNA segments in hypovirulent strains of *Endothia parasitica, Phytopathology,* 70, 1217, 1980.
29. **Dodds, J. A.,** Association of Type 1 viral-like dsRNA with club-shaped particles in hypovirulent strains of *Endothia parasitica, Virology,* 107, 1, 1980.
30. **Re, G. G., Morgan, E. M., and Kingsbury, D. W.,** Nucleotide sequences responsible for generation of internally deleted Sendai virus defective interfering genomes, *Virology,* 146, 27, 1985.
31. **Re, G. G. and Kingsbury, D. W.,** Nucleotide sequences that affect replicative and transcriptional efficiencies of Sendai virus deletion mutants, *J. Virol.,* 58, 578, 1986.
32. **Levis, R., Weiss, B. G., Tsiang, M., Huang, H., and Schlesinger, S.,** Deletion mapping of Sindbis virus DI RNAs derived from cDNAs defines the sequences essential for replication and packaging, *Cell,* 44, 137, 1986.
33. **Black, L. M.,** Wound tumor virus disease, in *Molecular Biology of Plant Tumors,* Kahl, G. and Schell, J., Eds., Academic Press, New York, 1982, 69.
34. **Matthews, R. E. F.,** Classification and nomenclature of viruses: fourth report of the International Committee on toxonomy of viruses, *Intervirology,* 17, 1, 1982.
35. **Black, L. M.,** Physiology of virus-induced tumors in plants, in *Handbuch der Pflanzenphysiologie,* Vol. 15 (Part2), W. Ruhland and A. Lang, Eds., Springer-Verlag, Berlin, 1965, 236.
36. **Black, L. M.,** Insect tissue cultures as tools in plant virus research, *Annu. Rev. Phytopathol.,* 1, 73, 1969.
37. **Black, L. M.,** Plant tumors of viral origin, *Prog. Exp. Tumor Res.,* 15, 110, 1972.
38. **Black, L. M.,** Vector cell monolayers and plant viruses, *Adv. Virus Res.,* 25, 191, 1979.
39. **Black, L. M.,** The controversy regarding multiplication of some plant viruses in their insect vectors, in *Current Topics in Vector Research,* Vol. 2, Harris, K. F., Ed., Praeger, New York, 1984, 1.
40. **Chiu, R. J. and Black, L. M.,** Monolayer cultures of insect cell lines and their inoculation with a plant virus, *Nature (London),* 215, 1076, 1967.
41. **Hsu, H. T., Nuss, D. L., and Adam, G.,** Utilization of insect tissue culture in the study of the molecular biology of plant viruses, in *Current Topics in Pathogen-Vector-Host Research,* Harris, K. F., Eds., Praeger, New York, 1983, 189.

42. **Nuss, D. L. and Peterson, A. J.,** Expression of wound tumor virus gene products in vivo and in vitro, *J. Virol.,* 34, 532, 1980.
43. **Nuss, D. L. and Peterson, A. J.,** Resolution and genome assignment of RNA transcripts synthesized in vitro by wound tumor virus, *Virology,* 114, 399, 1981.
44. **Nuss, D. L. and Peterson, A. J.,** In vitro synthesis and modification of mRNA by exvectorial isolates of wound tumor virus, *J. Virol.,* 39, 954, 1981.
45. **Asamizu, T., Summers, D., Motika, M. B., Anzola, J. V., and Nuss, D. L.,** Molecular cloning and characterization of the genome of wound tumor virus: a tumor-inducing plant reovirus, *Virology,* 144, 398, 1985.
46. **Peterson, A. J. and Nuss, D. L.,** Wound tumor virus polypeptide synthesis in productive, non-cytopathic infection of cultured vector cells, *J. Virol.,* 56, 620, 1985.
47. **Peterson, A. J. and Nuss, D. L.,** Regulation of wound tumor virus gene expression in persistently infected vector cells is related to change in translational activity of viral transcripts, *J. Virol.,* 59, 195, 1986.
48. **Reddy, D. V. R. and Black, L. M.,** Comparative infectivity of WTV isolates, in "Insect tissue cultures as tools in plant virus research," *Annu. Rev. Phytopathol.,* 1, 87, 1969.
49. **Liu, H. Y., Kimura, I., and Black, L. M.,** Specific infectivity of different wound tumor virus isolates, *Virology,* 51, 320, 1973.
50. **Shikata, E., Orenski, S. W., Hirumi, H., Mitsuhashi, J., and Maramorosch, K.,** Electron micrographs of wound-tumor virus in an animal host and in a plant tumor, *Virology,* 23, 441, 1964.
51. **Sinha, R. C.,** Sequential infection and distribution of wound-tumor virus in the internal organs of a vector after ingestion of virus, *Virology,* 26, 673, 1965.
52. **Skikata, E. and Maramorosch, K.,** Electron microscopy of wound tumor virus assembly sites in insect vectors and plants, *Virology,* 27, 461, 1965.
53. **Granados, R. R., Hirumi, H., and Maramorosch, K.,** Electron microscopic evidence for wound-tumor virus accumulation in various organs of an inefficient leafhoppers vector, *Agalliopsis novella, J. Invertebr. Pathol.,* 9, 147, 1967.
54. **Hirumi, H., Granados, R. R., and Maramorosch, K.,** Electron microscopy of a plant-pathogenic virus in the nervous system of its insect vector, *J. Virol.,* 1, 430, 1967.
55. **Maramorosch, K.,** Arthropod transmission of plant viruses, *Annu. Rev. Entomol.,* 8, 369, 1963.
56. **Maramorosch, K.,** Influence of temperature on incubation and transmission of the wound-tumor virus, *Phytopathology,* 40, 1071, 1950.
57. **Black, L. M.,** Viruses that reproduce in plants and insects, *Ann. N.Y. Acad. Sci.,* 56, 398, 1953.
58. **Black, L. M.,** Some viruses transmitted by agallian leafhoppers, *Proc. Am. Philos. Soc.,* 88, 132, 1944.
59. **Black, L. M., Wolcyrz, S., and Whitcomb, R. F.,** A vectorless strain of wound-tumor virus, in Abstr. 7th Int. Congr. Microbiology, Stockholm, 1958, 2255.
60. **Reddy, D. V. R. and Black, L. M.,** Specific infectivity of different isolates of wound-tumor virus, *Phytopathology,* 55, 1072, 1965.
61. **Black, D. R. and Knight, C. A.,** Ribonucleic acid transcriptase activity in purified wound tumor virus, *J. Virol.,* 6, 194, 1970.
62. **Reddy, D. V. R., Rhodes, D. P., Lesnow, J. A., MacLeod, R., Banerjee, A. K., and Black, L. M.,** *In vitro* transcription of wound tumor virus RNA by virion-associated RNA transcriptase, *Virology,* 80, 356, 1977.
63. **Nuss, D. L.,** Molecular biology of wound tumor virus transmissions: genome segment 5 and the loss of transmissibility, in *Double-Stranded RNA Viruses,* Bishop, D. H. L. and Compans, R. W., Eds., Elsevier, Amsterdam, 1983, 415.
64. **Wickner, R. B.,** Killer systems in *Saccharomyces cerevisiae,* in *Molecular Biology of the Yeast Saccharomyces: Life Cycle and Inheritance,* Strathern, J. N., Jones, E. W., and Broach, J. R., Eds., Cold Spring Harbor Laboratory, Cold Spring Harbor, N.Y., 1981, 415.
65. **Wickner, R. B.,** Genetic control of replication of the double-stranded RNA segments of the killer systems in *Saccharomyces cerevisiae, Arch. Biophys.,* 222, 1, 1983.
66. **Bussey, H.,** Physiology of killer factor in yeast, *Adv. Microb. Physiol.,* 22, 93, 1981.
67. **Tipper, D. K. and Bostian, L. A.,** Double-stranded ribonucleic acid killer systems in yeast, *Microbiol. Rev.,* 48, 125, 1984.
68. **Bostian, K. A., Hopper, J. F., Rogers, D. T., and Tipper, D. J.,** Translational analysis of the killer-associated virus-like particle dsRNA genome of *Saccharomyces cerevisiae:* M dsRNA encodes toxin, *Cell,* 19, 403, 1980.
69. **Bussey, H., Sacks, W., Galley, D., and Saville, D.,** Yeast killer plasmid mutations affecting toxin secretion and activity and toxin immunity function, *Mol. Cell. Biol.,* 2, 346, 1982.
70. **Bostian, K. A., Sturgeon, J. A., and Tipper, D. J.,** Encapsidation of yeast killer double-stranded ribonucleic acids: dependence of M on L, *J. Bacteriol.,* 143, 463, 1981.
71. **Bevan, E. A., Herring, A. J., and Mitchell, D. J.,** Preliminary characterization of two species of dsRNA in yeast and their relationship to the "killer" character, *Nature (London),* 245, 81, 1973.

72. **Harris, M. S.,** Virus-like particles and double-stranded RNA from killer and non-killer strains of *Saccharomyces cerevisiae, Microbiology,* 21, 161, 1978.
73. **Herring, A. J. and Bevan, E. A.,** Virus-like particles associated with the double-stranded RNA species found in killer and sensitive strains of the yeast *Saccharomyces cerevisiae, J. Gen. Virol.,* 22, 387, 1974.
74. **Hopper, J. F., Bostian, K. A., Rowe, L. B., and Tipper, D. J.,** Translation of the L-species dsRNA genome of the killer associated virus-like particles of *Saccharomyces cerevisiae, J. Biol. Chem.,* 252, 9010, 1977.
75. **Somers, J. M. and Bevan, E. A.,** The inheritance of the killer character in yeast, *Genet. Res.,* 13, 17, 1969.
76. **Furuichi, Y.,** Expression of eukaryotic viral genome dsRNAs, in *Annu. Rep. Roche Institute of Molecular Biology,* 1985, 102.
77. **Day, P. R., Dodds, J. A., Elliston, J. E., Jaynes, R. A., and Anagnostakis, S. L.,** Double-stranded RNA in *Endothia parasitica, Phytopathology,* 67, 1393, 1977.
78. **Anagnostakis, S. L.,** Biological control of chestnut blight, *Science,* 215, 466, 1982.
79. **Van Alfen, N. K.,** Biology and potential for disease control of hypovirulence of *Endothia parasitica, Annu. Rev. Phytopathol.,* 20, 349, 1982.
80. **L'Hostis, B., Hiremath, S. T., Rhoads, R. E., and Ghabrial, S. A.,** Lack of sequence homology between double-stranded RNA from European and American hypovirulent strains of *Endothia parasitica, J. Gen. Virol.,* 66, 351, 1985.
81. **Garrod, S. W., Fulbright, D. W., and Ravenscroft, A. V.,** Dissemination of virulent and hypovirulent forms of marked strain of *Endothia parasitica* in Michigan, *Phytopathology,* 75, 533, 1985.
82. **Fulbright, D. W., Weidlich, W. H., Haufler, K. Z., Thomas, C. S., and Paul, C. P.,** Chestnut blight and recovering American chestnut trees in Michigan, *Can. J. Bot.,* 61, 3164, 1983.
83. **Joklik, W. K.,** Reproduction of the reoviridae, in *Comprehensive Virology,* Vol. 2, Fraenkel-Conrat, H. and Wagner, R. R., Eds., Plenum Press, New York, 1974, 231.
84. **Tyler, K. L. and Fields, B. N.,** Reovirus and its replication, in *Virology,* Fields, B. N. et al., Eds., Raven Press, New York, 1985, 823.
85. **Shatkin, A. J. and Kozak, M.,** Biochemical aspects of reovirus transcription and translation, in *The Reoviridae,* Joklik, W. K., Ed., Plenum Press, New York, 1983, 79.
86. **Zarbl, H. and Millward, S.,** The reovirus multiplication cycle, in *The Reoviridae,* Joklik, W. K., Ed., Plenum Press, New York, 1983, 107.
87. **Bruenn, J. A.,** Virus-like particles of yeast, *Annu. Rev. Microbiol.,* 34, 49, 1980.
88. **Nemeroff, M. E. and Bruenn, J. A.,** Conservative replication and transcription of *Saccharomyces cerevisiae* viral double-stranded RNA *in vitro, J. Virol.,* 57, 754, 1986.
89. **Taniaguchi, T., Palmiere, M., and Weissman, C.,** Qβ DNA-containing hybrid plasmids giving rise to Qβ phage formation in the bacterial host, *Nature (London),* 274, 223, 1978.
90. **Racaniello, V. R. and Baltimore, D.,** Cloned poliovirus complementary DNA is infectious in mammalian cells, *Science,* 214, 916, 1981.
91. **Cress, D. E., Kiefer, M. C., and Owens, R. A.,** Construction of infectious potato spindle tuber viroid cDNA clones, *Nucleic Acids Res.,* 11, 6821, 1983.
92. **Semler, B. L., Dornea, A. J., and Wimmer, E.,** Production of infectious poliovirus from cloned cDNA is dramatically increased by SV40 transcription and replication signals, *Nucleic Acids Res.,* 12, 5123, 1984.
93. **Omata, T., Kohara, M., Sakai, Y., Kameda, A., Imura, N., and Nomoto, A.,** Cloned infectious complementary DNA of the poliovirus Sabin 1 genome: biochemical and biological properties of the recovered virus, *Gene,* 32, 1, 1984.
94. **Ahlquist, P., French, R., Janda, M., and Loesch-Fries, L. S.,** Multicomponent RNA plant virus infection derived from cloned viral cDNA, *Proc. Natl. Acad. Sci. U.S.A.,* 74, 4900, 1984.
95. **Dasmahapatra, B., Dasgupta, R., Saunders, K., Selling, B., Gallagher, T., and Kaesberg, P.,** Infectious RNA derived by transcription from cloned cDNA copies of the genomic RNA of an insect virus, *Proc. Natl. Acad. Sci. U.S.A.,* 83, 63, 1986.
96. **Mizutani, S. and Colonno, R. J.,** *In vitro* synthesis of an infectious RNA from cDNA clones of human rhinovirus type 14, *J. Virol.,* 56, 628, 1985.
97. **Van Der Werf, S., Bradley, J., Wimmers, E., Studier, W. F., and Dunn, J. J.,** Synthesis of infectious poliovirus RNA by purified T7 RNA polymerase, *Proc. Natl. Acad. Sci. U.S.A.,* p. 2330, 1986.
98. **Ahlquist, P. and Janda, M.,** cDNA cloning and *in vitro* transcription of the complete brome mosaic virus genome, *Mol. Cell. Biol.,* 4, 2876, 1984.
99. **Chamberlin, M., Kingston, R., Gilman, M., Wiggs, J., and deVera, A.,** Isolation of bacterial and bacteriophage RNA polymerases and their use in synthesis of RNA *in vitro, Methods Enzymol.,* 101, 540, 1983.
100. **Chamberlin, M., McGrath, J., and Waskell, L.,** New RNA polymerase from *Escherichia coli* infected with bacteriophage T7, *Nature (London),* 228, 227, 1970.

101. **Melton, D. A., Krieg, P. A., Rebagliati, M. R., Maniatis, T., Zinn, K., and Green, M. R.,** Efficient *in vitro* synthesis of biologically active RNA and RNA hybridization probe from plasmids containing a bacteriophage SP6 promoter, *Nucleic Acids Res.*, 12, 7035, 1984.
102. **Oka, A., Sugisaki, H., and Takanami, M.,** Nucleotide sequence of the kanamycin resistance taumsposon Tn903, *J. Mol. Biol.*, 147, 217, 1981.
103. **Holland, J., Spindler, K., Horodyski, F., Grabau, E., Nichol, S., and VandePol, S.,** Rapid evolution of RNA genomes, *Science*, 215, 1577, 1982.
104. **Perrault, J.,** Origin and replication of defective interfering particles, *Curr. Top. Microbiol. Immunol.*, 93, 151, 1981.
105. **Pedley, S., Hundley, F., Chrystie, I., McCrae, M. A., and Desselberger, U.,** The genomes of rotaviruses isolated from chronically infected immunodeficient children, *J. Gen. Virol.*, 65, 1141, 1984.
106. **Allen, A. M. and Desselberger, U.,** Reassortment of human rotaviruses carrying rearranged genomes with bovine rotavirus, *J. Gen. Virol.*, 66, 2703, 1983.
107. **Hundley, F., Biryahwaho, B., Gow, M., and Desselberger, U.,** Genome rearrangements of bovine rotavirus after serial passage at high multiplicity of infection, *Virology*, 143, 88, 1985.
108. **Reddy, D. V. R. and Black, L. M.,** Electrophoretic separation of all components of the double-stranded RNA of wound tumor virus, *Virology*, 54, 557, 1973.
109. **Salser, W.,** Globin mRNA sequencies: analysis of base pairing and evolutionary implications, *Cold Spring Harbor Symp. Quant. Biol.*, 42, 985, 1977.
110. **Hansen, D. R., Van Alfen, N. K., Gillies, K., and Powell, W. A.,** Naked dsRNA associated with hypovirulence of *Endothia parasitica* is packaged in fungal vesicles, *J. Gen. Virol.*, 66, 2605, 1985.
111. **McCrae, M. A. and Joklik, W. F.,** The nature of the polypeptide encoded by each of the 10 double-stranded RNA segments of reovirus type 3, *Virology*, 89, 578, 1978.
112. **Xu, Z., Anzola, J. V., and Nuss, D. L.,** Tailored removal of flanking homopolymer sequences from cDNA clones, *DNA*, 6, 505, 1987.

Chapter 10

EVOLUTION OF RNA VIRUSES

David Zimmern

TABLE OF CONTENTS

I. Introduction 212
  A. RNA as Genetic Material 212
  B. Mechanisms and Routes of RNA Genome Variation 212

II. Specific Mechanisms 213
  A. Point Mutation 213
  B. Recombination 214

III. Interviral Homologies 215
  A. General Routes of Variation in Nature are Deducible From Their End Products 215
  B. Modular Construction of RNA Virus Genomes 216
    1. Homologies Between Diverse Virus Groups 216
    2. Selection Pressure: Conservation of Amino Acid Sequence Despite High Mutation Rates 220
    3. Recombination of Functional Modules 223

IV. Support for the Protovirus Hypothesis? 225

V. Addendum 227

Acknowledgments 228

Appendix 1: Alignment of Conserved Core of Picornaviral P2C Proteins 228

Appendix 2: Alignment of Residues Around Active Site of Picornaviral P3C Protease 229

Appendix 3: Alignment of Picornaviral P3D Polymerase Proteins 230

Appendix 4: Alignment of the Conserved N Terminal Domain of Alphavirus nsP1 Proteins and Their Plant Viral Homologues Listed in Table 1 231

Appendix 5: Alignment of the Conserved Core of Alphavirus nsP2 Proteins and Their Plant Viral Homologues Listed in Table 1 232

Appendix 6: Alignment of the Alphavirus nsP4 Proteins and Their Plant Viral Homologues Listed in Table 1 234

References 235

I. INTRODUCTION

A. RNA as Genetic Material

It is thought that RNA is only used as the primary repository of genetic information by a limited and heterogeneous assortment of viruses. This characteristic sets the relevant groups of viruses apart from all other living forms, and raises the evolutionary question of how such a distinction arose. Despite its intrinsic interest, this large question has attracted only passing interest from virologists because, in common with most questions of ultimate origins, it seemed unlikely that we would ever know. Nevertheless, the question remains an intriguing one and has attracted a certain amount of speculation.

Although the answer to this large problem remains elusive, it incorporates a number of other questions to which the answers, either for reasons of urgency, opportunity, or pure accident, are beginning to emerge. Foremost among these is the question of the mechanism(s) of variation among existing viruses. This is of obvious practical importance in understanding the emergence of new pathogenic strains of medically or agriculturally troublesome viruses, besides having an academic interest to those involved in various aspects of the molecular biology of RNA. In considering this question, the very uniqueness of RNA viruses which makes their remoter origins so hard to fathom can be turned to advantage because the group of RNA viruses is well defined and its members, although diverse, must have evolved under very similar constraints on sources of new variation. This chapter will attempt to review and collate the available information concerning the mechanisms of such variation, returning at the end to ask if this new information sheds any additional light on the theoretical question of how RNA-based genetic elements arose (or became autonomous) in the first place.

B. Mechanisms and Routes of RNA Genome Variation

Genetic elements made of DNA have two major mechanisms of variation: point mutation and recombination, the latter mostly involving breakage and rejoining of DNA molecules, but including also an assortment of more specialized variations on the theme of replicative recombination such as certain kinds of transposition.[1] Very early work on RNA virus genetics, done in some cases even before the virus in question had been identified as having an RNA genome, initially made no distinction regarding mechanisms of variation between RNA viruses and those with DNA genomes. As it became clear that (1) specialized enzymes deal with DNA recombination which do not recognize RNA, (2) some well-characterized cases of apparent recombination in RNA viruses (particularly flu) were actually due to reassortment of separate genome segments,[2] and (3) RNA recombination was undetectable in all but one of the best experimental systems,[2] and specifically not in the one most extensively studied (RNA phage in *Escherichia coli)*[3] this undifferentiated view was displaced by the conviction that with possible rare exceptions recombination did not occur in RNA. One major current in contemporary research is the reexamination of the generality of this assumption, which in hindsight can be clearly seen to represent a prejudice based on overreliance on the *E. coli* phage model.

Although (with the notable exception of poliovirus, and disregarding rearrangements in defective interfering RNAs) little evidence for true recombination could be found in early work, it was quickly established that RNA viruses undergo mutation at very high rates. It is commonly thought, again, based principally on the *E. coli* phage model, that this high mutation rate represents the raw rate of polymerase errors because there are no mechanisms for proofreading in RNA virus replication. The classical work on which this view is based represents a second major research theme in this area.

Finally, while attempts to define the general mechanisms of variation have necessarily relied to varying degrees on studies with model systems, it is evident that they only acquire relevance in relation to real viruses as functional entities. In order to understand change and variation in its natural context of viral evolution, one must first understand static relationships between different virus groups as they now exist. With the advent of nucleotide sequencing techniques that make it feasible to determine complete genomic sequences routinely, a rational taxonomy based on nucleotide sequence comparisons has become possible which has revolutionized our understanding of viral relationships. This represents a third important strand of contemporary research.

II. SPECIFIC MECHANISMS

A. Point Mutation

Evidence for single base changes in RNA viruses is so widespread and includes so many sporadic instances that it would be superfluous to document every case. Systematic studies have been made by several groups, dating from historical studies on the effects of mutagenic agents. They include classic studies on the genetic code using TMV coat protein as a model system[4,5] and on specific mutations in Qβ phage that introduced the concept of site-directed mutagenesis.[6,7] From the point of view of understanding viral variation, the most important of these studies were those responsible for establishing our current view of the high rate of mutation in RNA virus genomes. Notable among these were formal genetic studies by Zinder and other early workers on RNA phages (reviewed in Reference 3), culminating in the studies of Weissmann and colleagues on Qβ phage, and both formal genetic and serial passaging studies on a variety of RNA viruses of eukaryotes (reviewed in the context of the earlier work in Reference 8). The major points emerging from all this are summarized here (and are also the subject of the chapter by Domingo and Holland elsewhere in this book):

1. Mutation rates are always very high, although the polymerase intrinsic error rate may not be quite as high as the impression sometimes given by spectacularly high rates of reversion in cases where selection may have intervened. One of the earliest, and still among the best, estimates was on the order of 10^{-4} per base per round of replication for reversion of a viable site-specific mutant of Qβ.[7] Since exact rates can only be measured for proven single base changes growing to a measurable final titer in the absence of competition, such rigorous determinations cannot usually be made, but most other estimates have been of the same order of magnitude for all RNA viruses studied.[8] (One noteworthy exception to the prevailing finding of high mutation rates is an E2 envelope glycoprotein mutant of Sindbis virus where a reversion rate of $<7 \times 10^{-7}$ has recently been reported).[9] These rates apply to individual sites, and corresponding mutation rates for gross phenotypic characters (e.g., temperature-sensitive RNA synthesis) where selection is applied to whole genes may be several percent. Some variation in reported rates is to be expected since mutation rates are site dependent, and apparent rates may be subject to selection bias within a single growth cycle, perhaps accounting for reversion rates as high as 10^{-2} which have been reported.[3] Notwithstanding such uncertainties, there is a clear consensus that rates are several orders of magnitude higher than for DNA (10^{-6} in prokaryotes).

   These high mutation rates can be tolerated only because the genomes of RNA viruses are small, since there is an inverse relationship for any genome between size and acceptable mutational load (Equation 28 in Reference 10). Even so, RNA viruses may be approaching the limits of size imposed by this constraint, since their typical size range of 3×10^3 to 2×10^4 residues approaches or equals the inverse of the mutation rate. It follows that:
2. There is a fairly high probability that any given progeny RNA molecule in a virus population will carry a point mutation, although many such changes (e.g., at codon third positions) are expected to be phenotypically silent. The fact that consensus

sequences can be derived at all for RNA virus genomes presumably testifies to the rapidity with which selection homogenizes the population and suggests that the majority of point mutations are not phenotypically silent, since alleles of equal fitness are expected in the population at equal frequency.[11,12] Experimental support can be found for this conclusion in studies of persistent virus infections in vivo where the functions of certain viral genes are by-passed. The redundant genes accumulate mutations in a way analogous to the accumulation of mutations in cellular pseudogenes, without the bias seen in viable stocks towards silent mutations rather than those which change the amino acid sequence.[13]

Classical population genetics suggests that in such a situation only those mutations which are not selectively disadvantageous will be carried in a population and hence major changes in a genome by progressive accumulation of point mutations will be constrained in spite of the high mutation rate. The importance of the high mutation rate may instead lie in providing a pool of variation to respond rapidly to marginal selective advantages and disadvantages arising from a changing environment. The dynamic nature of the resulting equilibrium is nonetheless evident from the large numbers of variants which can usually be found in the population at low frequency by cloning either the virus itself[11,14,15] or its cDNA[16,17] (a fact which should be kept in mind more firmly than it sometimes is when interpreting cDNA cloning experiments). Thus, the biology of RNA viruses is the biology of the population consensus, and their genetics are essentially population genetics in which the relative proportion of any particular mutant is a sensitive function of its growth rate.[12]

3. There is no evidence for error correction by proofreading of progeny RNA sequences, and this probably accounts for the difference in apparent mutation rates compared to DNA. However, the effectiveness of selection in maintaining a relatively uniform population suggests that RNA viruses may circumvent this limitation in a different way by simply discarding nonviable variants as soon as possible by competition during a single growth cycle, and perhaps sometimes more actively. For example, it has been suggested that each poliovirus replicase molecule acts only once in *cis*, using the RNA from which it was translated as a template, and that subsequent reinitiation is prevented by the cleavage of VPg (reviewed in Reference 18). In genetic terms, this manifests itself by the absence of complementation between appropriate mutants, a phenomenon which also occurs in other RNA viruses, though not necessarily for the same reason. If a replicase worked only in *cis,* mutants producing nonviable replicase would be instantly aborted. Such options are, presumably, impossibly wasteful for a cellular chromosome, but would be quite efficient as a feedback control for a molecule devoted largely to coding for its own replication. While the generality of such mechanisms has yet to be established, and the poliovirus data are not definitive, it is possible to see from this model that the lack of proofreading need not entirely prevent error correction.

In its reliance on selection, rather than proofreading, to establish a population consensus, it could be argued that RNA virus replication was more truly Darwinian than DNA replication, because proofreading increases the probability of propogation of the parental sequence regardless of its fitness.[19] Speculations about the possible adaptive advantages which these characteristics may confer on viruses, and perhaps on some cellular genetic elements, have stimulated a large enough body of theory to constitute a literature of their own (see, for example, Reference 19a).

B. Recombination

Amid the run of early experimental evidence discounting the existence of general recombination in RNA viruses was one counterexample that obstinately refused to go away. This was the genetic evidence for recombination in picornaviruses originally discovered by Hirst and Ledinko.[2] Extensive studies of this phenomenon by a generation of picornavirologists eventually demonstrated beyond reasonable doubt the existence of a linear genetic map approximately colinear with the viral polyprotein[20] and ultimately provided nucleotide sequences spanning crossover points (see chapter by King, this volume). It is debatable whether picornaviruses are intrinsically unique in their ability to recombine. The very high rate of reversion commonly observed with RNA virus mutants, combined with the frequent need to use conditional lethal mutants that may be leaky, may make it very difficult to detect recombinants above the background in classical two- or three-factor crosses. Picornavirus recombination occurs with high frequency (around 1%) and good selectable markers are available; in less favorable cases recombination might be overlooked. Classical methods have proved sufficient; however, for the recent recombination in coronaviruses,[21] while with the demonstration of availability of high-level transcription vectors, evidence for general recombination is now being found even in systems that are notoriously unfavorable for genetic studies, such as plant viruses,[22] suggesting that additional cases may await discovery.

Although genetic tests for recombination were usually inconclusive, accidental evidence for a more specialized kind of recombination was found in laboratory stocks of many kinds of RNA virus, in the form of extensive rearrangements which crippled the virus as a functional whole. A tendency to spawn variants with deletions (often multiple) and/or partial duplications is characteristic of many RNA viruses serially passaged at high multiplicity. These RNAs are known as defective interfering (DI) RNAs (see chapters by Schlesinger and Huang in this book).

DI RNAs are thought to be random variants which overgrow the population because they are truncated templates which are copied faster than their parental RNAs. These truncated templates are thought in turn to originate in strand switching or slipping errors made by viral RNA polymerase during replication (reviewed in Reference 23). Formally, this amounts to copy choice recombination, whose possible existence was thus tacitly conceded, although for many years DI RNAs were thought, for no convincing reason other than the lack of evidence to the contrary, to be formed only by intramolecular events. Intermolecular DI RNAs combining sequences from different genome segments of influenza virus have now been clearly demonstrated,[24] besides bizarre Sindbis virus DI RNAs which have covalently attached cellular tRNA sequences.[25] Thus, it is clear that whatever mechanism operates in these instances, it must be capable of acting in *trans* and even on unrelated sequences. While there is a precedent for transfer between templates by a nucleic acid polymerase, namely, proviral DNA synthesis by reverse transcriptase,[26] there is still no direct evidence either for or against the copy choice model in DI RNA formation (see Addendum for more recent information). However, it is clear in certain cases (e.g., Reference 27) that little or no base pairing is required at the crossover point, in contrast to the classically defined recombination in picornaviruses where the sequenced examples appear to involve recombination within regions of more extended homology. Thus, more than one recombination mechanism may operate in different cases. (See the recent review by King et al.[28] for further discussion of homologous versus nonhomologous RNA recombination.) The detailed mechanism(s) of recombination await(s) further study, and since a variety of recent work has emphasised the generality and importance of the phenomenon, these various kinds of model systems will probably receive much closer attention in the future than has hitherto been the case.

III. INTERVIRAL HOMOLOGIES

A. General Routes of Variation in Nature are Deducible From Their End Products

The major difficulty with mechanistic studies that necessarily use model systems is that generalizations based on them may be inappropriate or biased. As was argued in Section

II, it may be useful to look for the complementary information already implicit in the variety of different viruses, their natural groupings, and the relationship of these to each other, in order to get a wider perspective. Although the exact route of evolution by which viruses arrived at their current state as we observe them may be inaccessible to experiment, their structures must ultimately be explicable in terms of experimentally observable processes. The twin surprises that have emerged so far from such comparisons are the obvious importance of recombination, which was only detected with difficulty in model systems, and, conversely, the relative stability (reflected in recognizable sequence homology) of key amino acid coding regions of otherwise diverse RNA genomes despite their well-documented high mutation rates. The latter observation suggests the strong selection and/or a recent origin have been important in the evolution of existing RNA viruses.

B. Modular Construction of RNA Virus Genomes

1. Homologies Between Diverse Virus Groups

While viruses with RNA genomes exhibit great variety in morphology and genetic organization, it has emerged that this diversity may be elaborated around a limited pool of common genes. The variety arises from the association of these genes with different ancillary genes, their assembly on one or more RNA molecules, and in some cases, expression of related genes by different mechanisms, in individual cases. In the past few years, nucleotide sequences representative of many of the major groups of RNA viruses have been completed, and comparisons of amino acid sequences predicted from these nucleic acid sequences suggest that some surprisingly diverse groups of viruses might be related. These relationships cut across conventional groupings based on virion structure, and whether the virus infects an animal or a plant.

Sequence homologies between functionally equivalent proteins in different virus groups first came to light when it was found that the matrix proteins of VSV (a rhabdovirus) and influenza were distantly related.[29] Matrix proteins are structural proteins found in the layer between a helical nucleocapsid and the membrane in several classes of enveloped animal viruses; therefore, a structural relationship might reasonably have been expected. Nevertheless, the relationship presented a challenge to virologists' preconceptions. Although both groups are negative strand viruses, meaning that the mRNA is complementary to the packaged RNA and is transcribed from it by a virion-associated polymerase, VSV has a genome comprising a single RNA molecule carrying all the viral genes, while influenza virus has a genome comprising eight RNA molecules, most of which carry only a single gene. These differences in genetic structure are associated with substantial differences in the mechanism of gene expression.[18]

Observations such as those of Rose et al.[29] may be explained in terms of descent from a common ancestor, if the different gene arrangements, and associated differences in the mechanisms of gene expression, arose as a secondary characteristic. From the fact that almost identical transcriptional control sequences were reiterated in each intercistronic gap between genes on the polycistronic genomic (−) RNA of paramyxoviruses (rhabdoviruses are similar in this regard), Giorgi et al.[30] suggested that viruses with multiple RNAs might give rise to those with one RNA by segment fusion. Subsequently, as a range of complete genomic sequences became available, this idea received much stronger support, since the occurrence of genes that are common to viruses with multiple RNAs, and those with single RNAs, emerges as a frequent theme of amino acid sequence comparisons. Such comparisons, while identifying relationships, do not indicate which genomic arrangement was derived from the other and it is equally possible to argue that segmented genomes were derived from single RNAs by fission. Since these comparisons also show that "mixing and matching" may associate genes that are unique to one or a subset of viruses, besides simply providing the option of expressing identical sets of genes on one or a number of RNAs, such ideas are more usefully generalized to the notion that it is necessary to postulate recombination during evolution to relate such pairs of virus groups,[31,32] a conceptual framework in which it is also possible to consider events such as the formation of the Sindbis DIs mentioned in Section IV.

These arguments are most clearly visualized in relation to concrete examples where detailed sequence comparisons have been made over entire genomes from several related viruses. While certain negative strand virus structural proteins have been shown to be related, the nonstructural proteins of this group show an unexpected diversity, at least based on sequence information currently available. Consequently, although several lines of evidence going back over a number of years suggest that (for example) the rhabdovirus and paramyxovirus groups must be related (reviewed in Reference 18), no clear similarities are detectable in their putative polymerase (L gene) protein sequences (but see Addendum).[17,33,34] More sense can be made of the sequenced nonstructural proteins of positive strand viruses, although the evidence that structural relationships correspond to related virally encoded functions is somewhat fragmentary as yet, because the exact functions of many of the proteins in question have not been established. Insofar as these functions are known, however, the structural relationships make excellent sense, and they also correlate well with features of the respective viral genomes that are related to, or constrain, their mode of replication. Indeed, they seem to reflect fundamental similarities in the way that apparently diverse groups of viruses replicate. Accordingly, the following discussion concentrates on two groups of (+) strand viruses. It seems likely that other similar groupings should emerge eventually, revealing additional major families of RNA viruses. Extensive similarities which have already been found among widely divergent viruses which use reverse transcription are discussed in Section IV.

Consider, in the first instance, picornaviruses, like polio, which make a large polyprotein translation product that is subsequently cleaved by a virally encoded protease. The protease maps in a group of genes involved in replication where its 5′ neighbor is a small protein called VPg, which eventually becomes attached to the 5′ end of the progeny RNA in the course of replication.[18] A similar arrangement of genes has been established in the plant virus cowpea mosaic virus (CPMV), whose resemblance to picornaviruses was first noticed when it, too, was found to have a VPg[35,36] and a virally encoded protease.[37] Neither of these proteins is particularly well conserved, except in a limited area around the protease active site, but they are flanked by two other genes which are clearly related structurally when polio is compared with CPMV, one of these being the probable replicase, or a subunit thereof[31,38] (see Chapters 2 and 3 in Volume I and Appendixes 1 to 3). The entire set of four genes is thus conserved in arrangement, and at least partly in sequence, and may constitute a module of replicative genes which function together[31,38] (Figure 1).

Although a relationship in mode of replication between these two groups of viruses was already suspected before the nucleotide sequences were available, there is one obvious and major difference: the plant picornaviruses, in contrast to the animal ones, have a genome which is divided into two separate RNA molecules. Not counting the capsid protein genes (whose phylogenetic relationships are complicated to assess since they share a core tertiary structure universally found in icosahedral viruses, but have little or no amino acid sequence homology and a different pattern of proteolytic cleavage),[31,38,39] it is evident that the CPMV genome carries at least one gene on the smaller RNA that has no counterpart in animal picornaviruses (48-kdalton protein — see Figure 1). Thus, in order to transform the picornaviral gene arrangement into that of CPMV or vice versa, extensive rearrangements would be necessary involving, at a minimum, the deletion of the structural protein genes at the 5′ end of picornaviral RNAs and their coupling to the 48-kdalton protein gene to form a second genomic RNA or, conversely, the deletion of the CPMV-unique gene(s) from the smaller RNA and the integration of the capsid protein gene(s) into the larger one.

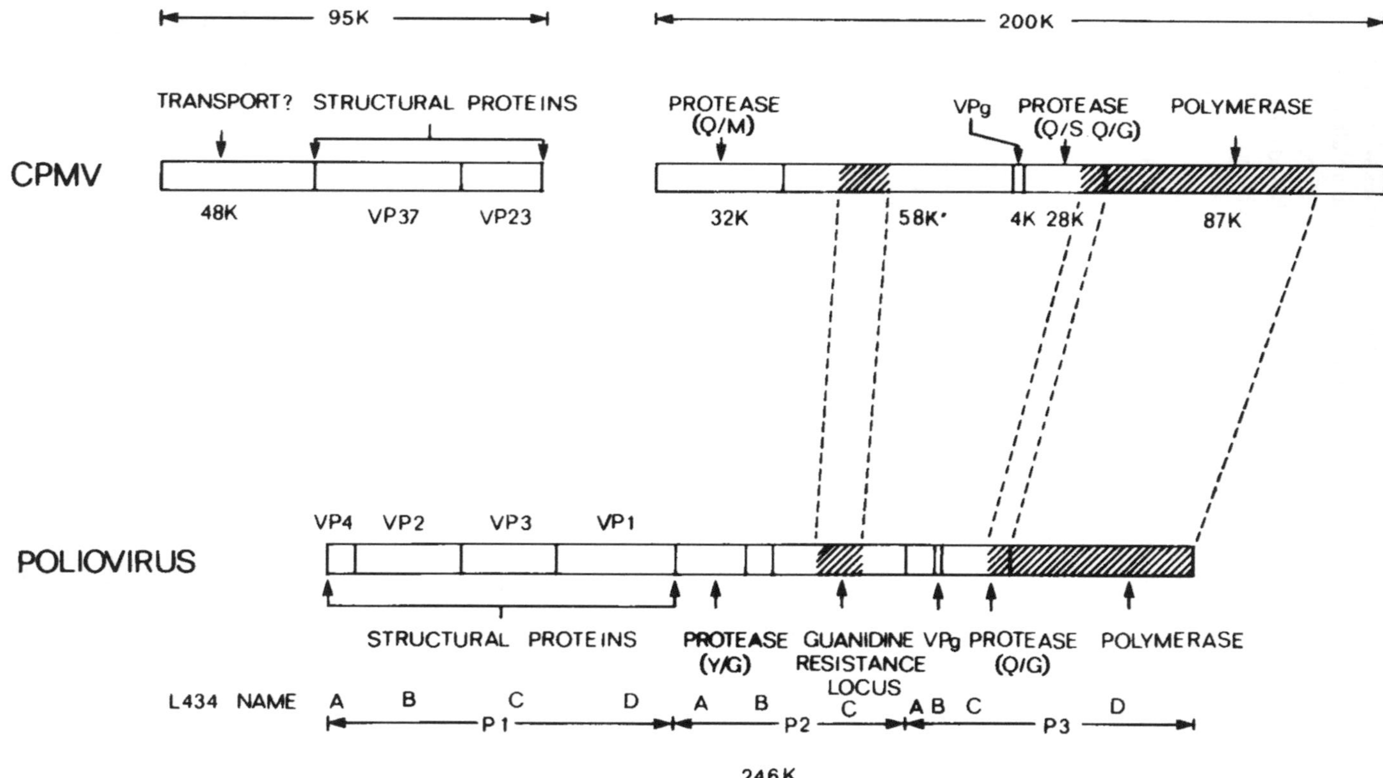

FIGURE 1. Comparison of the genetic maps of CPMV and picornaviruses. Picornaviral proteins are synthesised as an approximately 250-kdalton precursor which is initially cleaved into three domains denoted by convention P1 (the structural protein precursor), P2, and P3 (precursors to two proteases, polymerase, and other proteins involved in replication). The similarity to the CPMV gene arrangement is evident if the structural protein genes on the smaller CPMV RNA (M RNA) are drawn on the left of the replicative protein genes carried on the larger CPMV RNA (B RNA) (top of figure). Crosshatched regions of protein correspond to stretches of significant amino acid sequence homology (see Appendixes 1 to 3). Additional examples of homology in more divergent viruses are also known (see References 40 to 42 and Appendixes 2 and 3). (Redrawn and updated from References 31 and 58.)

Following the identification of a related set of genes common to both picornaviruses and CPMV, more distantly related homologues of the protease and polymerase genes have been identified in another group of plant viruses, the potyviruses[40,41] (see Appendixes 2 and 3), and there are preliminary reports of a third group of plant viruses with a related polymerase gene.[42] All these groups of viruses have RNAs with a VPg at the 5′ end and poly(A) at the 3′ end, supporting the idea of a functional correlation between RNA terminal structures and mode of replication in this group of viruses. However, both of these more recently characterized groups of plant viruses must diverge even more from the canonical picornavirus gene arrangement than CPMV since their genetic maps, which are currently at various stages of completeness, suggest each group carries additional ancillary genes.

The appearance of related genes in different contexts (i.e., in association with different ancillary genes) is a recurring theme which is even more clearly seen in the second major virus grouping which has emerged. This second group contains a more diverse array of different viruses, from Sindbis virus (an icosahedral enveloped animal virus with a 12-kilobase genome) to TMV (a rod-shaped plant virus with a 6.4-kilobase genome) (Table 1). In these two cases, there is little doubt that the capsid protein genes are unrelated, despite the fact that no structures have yet been determined for icosahedral viruses in the group. Aside from the gross differences in quaternary structure between helical and icosahedral particles, the crystal structure of the TMV coat protein subunit is predominantly α-helical[43] and completely different to the β-barrel structure common to all icosahedrally symmetrical viruses so far examined.[39,44,45] Also, unlike the first group, only the 5′ end group of the RNA (a cap) is conserved; the 3′ end structures are quite variable (Table 1). Nevertheless, sequence comparisons show evidence for relatedness of three Sindbis virus nonstructural proteins to corresponding nonstructural genes found in TMV, as well as in four other plant viruses (AlMV, BMV, CMV, and TRV).[32,46-50] This group therefore provides a particularly clear example of the coupling of related nonstructural genes to unrelated structural genes. AlMV, BMV, and CMV provide a further contrast since the genes they have in common with the other viruses are carried on two separate RNAs, whereas in Sindbis, TMV, and TRV they are fused on one RNA and separated only by a translational termination codon (Figure 2). The gene downstream of the terminator is expressed in both cases by translational suppression (at an opal codon in Sindbis and an amber codon in TMV).

Detailed information regarding the functions of the nonstructural genes in this group is much less extensive than in picornaviruses (see Chapters 4 and 5 in Volume I). However, in contrast to picornaviruses, all the viruses in question make separate mRNAs of less-than-full genomic length (subgenomic RNAs) for their capsid proteins, and there is evidence in alphaviruses such as Sindbis virus that subgenomic RNA synthesis requires the products of genes in two early complementation groups.[51] This may represent a common mode of replication reflected in protein homologies.

The differences in detailed strategy of replication between the two major virus groups presumably necessitate a somewhat different array of genes. However, since RNA viruses are all expected to encode RNA-dependent RNA polymerases, there are clearly grounds for believing that at least one of the related proteins in each case is such an enzyme, or its virally encoded subunit. In polio and CPMV, this viewpoint is supported by substantial evidence.[52,53] In the second group there is only limited and conflicting evidence to identify the relevant protein, but a variety of experiments in Sindbis virus suggest that it might be the protein downstream of the translational terminator.[51] Marginal homology can be detected between this protein and the picornavirus replicase, with stronger local conservation around a Gly–Asp–Asp sequence, which may even be common to bacteriophage replicases.[54] In the quest for unifying themes in RNA virus replication, this observation is obviously a striking one. However, the Gly–Asp–Asp motif does not seem to be universal among RNA-dependent RNA polymerases as was at first thought possible, because it is absent from the

Table 1

Virus group	Example	Host	Genomic RNAs (kilobases)	5′ end	3′ end	Capsid morphology
Alphaviruses	Sindbis virus	Mammals/ insects	1:11.7	Cap	Poly(A)	Icosahedral, enveloped
Tobamoviruses	TMV	Plants	1:6.4	Cap	Pseudo tRNA	Helical
Tobraviruses	TRV	Plants	2:total 8.1—12.3	Cap	C_{OH}	Helical
Bromoviruses/ cucumoviruses	BMV CMV	Plants	3:total 8.2	Cap	Pseudo tRNA	Icosahedral
Alfalfa mosaic/ ilarviruses	AlMV TSV	Plants	3:total 8.3	Cap	C_{OH}	Bacilliform or icosahedral

rhabdo- and paramyxovirus L proteins, which, as was noted earlier, are unexpectedly divergent.[17,33,34] The next episode of this saga is awaited with great interest.

Assuming that all the various structural relationships which have been inferred are confirmed in functional assays, it emerges that the distinct virus groups are related by the association of shared common components with unique components that define the features of replication and structure characteristic of each group. Thus, individual virus families can be represented as different permutations of genes drawn from functionally distinct sets (Figure 3). In essence, one can regard each virus as being built from modules of genetic information and the evolutionary process as one of mixing and matching these modules. This would not be possible without recombination nor recognizable without conservation within the modules.

2. Selection Pressure: Conservation of Amino Acid Sequence Despite High Mutation Rates

Implicit in the foregoing discussion was the observation that similar genes are used by both plant and animal viruses. This is true not only in the above two cases, but also for reverse transcriptase, which has been found in the plant virus cauliflower mosiac virus as well as in retroviruses, hepadnaviruses, and *Drosophila* and yeast transposable elements.[55-57] Plants and animals diverged about 10^9 years ago, so this conservation is remarkable when contrasted with the high rate of point mutation expected from in vitro studies. It raises the issue of whether the genes in their viral context could conceivably have predated the eukaryotic radiation. The available evidence (Section II.A) suggests that the major determinant of protein sequence conservation on an experimental time scale is selection pressure, so the question resolves itself into that of whether selection pressure would be adequate to conserve sequence similarity for so long. If not, some mechanism of interphyletic dispersal must be at work. Although at first sight this also seems unlikely, the phylogenetic distance between plants and animals need not be a barrier in this respect since insect vectors are common for both plant and animal viruses, and examples are known of plant viruses which are capable of multiplying in their insect vectors (see Reference 58 for further discussion). It is also possible that both these factors are operating, and that the genes themselves are very old, but their reassortment and dispersal on viral genomes are more recent.

There is certainly evidence that genes involved in replication are well conserved compared with other viral genes. In the most extreme examples, replicative functions (reverse transcriptase, for example) may survive in identifiable form in viruses so divergent that their genetic maps are not directly comparable, whereas the structural proteins can diverge in sequence to the extent that sequence homologies may not be detectable even between isogenic viruses.[59,60] In such extreme cases, recombination or genome segment reassortment may well have intervened. A clear pattern of selection can be seen; however in groups of relatively closely related viruses with similar gene arrangements (for example, the picornaviruses or the alphaviruses, where to a first approximation recombination can be neglected), this results in the commonplace observation of higher percentage homology between replicative functions

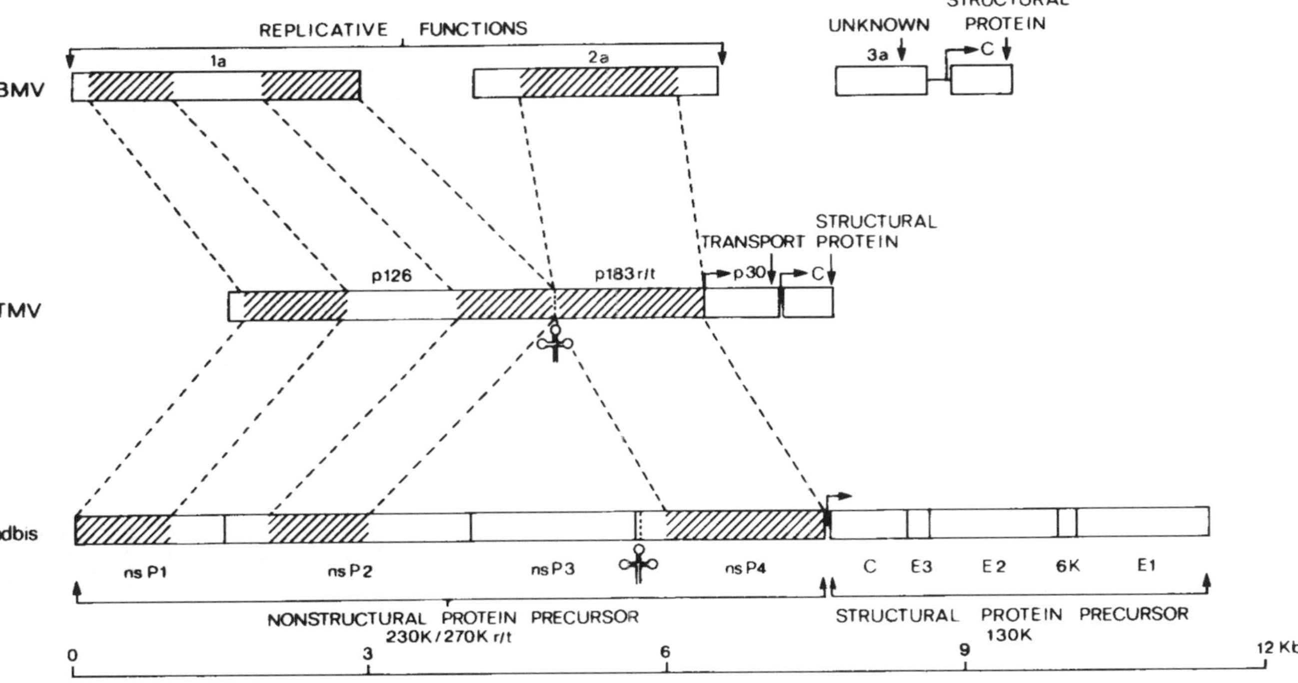

FIGURE 2. Comparison of the genetic maps of BMV, TMV, and alphaviruses. The BMV genome is divided between three RNAs (top). The two largest are monocistronic mRNAs; the smallest is dicistronic with the distal coat protein gene expressed via a subgenomic RNA (denoted with an angle arrow, bent right, here and in all other cases). Gene arrangements of AlMV and CMV (Table 1) are similar to BMV. Three domains (crosshatched) encoded by the single TMV RNA have related equivalents in the two largest BMV proteins (see Appendixes 4 to 6). Domains encoded in the two separate BMV proteins 1a and 2a are instead separated only by a suppressible termination codon (denoted with a tRNA symbol) in TMV. Sindbis virus also has these three domains encoded on a single RNA, but the resulting protein is a precursor which is proteolytically cleaved into four products, one of which (nsP3) is not detectably related to any of the plant viral proteins. Not all alphaviruses have the suppressible opal codon between nsP3 and nsP4. (Redrawn and updated from References 32 and 46.)

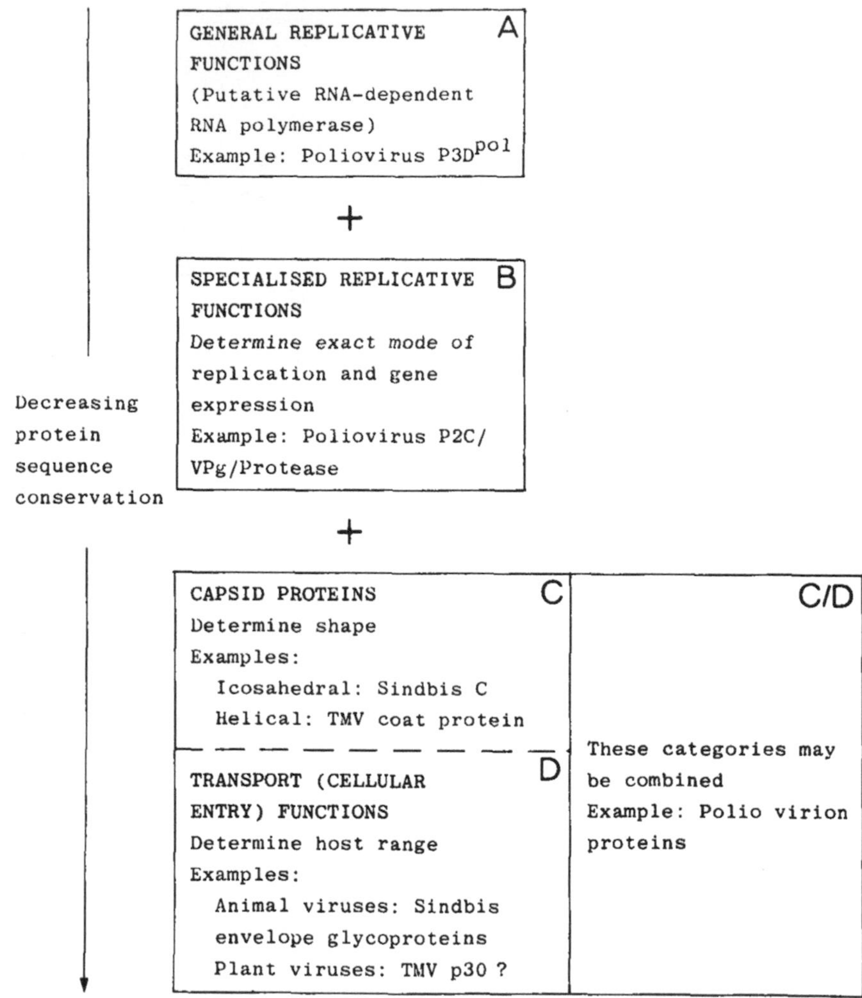

FIGURE 3. Modular construction of RNA viral genomes. Viral genomes can be thought of as individual permutations of genes belonging to a restricted number of sets of functions. The basic three or four sets shown here turn up in most if not all RNA viruses that have been examined — the repertoire could be expanded by adding members of optional extra sets, e.g., for vector transmission.

than between structural genes.[38,51] Assuming in such cases that the underlying point mutation rate is the same for all the viral genes, it follows that selection favors conservation of the genes involved in replication. Conversely, it also follows that variation in structural genes is tolerated and may even be encouraged since it may assist in "trying out" surface variants capable of binding to new receptors or (in the case of vertebrates) escaping immune surveillance.

Clearly, these observations provide in principle the basis for understanding the rate of drift in either class of sequence as a function of point mutation and selection. Unfortunately, they cannot be used to date the genes since, unlike the higher organisms which they infect, no viral fossil record is available with which to calibrate the molecular clock. Thus, protein sequence comparisons cannot settle the question of antiquity of genes now found on viral genomes, relative to the phylogenetic distances between their respective hosts.

Although the molecular palaeontological approach cannot be applied to viral proteins, the sequence alignments contain another kind of information which is almost equally informative. In relatively well-studied families of proteins whose crystal structures are known, it can be seen that the most highly conserved residues include those required for proper folding of the protein and for interaction with various ligands and with other proteins. (The degree of latitude allowed in less well-conserved positions is incompletely understood.) The relative rates of drift in different classes of viral genes would therefore follow the pattern outlined above if the interactions of replicative proteins with ligands such as template RNA and nucleoside triphosphates, as well as with each other, and perhaps host proteins as well, were invariant or very slowly changing in evolution, whereas in structural genes more rapid variation was dictated by changing or evolving substrates. Information regarding substrate interactions can sometimes be extracted from the sequence alignments in the following way.

In a sufficiently large sample of related protein sequences, clusters of invariant residues will become apparent, some of which will be involved in ligand binding (a positive identification of the relevant residues can be made from crystallographic analysis). Several examples are now known, including binding sites for some particularly common ligands, which turn out to be formed by conserved amino acid motifs that are found repeatedly in various families of otherwise unrelated proteins. Such motifs may constitute semi-independent domains of protein folding[61] which can be recombined with a variety of protein secondary structural elements in a manner analogous to intact genes. A gratifying connection between this classical protein structural work and the RNA viral protein families established from sequence comparisons has now emerged with the recent recognition within the putative replicase genes of both the TMV-alphavirus family and the picornaviruses of a sequence motif resembling that previously identified in a wide range of purine nucleoside triphosphate binding proteins[62-64] (Figure 4). In the former case, this motif is the most highly conserved part of the viral sequences in which it is embedded,[32,46] which is consistent with its known importance in other proteins where it has been found and subjected to site-directed mutagenesis (such as p21 *ras*).[63,65] Although the functional identity of the viral sequence has not yet been experimentally confirmed, it seems reasonable, based on our prior notions of the structural and functional basis for replicase conservation outlined above, since nucleoside triphosphates are clearly prime candidates for replicase substrates and have not changed in evolution. Assuming experimental confirmation is forthcoming, the presence of this motif in viral proteins leads to the important conclusion that the basic building blocks of RNA viral genes at the level of conserved protein tertiary structure are the same as those used in cellular genes, which provides a glimpse of a possible answer to the large question of ultimate origins with which we began.

If corroborated, RNA viral protein motifs resembling cellular ones would add weight to the persuasive, if so far circumstantial, case for the ultimate derivation of viral genes from cellular genes (see Section IV). However, a cautious interpretation seems advisable at present in the care shown in Figure 4, because the motif is so common that it could arguably represent a uniquely favorable fold that may have evolved repeatedly (convergent evolution). More generally, apart from the question of whether both viral and cellular examples did evolve from a common ancestor, such highly conserved sequences further complicate efforts to put an absolute time scale on the processes involved, because such motifs may be so highly constrained that they hardly change at all in evolution. If this were true, selection pressure might have been enough to conserve the sequence essentially indefinitely, even in the context of a viral high mutation rate.

3. *Recombination of Functional Modules*

The evidence for modular construction of RNA viruses (or "mix and match" — see Figure 3) elaborated above, raises two central mechanistic problems: whether the underlying recombination requires viral or cellular functions, and, if viral functions are required, how did the whole cycle get started? It should be noted that the idea of modular construction serves only to emphasise the opportunism of evolution in adapting preexisting modules for

ATP binding:

Protein	Residues	β sheet		Loop		α helix
Adenylate kinase	9–29	K I I F V V	G	G P G S	G K G	T Q C E K I V
Rec. A	60–80	K I V E I Y	G	P E S S	G K T	T L T L Q V I
Bovine ATPase β	152–171	K I G L F –	G	G A G V	G K T	V F I M E L I
E. coli ATPase β	145–164	K G V L F –	G	G A G V	G K T	V N M M E L I
E. coli ATPase α	164–183	R E L I I –	G	D R G T	G K T	A L A I D A I
Myosin, C. elegans	164–183	S M L I T –	G	E S G A	G K T	E N T K K V I
Myosin, rabbit	173–192	S I L I T –	G	E S G A	G K T	V N T K R V I

GTP binding:

Protein	Residues	β sheet		Loop (* *)		α helix
EF-Tu	13–31	N V G T I –	G	H V D H	G K T	T L T A A I
EF-G	11–29	N I G I S –	A	H I D A	G K T	T T T E R I
p21 Ha-*ras*	5–23	K L V V V –	G	A G G V	G K S	A L T I Q L
Yeast $YP2_{sc}$	10–28	K L L L I –	G	N S G V	G K S	C L L L R E
$c\text{-}ras^{sc}\text{-}1$	12–30	K I V V V –	G	G G G V	G K S	A L T I Q F
Transducin α s/u	31–49	K L L L L –	G	A G E S	G K S	T I V K Q M
G_i α s/u	35–53	K L L L L –	G	A G E S	G K S	T I V K Q M

Viral:

Protein:	Residues					
Polio P2c	123–143	V C L L V H	G	S P G T	G K S	V A T N L I A
Rhinovirus 14 P2c	1223–1243	V C V L I H	G	T P G S	G K S	L T T S I V C
FMDV A10 P2c	1211–1231	V V V C L R	G	K S G Q	G K S	F L A N V L A
EMC P2c	1305–1325	V V I V L R	G	D A G Q	G K S	L S S Q V I A
Hepatitis A P2c	1224–1244	V V C Y L Y	G	K R G G	G K S	L T S I A L A
CPMV 58K	488–508	F T I F F Q	G	K S R T	G K S	L L M S Q V T
AlMV 1a	819–839	S V T I V D	G	V A G C	G K T	T N I K Q I A
BMV 1a	679–699	D I S M V D	G	V A G C	G K T	T A I K D A F
CMV 1a	708–728	T I S Q V D	G	V A G C	G K T	T A I K S M F
TMV p126	827–847	K V V L V D	G	V P G C	G K T	K E I L S R V
SFV nsP2	180–200	T V V G V F	G	V P G S	G K S	A I I K S L V
Sindbis nsP2	180–200	E T I C V I	G	T P G S	G K S	A I I K S T V

FIGURE 4. Conserved amino-acid sequence motif around the phosphate binding pocket in selected purine nucleoside triphosphate binding proteins, with some viral proteins for comparison. Alignments are based on those in References 62 (ATP binding proteins) and 63 (GTP binding proteins) with additional G protein sequences from Reference 90. Viral protein alignment is based on Reference 64. Apparently homologous Gly and Gly–Lys–Thr/Ser sequences are boxed. Correlation with protein secondary structure is based on adenylate kinase,[91] where Lys 21 is thought to neutralize the charge on the ATP α-phosphate, and on EF-Tu, where Lys 24 may interact with the GTP β-phosphate.[92] For additional information see Reference 93. The positions indicated by asterisks above the sequence corresponding to Gly 12 and Gly 13 in p21 *ras* are those at which point mutations have been found to alter activity. For additional information see Reference 65.

use in a variety of contexts and does not help to identify the original source of such modules or the mechanism of recombination. It simply suggests that it is much easier to create new combinations of functional units once they have arisen then it is to create those functional units in the first place.

A retrospective view of the mechanism of viral recombination on an evolutionary time scale is obviously ruled out, but various lines of experimental work, including those discussed in Section II.B, suggest several possibilities. First, there is the conservative possibility of recombination at the DNA level, if viral RNAs can occasionally be copied into DNA in the manner known to occur with cellular mRNAs during pseudogene formation.[66,67] This would allow the possibility of recombination with a wide range of cellular genes and their subsequent recruitment to perform viral functions, in line with the existence of structural features common to both viral and cellular proteins (see above) and as predicted by Temin's protovirus hypothesis (see next section). However, natural pseudogenes of RNA viruses have never been described, and in the light of the recent direct evidence for RNA recombination already discussed, compulsory reverse transcription would appear unnecessary.

A second possibility, perhaps the most tenable at the moment, is suggested by DI RNAs and could operate independently of DNA. It would involve the recruitment of molecules rearranged by normal low-level errors of replication as a source of variation leading to new consensus genomic arrangements. The example of Sindbis DI RNAs with covalently attached cellular tRNAs shows that the recombinations involved in DI RNA formation need not be confined to preexisting viral RNAs. Direct RNA recombination alone may be capable of drawing on a range of cellular genes, besides viral sequences, as a source of new viral variation. For purely viral gene reassortment, the attractions of the model are obvious. However, assuming the copy choice model of recombination applies in most cases of DI RNA formation, this model leaves us with the chicken and egg problem that a suitable polymerase must be available before the cycle of modular reassortment can begin.

A more radical alternative for an RNA-level mechanism of recombination which could avoid the need for preexisting viral functions is suggested by the involvement of splicing in the synthesis of at least some plant viral satellite RNAs.[68-73] Cleavage and ligation, i.e., splicing (either autocatalytic or using cellular enzymes), might offer an alternative to copy-choice recombination by RNA polymerase slippage in certain cases. Since splicing is normally extremely site specific, such a mechanism would in principle be somewhat less general than copy choice. Nevertheless, such a mechanism could be capable of exchanging information between viruses or between a virus and a cellular RNA via *trans*-splicing events, the existence of which is by now fairly well established.[74-77] (For readers not averse to speculation, these ideas have been discussed at greater length elsewhere.[78])

Both the *trans*-splicing and copy choice models require systems that are already quite sophisticated and highly evolved. The contrast between the probable antiquity of some of the protein secondary structural elements found as viral components and the sophistication of even the simplest mechanisms we can find to put them together, only serves to emphasize the fact that we are still several steps away from being able to relate all these observations into a coherent overall model of viral evolution. Although a final synthesis may still be elusive, the next and final section will attempt to give a progress report.

IV. SUPPORT FOR THE PROTOVIRUS HYPOTHESIS?

From all the various lines of evidence considered in the preceding sections, several conclusions emerge quite clearly which are summarized below:

1. Modular construction of RNA viruses is evident from amino acid sequence comparisons. Viral genes can almost certainly be reassorted by recombination.
2. Viral replicative genes are relatively well conserved, while genes involved in transmission and for capsid formation are more variable. These different rates of accepted point mutation must result from the differing effects of selection on different genes working against a constantly high rate of mutation.
3. Conserved replicative functions are found in viruses infecting diverse kinds of host, and are related to the mode of viral replication, not the host species. There seems to be relatively little constraint on their free reassortment with a variety of structural or host range genes.
4. Although there is as yet no evidence that any whole genes from RNA viruses (excluding retroviruses) have cellular counterparts, there is evidence, although limited so far, that

parts of their tertiary structure may be folded into motifs that they share with a variety of cellular proteins.

If the last conclusion receives more support, as seems entirely plausible, it suggests that the basic building blocks of RNA viruses are the same as those of cells, although at a very remote level. It is consistent with the accepted notion that protein structures are selected at the tertiary structural level, which in turn is responsible for observable primary sequence homology if conservation is sufficient.

It should be emphasized in this regard that most of the evidence so far involves proteins implicated in viral replication, although it has also been argued that icosahedral virus capsid proteins resemble carbohydrate binding proteins such as concanavalin A in their tertiary structures, even though their primary sequences are too divergent to align.[79] Such structural homologies may eventually suffice to discard any ideas of a separate prebiotic origin for RNA viruses, along with the fact that it is difficult to imagine any functional virus remotely resembling current ones without a cell for the virus to infect, or, at the very least, a translation apparatus. Given a cell to start with, one can see in outline how a virus could be put together using only routes consistent with the four rules outlined above.

These conclusions are essentially the same as those put forward over many years by Howard Temin to account for retrovirus origins, which have come to be known in aggregate as the protovirus hypothesis.[80] Although retrovirology is somewhat outside the scope of this chapter, it is relevant in this context to note that many of the main conclusions outlined above apply equally to retroviruses. Reverse transcriptase has been found in two kinds of viruses that are substantially different from the retroviruses where it was first described.[55,56] The gene occurs in different contexts both with or without associated protease and DNA endonuclease genes, and with or without LTRs, while within the retroviruses, env genes may be reassorted independently of pol.[81] The main distinction dictated by the DNA proviral form of retroviruses is that DNA recombination is obviously a much more likely route of reassortment than it is with "conventional" RNA viruses. Nevertheless, the evidence also favors recombination in the evolution of RNA viruses without a DNA provirus, and we have seen that mechanisms for RNA recombination exist that dispense with the need for a DNA copy. (In fact, copy choice recombination by reverse transcriptase on RNA templates may also be involved in some steps of retroviral recombination).[82,83] Thus, it seems likely that the same general principles have operated during the recent evolution of both kinds of virus, and there seems to be no reason why the protovirus hypothesis should not also be applicable to conventional RNA viruses.

The major conceptual barrier still remaining is the apparent uniqueness of RNA replicases to viruses. In contrast to the idea embodied in the protovirus hypothesis that the enzyme at the core of the protovirus was originally a cellular enzyme, which contributed to the original attractiveness of the protovirus hypothesis in explaining how closely retrovirus biology was integrated into that of the host cell, RNA virus biology is much more conveniently conceptualized by assuming that RNA replicases are unique to viruses.[84] However, this is a problem in every theory — if the enzyme did not already exist, evolution would have to invent it.

The protovirus theory postulates that the proto reverse transcriptase is (or was) an enzyme involved in development, specifically as a component of transposible elements. The latter conjecture is supported by recent evidence associating pol genes with a number of cellular elements that are either known to have or are suspected of an ability to transpose[57,85-87] (see also chapter by Boeke, this volume). RNA replicases, on the other hand, are not well characterized in any context other than viruses. Although there is quite respectable evidence for the existence of an enzyme having RNA-dependent RNA polymerase activity in plants,[88] the natural function of this enzyme remains a mystery, and hence it is not entirely clear whether the activity measured in the standard in vitro assays is genuine or artifactual, or how it relates to the role of the enzyme in vivo. Nevertheless, it remains possible that such an enzyme could be the original RNA replicase or have shared a common ancestor with the viral enzyme.

However, even if cellular RNA replicases as such do not exist, this is not a fatal blow to the protovirus hypothesis. It is conceivable that such an enzyme could be the product of a recombination event which altered the substrate specificity of another polymerase altogether by attaching an unrelated substrate binding domain. A short sequence motif has recently been recognized in one of the tobamovirus/alphavirus family of replicative genes which is also found in DNA polymerases from herpes, vaccinia, and adenoviruses.[89] Because such short motifs usually delineate conserved folds involved in substrate binding and may turn up in otherwise unrelated proteins, this observation does not necessarily mean that the proteins as a whole are directly related, but it does add substance to the idea that viral RNA polymerases may ultimately derive from (a) preexisting polymerase(s) with a different specificity.

Whatever their source, the limited number of core replicative genes found in RNA viruses suggests that the postulated reservoir of potential precursor genes is rather limited. The conservation of recognizably similar genes as modular units of recombination and the already low frequency of successful reassortments of these modules suggests that if any of the modules themselves arose from combinations of other genes by nonhomologous recombination, the odds against the product being functional must have been astronomical. However, evolution deals in precisely such large numbers, and some such rare event must have occurred to give rise to the ancestral RNA replicase gene at whatever stage it occurred. Either a recent or more remote rearrangement model can be accommodated within the protovirus hypothesis, the only distinction being the stage at which an RNA replicative enzyme arose in the form in which it is found in current viruses. Either way, the original derivation of viral enzymes from preexisting gene(s) seems more consistent with the apparent pattern of recent virus evolution than any other of the various alternative suggestions that have been made from time to time. Thus, studies of viral evolution seem to reconverge upon the cell, and it seems probable to this author that future insights into the ultimate origins of RNA viruses are more likely to come from this quarter than from any study of the viruses themselves.

V. ADDENDUM

Since this chapter was written several important new facts have been established, principally:

1. Kirkegaard and Baltimore[113] have shown in an elegant experiment that poliovirus recombination proceeds by a copy choice mechanism.
2. Additional paramyxovirus L gene sequences have now been completed,[114] allowing a clear demonstration of their relatedness to the VSV L gene amino acid sequence and the absence of the Gly–Asp–Asp motif.
3. Retroviral proteases are known to have amino acid sequence homology to cellular acid proteases in a restricted region around the active site,[57,115] similarly picornaviral proteases resemble cellular cysteine proteases.[38] 3-D image reconstruction and model building studies of the alphavirus capsid protein,[116] which doubles as a serine autoprotease, have now shown that a chymotrypsin-like active site can be formed by residues which are part of an otherwise typical β-barrel-based icosahedral capsid subunit. This provides a dramatic illustration of a phenomenon well known to crystallographers in other contexts, that of unrelated proteins with similar active sites. This example once again underlines the difficulty of distinguishing between convergent and divergent evolution in sequence comparisions based only on short motifs (Section III. B. 2) and the dangers of arguing the overall relatedness of proteins in such cases.

4. More extensive homology searches[117] suggest that the motif mentioned in Reference 89 derives from a large family of ATP and nucleic acid binding proteins and is not unique to polymerases (see also Appendix 5).

ACKNOWLEDGMENTS

Obviously this review could not have been written without drawing on the accumulated wisdom of many people, especially the virologists and structural molecular biologists with whom I have been fortunate enough to work at various times. I should particularly like to thank Jim Haseloff, who found the AlMV/BMV/TMV homologies shown in Figure 2, together with the many friends and colleagues whose hard work provided the data on which it was based, and John Rogers for helpful comments on the manuscript.

APPENDIX 1*
ALIGNMENT OF CONSERVED CORE OF PICORNAVIRAL P2C PROTEINS

```
                  *  *      *    ***                                 **  *
                  * ***     * * ***                      *  **    ****  *
Polio    # 117  KHRIEPVCLLVHGSPGTGKSVATNLIARAIAE----RENTSTYSLPPDPSHFDGYKQQGV
Rhino 14 #1217  KGRIEPVCVLIHGTPGSGKSLTTSIVGRAIAEHF----NSAVYSLPPDPKHFDGYQQQEV
Rhino  2 #1199  KKRCEPVAIVIHGPPGAGKSITTNFLAKMITN------DSDIYSLPPDPKHFDGYDQQSV
FMDV A10 #1205  KSRPEPVVVCLRGKSGQGKSFLANVLAQAISTHFTGRID-SVWYCPPDPDHFDGYNQQTV
EMC      #1299  TARCEPVVIVLRGDAGQGKSLSSQVIAQAVSKTIFGRQ--SVYSLPPDSDFFDGYENQFA
Hep A    #1218  VTRCEPVVCYLYGKRGGGKSLTSIALATKICKHYGVEPEKNIYTKPVASDYWDGYSGQLV
CPMV     # 482  GVRKMPFTIFFQGKSRTGKSLLMSQVTKDFQDHYGLGGE-TVYSRNPCDQYWSGYRRQPF
                            ↑    ↑↑↑

                   ** G                **       ** *  **  *         *      *
                  *** R**     *   * ****     **** *  **  *    *   **      *
Polio    # 173  VIMDDLNQNP-DGADMKLFCQMVSTVEFIPPMASLEEKGILFTSNYVLASTNS-SRISPP
Rhino 14 #1273  VIMDDLNQNP-DGQDISMFCQMVSSVDFLPPMASLDNKGMLFTSNFVLASTNS-NTLSPP
Rhino  2 #1253  VIMDDIMQNP-AGDDMTLFCQMVSSVTFIPPMADLPDKGKAFDSRFVLCSTNH-SLLTPP
FMDV A10 #1264  VVMDDLGQNP-DGKDFKYFAQMVSTTGFIPPMASLEDKGKPFNSKVIIATTNLYSGFTPR
EMC      #1357  AIMDDLGQNP-DGSDFTTFCQMVSTTNFLPNMASLERKGTPFTSQLVVATTNL-PEFRPV
Hep A    #1278  CIIDDIGQNTTD-EDWSDFCQLVSGCPMRLNMASLEEKGRHFSSPFIIATSNW-SNPSPK
CPMV     # 541  VLMDDFAAVVTEPSAEAQMINLISSAPYPLNMAGLEEKGICFDSQFVFVSTNF-LEVSPE


                       *   *
                *      *  **      *
Polio    # 231  TVAH-SDALARRFAFDMDIQV
Rhino 14 #1331  TILN-PEALVRRFGFDLDICL
Rhino  2 #1311  TITS-LPAMNRRFFLDLDIIV
FMDV A10 #1323  TMVC-PDALNRRFHFDIDVSA
EMC      #1415  TIAH-YPAVERRITFDYSVSA
Hep A    #1336  TVYV-KEAIDRRLHFKVEVKP
CPMV     # 600  AKVRDDEAFKNRRHVIVQVSN
```

Protein sequences are translated from nucleotide sequences of the viruses indicated on the left of the alignment as determined by the following authors: poliovirus (type 1, strain Mahoney;[94,95] rhinovirus 14;[96,97] Rhinovirus 2;[98] foot-and-mouth disease virus, strain A [10]61;[99] encephalomyocarditis virus;[100] hepatitis A virus;[101] and cowpea mosaic virus B RNA.[102] The standard one-letter amino acid code is used. For genetic map locations of the

* The appendixes were authored by T. C. Hodgman and D. Zimmern.

conserved domain, see text, Figure 1 (all except CPMV resemble polio). Residue numbers refer to the first amino acid on each line of the alignment, with respect to the P2C N terminus (polio), or to the inferred polyprotein initiator methionine (all the rest). Conserved Gly and Gly–Lys–Ser residues in the putative nucleoside phosphate binding site (see text, Figure 4) are marked ↑. (Note: Protein P2C was formerly known as VP-X or P2-X).

Key to symbols: double asterisk, strictly conserved amino acid; single asterisk, highly constrained amino acid (only two alternatives); GR, poliovirus mutants resistant to 2 m*M* guanidine have a point mutation at this position[103] (see also Reference 28).

APPENDIX 2
ALIGNMENT OF RESIDUES AROUND ACTIVE SITE OF PICORNAVIRAL P3C PROTEASE

```
                                          **               * *  *
                          *      *   *  * *** *    *     * * ** *  *
Polio     #  123 GYLNLGGRQTAR-TLMYNFPTRAGQCGG-VITCTGK----VIGMHV-GGNGSHGFAAALKRS
Rhino 14  # 1659 GLINLSSTPTNR-MIRYDYATKTGQCGG-VLCATGK----IFGIHV-GGNGRQGFSAQLKKQ
Rhino  2  # 1630 GNILLSGNQTAR-MLKYSYPTKSGYCGG-VLYKIGQ----VLGIHV-GGNGRDGFSAMLLRS
FMDV A10  # 1788 DIVVCMDGDTMPGLFAYKAATRAGYCGGAVLAKDGADTF-IVGTHSAGGNGVGYCSCVSRSM
EMC       # 1759 VSVPVETGQTFNHCIHYKANTRKGWCGSALLADLGGSKK-ILGIHSAGSMGIAAASIVSQEM
Hep A     # 1666 KNDGTTVDLTVDQAWRGKGEGLPGMCGGALVSSNQSIQNAILGIHVAGGNSILVAKLVTQEM
CPMV      # 1090 QSGNYVNKVSR--YLEYEAPTIPEDCGSLVIAHIGGKHK-IVGVHVAGIQGKIGCASLLPPL
                                          ↑                  ↑
TEV       # 2175 SDTSCTFPSSDGIFWKHWIQTKDGQCGSPLVSTRDGF---IVGIHSASNFTNTNNYFTSV
TVMV      # 2109 SESSHIVHKEDTSFWQHWITTKDGQCGSPLVSIIDGN---ILGIHSLTHTTNGSNYFVEF
```

Protein sequences are translated from nucleotide sequences of the viral RNAs indicated on the left of the alignment. For original references and key to symbols, see legend to Appendix 1, and for genetic map locations, see text, Figure 1. Residues are numbered with respect to the protease N terminus (polio) or the polyprotein initiator methionine (all the rest). Homologous sequences from two plant potyviral nuclear inclusion A proteins are shown below the others (from References 40 and 41).

Conserved Cys and His residues believed to form the active site by analogy with other cysteine proteases are marked ↑ (see Reference 38).

APPENDIX 3
ALIGNMENT OF PICORNAVIRAL P3D POLYMERASE PROTEINS

```
                       *      *            *                        **
                       ** *   *            ** *   ***               ***
Polio     #   16 IINAPSKTKLE--PSAFHYVFEGVKEPAVLTKNDPRLKTN-----FEEAIFSKYVGNKIT
Cox       #   16 VINTPSKTKLE--PSVFHQVFVGNKEPAVLRSGDPRLKAN-----FEEAIFSKYIGNVNT
Rhino 14  # 1735 PVNTPTKSKLH--PSVFYDVFPGDKEPAVLSDNDPRLEVK-----LTESLFSKYKGNVNT
Rhino  2  # 1706 SIHTPCKTKLQ--PSVFYDVFPGSKEPAVLSEKDARLQVD-----FNEALFSKYKGNTDC
FMDV A10  #   12 RVHVMRKTKLA--PTVAYGVFNPEFGPAALSNKDPRLNEG---VVLDDVIFSKHKGDAKM
EMC       # 1841 RIHVPRKTALR--PTVARQVFQPAYAPAVLSKFDPRTEAD-----VDEVAFSKHTSNQES
Hep A     # 1749 SMNVVSKTLFRKSPIHHHIDKTMINFPAAMPFSKAEID-------PMAMMLSKYSLPIVE
CPMV      # 1186 YIPLPTKTALVETPSEWHLDTPCDKVPSILVPTDPRIPAQHEGYDPAKSGVSKYSQPMSA
```

```
                                                        *  *             * *
                               *         *           *  *  * *        ** * *
Polio     #   69 EVDEH--MKEAVDHYAGQL---MSLDINTEQMCLEDAMYGTDGLEALD---LSTSAGYPY
Cox       #   69 HVDE--YMLEAVDHYAGQL---ATLDISTEPMKLEDARYGTEGLEALD---LTTSAGYPY
Rhino 14  # 1788 EPTEN--MLVAVDHYAGQL---LSLDIPTSELTLKEALYGVDGLEPID---ITTSAGFPY
Rhino  2  # 1759 SINDH--IRIASSHYAAQL---ITLDIDPKPITLEDSVFGTDGLEALD---LNTSAGFPY
FMDV A10  #   67 TEEDKALFRRCAADYASRLH--SVLGTANAPLSIYEAIKGVDGLDAME---PDTAPGLPW
```

```
EMC        # 1894 LP---PVFRMVAKEYANRVF--TLLGKDNGRLTVKQALEGLEGMDPMD---RNTSPGLPY
Hep A      # 1802 EPED---YKEASVFYQNKIVGKTQLVDDF--LDLDMAITGAPGIDAIN---MDSSPGFPY
CPMV       # 1246 LDPE--LLGEVANDVLELWHDCAVDWDDFGEVSLEEALNGCEGVEYMERIPLATSEGFPH

                                                                          ***
                     *  *                                                 ***
Polio      #  121 VAM------GKKKRDILN--KQTRDT-----------KEMQKLLDTYGINLPLVTYVKDE
Cox        #  121 VAL------GIKKRDILS--KKTKDL-----------TKLKERMDKYGLNLPMVTHVKDE
Rhino 14   # 1840 VSL------GIKKRDILN--KETQDT-----------EKMKFYLDKYGIDLPLVTYIKDE
Rhino  2   # 1811 IAM------GVKKRDLIN--NKTKDI-----------SKLKEAIDKYGVDLPMVTFLKDE
FMDV A10   #  122 ALQ------GKRRGALIDFENGTVGP---------EVEAALKLMEKREYKFACQTFLKDE
EMC        # 1946 TAL------GMRRTDVVDWESATLIP--------FAAERLRKMNEGDFSEVVYQTFLKDE
Hep A      # 1854 VQE------KLTKRDLIWLDENGLLLGVHPRLAQRILFNTVMMENCSDLDVVFTTCPKDE
CPMV       # 1304 ILSRNGKEKGKRRFVQGDDCVVSLIPGTT---VAKAYEELEASAHRFVPALVGIECPKDE

TEV        # 2413          GKKKEALSE---LTLDE-------QEAMLKASCLRLYTGKLGIWNGSLKAE
TVMV       # 2358          GKKKQYFED---LSDDA-------VANLVQKSCLRLFKNKLGVWNGSLKAE

                   **    * *                                    *  *    *
               *   **    ***     *       *    *     *   **   * ******   *
Polio      #  162 LRSKTKV-EQGKSRLIEASSLNDSVAMRMAFGNLYAAFHKNPGVITGSAVGCDP-DLFWS
Cox        #  162 LRSIEKV-AKGKSRLIEASSLNDSVAMRQTFGNLYKTFHLNPGVVTGSAVGCDP-DLFWS
Rhino 14   # 1881 LRSVDKV-RLGKSRLIEASSLNDSVNMRMKLGNLYKAFHQNPGVLTGSAVGCDP-DVFWS
Rhino  2   # 1852 LRKHEKV-IKGKTRVIEASSVNDTLLFRTTFGNLFSKFHLNPGIVTGSAVGCDP-EVFWS
FMDV A10   #  167 IRPMEKV-RAGKTRIVDVLPVEHILYTKMMIGRFCAQMHSNNGPQIGSAVGCNP-DVDWQ
EMC        # 1992 LRPIEKV-QAAKTRIVDVPPFEHCILGRQLLGKFASKFQTQPGLELGSAIGCDP-DVHWT
Hep A      # 1908 LRPLEKV-LESKTRAIDACPLDYTILCRMYWGPAISYFHLNPGFHTGVAIGIDP-DRQWD
CPMV       # 1361 KLPMRKVFDKPKTRCFTILPMEYNLVVRRKFLNFV-RFIMANRHRLSCQVGINPYSMEWS

TEV        # 2454 LRPIEKV-ENNKTRTFTAAPIDTLLAGKVCVDDFNNQFYDLN-IKAPWTVGMTKFYQGWN
TVMV       # 2399 LRPFEKL-IENKTRTFTAAPIETLLGGKVCVDDFNNHFYSKH-IQCPWSVGMTKFYGGWN

                            *    *
                           *** **                   *                *
Polio      #  220 KIPVLMEEK---LFAFDYTGYDASLSPAWFE-----ALEMVLEKIGFGDRVD-YIDYLNH
Cox        #  220 KIPVMLDGH---LIAFDYSGYDASLSPVWFA-----CLKMLLEKLGYTHKETNYIDYLCN
Rhino 14   # 1939 VIPCLMDGH---LMAFDYSNFDASLSPVWFV-----CLEKVLTKLGFAGSS--LIQSICN
Rhino  2   # 1910 KIPAMLDDKC--IMAFDYTNYDGSIHPIWFE-----ALKQVLVDLSFNPT---LIDRLCK
FMDV A10   #  225 RFGTHFAQYRN-VWDVDYSAFDANHCSDAMNI---MFEEVFRTDFGFHPNAEWILKTLVN
EMC        # 2050 AFGVAMQGFER-VYDVDYSNFDSTHSVAMFRL---LAEEFFTPENGFDPLTREYLESLAI
Hep A      # 1966 ELFKTMIRFGDVGLDLDFSAFDASLSPFMIREAGRIMSELSGTPSHFGTA---LINTIIY
CPMV       # 1420 RLAARMKEKGNDVLCCDYSSFDGLLSKQVMDVIASMINELCGGEDQLKNARRNLLMACCS

TEV        # 2512 ELMEALPS-GWVYCDADGSQFDSSLTPFLINAVLKVRLAFMEEWDIGEQMLRNLYTEIVY
TVMV       # 2457 ELLGKLPD-GWVYCDADGSQFDSSLSPYLINAVLRLRLSSMEEWDVGQKMLQNLYTEIVY

                            ***   *   *   *
                            ***   *   *** ***                       *
Polio      #  271 SHHL-YKNKTYCVKGGMPSGCSGTSIFNSMINNLIIRTLLLKTYKGID--------LDHL
Cox        #  272 SHHL-YRNKHYFVRGGMPSGCSGTSIFNSMINNIIIRTLMLKVYKGID--------LDQF
Rhino 14   # 1989 THHI-FRDEIYVVEGGMPSGCSGTSIFNSMINNIIIRTLILDAYKGID--------LDKL
Rhino  2   # 1960 SKHI-FKNTYYEVEGGVPSGCSGTSIFNTMINNIIIRTLVLDAYKNID--------LDKL
FMDV A10   #  281 TEHA-YENKRITVEGGMPSGCSATSIINTILNNIYVLYALRRHYEGVE--------LDTY
EMC        # 2106 STHA-FEEKRFLITGGLPSGCAATSMLNTIMNNIIIRAGLYLTYKNFE--------FDDV
Hep A      # 2023 SKHL-LYNCCYHVCGSMPSGSPCTALLNSIINNINLYYVFSKIFGKSP-----VFFCQAL
CPMV       # 1480 RLAI-CKNTVWRVECGIPSGFPMTVIVNSIFNEILIRYHYKKLMREQQAPELMVQSFDKL
                                       ↑   ↑   ↑
TEV        # 2571 TPILTPDGTIIKKHKGNNSGQPSTVVDNTLM--VIIAMLYTCEKCG-------INKEEIV
TVMV       # 2516 TPISTPDGTIVKKFKGNNSGQPSTVVDNTLM--VVLAMYYALSKLGVD-----INSQEDV

                   ****            *              *  *
                   ****            *              *  *   **
Polio      #  322 -KMIAYGDDVIASYPHE-----VDASLLAQS--GKDYGLTMTPA-DKSAIFETV--TWEN
Cox        #  323 -RMIAYGDDVIASYPWP-----IDASLLAEA--GKGYGLIMTPA-DKGECFNEV--TWTN
```

```
Rhino 14 # 2040 -KILAYGDDLIVSYPYE-----LDPQVLATL--GKNYGLTITPP-DKSETFTKM--TWEN
Rhino  2 # 2011 -KIIAYGDDVIFSYIHE-----LDMEAIAIE--GVKYGLTITPA-DKSNTFVKL--DYSN
FMDV A10 #  332 -TMISYGDDIVVASDYD-----LDFEALKPH--FKSLGQTITPA-DKSDKGFVLGQSITD
EMC      # 2157 -KVLSYGDDLLVATNYQ-----LDFDKVRAS--LAKTGYKITPA-NTTSTFP-LNSTLED
Hep A    # 2077 -RILCYGDDVLIVFSRDVQIDNLDLIGQKIVDEFKKLGMTATSA-DKNVPQLKP---VSE
CPMV     # 1539 IGLVTYGDDNLISVNAVVTP-YFDGKKLKQS--LAQGGVTITDGKDKTSLELPF-RRLEE
                      ↑↑↑
TEV      # 2621 --YYVNGDDLLIAIHPDKA-ERLSRFKES----FGELGLKYEF
TVMV     # 2569 CKFFANGDDLIIAISPELE-HVLDGFQQH----FSDLGLNYDF

                  **** *           *                                      *
                  **** *           *      *                           *   *
Polio    #  371 VTFLKRFFRADEKYPFLIHPVMPMKEIHESIRWTKDPRNTQDHVRSLCLLAWHNGEEEYN
Cox      #  372 VTFLKRYFRADEQYPFLVHPAMPMKDIHESIRWTKDPKNTQDHVRSLCLLAWHNGEHEYE
Rhino 14 # 2089 LTFLKRYFKPDQQFPFLVHPVMPMKDIHESIRWTKDPKNTQDHVRSLCMLAWHSGEKEYN
Rhino  2 # 2060 VTFLKRGFKQDEKYNFLIHPTFPEDEIFESIRWTKKPSQMHEHVLSLCHLMWHNGRDAYK
FMDV A10 #  383 VTFLKRHFHMD-YGTGFYKPVMASKTLEAILSFARR-GTIQEKLISVAGLAVHSGPDEYR
EMC      # 2207 VVFLKRKFKKE---GPLYRPVMNREALEAMLSYYRP-GTLSEKLTSITMLAVHSGKQEYD
Hep A    # 2132 LTFLKRSFNL---VEDRIRPAISEKTIWSLMAWQRSNAEFEQNLENAQWFAFMHGYEFYN
CPMV     # 1595 CDFLKRTFVQRS-STIWDAPED-KASLWSQLHYVN--CNNCEKEVAYLTNVVNVLRELYM

                                          *
                                          *
Polio    #  431 KFLAKIRSVPIGRALLLPEYSTLYRRWLDSF*
Cox      #  432 EFIRKIRSVPVGRCLTLPAFTTLGRKWLDSF*
Rhino 14 # 2149 EFIQKIRTTDIGKCLILPEYSVLRRRWLDLF*
Rhino  2 # 2120 KFVEKIRSVSAGRALYIPPYDLLLHEWYEKF*
FMDV A10 #  441 RLFEPFQGLFE-----IPSYRSLYLRWVNAVCGDA*
EMC      # 2263 RLFAPFREVGV----VVPSFESVEYRWRSLFW*
Hep A    # 2189 KFYYFVQSCLEKE---MIEYRLKSYDWWRMRFYDQCFICDLS*
CPMV     # 1651 HSPREATEFRRK--------VLKKVSWITSGDLP 189 residues to C term-->
```

Protein sequences are translated from nucleotide sequences of the viral RNAs indicated on the left of the alignment. For original references and key to symbols, see legend to Appendix 1 and Reference 104 (Coxsackie virus polymerase). For genetic map locations, see text, Figure 1. Residue numbers refer to the first amino acid on each line of the alignment, with respect to the known N terminus (poliovirus, Coxsackie, and FMDV), or to the inferred polyprotein initiator methionine. Homologous sequences from two plant potyviral nuclear inclusion B proteins are also shown (references in Appendix 2). Residues thought to be homologous to those in alphaviral readthrough and related proteins (see Appendix 6 and Reference 54), including the Gly–Asp–Asp sequence common to many other polymerases, are marked ↑.

APPENDIX 4
ALIGNMENT OF THE CONSERVED N TERMINAL DOMAIN OF ALPHAVIRUS nsP1 PROTEINS AND THEIR PLANT VIRAL HOMOLOGUES LISTED IN TABLE 1

```
                                    *  *                          * *
                    **              *  * * *                      * **
AlMV 1a  #   67 LRANFPG-RRTVFSNSSSSSHCFAAAHRLLETDFVYRCFGNT---VDSIIDLGGNFVSHM
BMV  1a  #   60 FRDRYGGAFDLNLTQQYHAPHSLAGALRVAEHYDCLDSFPP----EDPVIDFGGSWWHHF
CMV  1a  #   61 IRTRYGGKYDLHLAQQELAPHGLAGALRLCETLDCLDFFPRSGLRQDLVLDFGGSWVTHY
TMV p126 #   62 ATRAYP-EFQITFYNTQNAVHSLAGGLRSLELEYLMMQIPY----GSLTYDIGGNFASHL
TRV p134 #   76 LMEIYP-EFNIVFKDDKNMVHGFAAAERKLQALLLLDRVPA----LQEVDDIGGQWSFWV
SFV nsP1 #   19 LQKAFP-SFEVESLQVTPNDHANARAFSHLATKLIEQETDK----DTLILDIGSAPSRRM
Sin nsP1 #   19 LQKSFP-QFEVVAQQVTPNDHANARAFSHLASKLIELEVPT----TATILDIGSAPARRM
```

```
                         *   *     *  *
                         *** * *  **  *
A1MV 1a   #  123  KVKRHNVHCCCPILDARDGARLTERILSLKSYVRKHPE--------------------IV
BMV  1a   #  116  SRRDKRVHSCCPVLGVRDAARHEERMCRMRK------------ILQESDDFD--------
CMV  1a   #  121  LRG-HNVHCCSPCLGIRDKMRHTERLMSMRK-----------VILNDPQQFDG-------
TMV p126  #  117  FKGRAYVHCCMPNLDVRDIMRHEGQKDSIELYLSRLERGGKTVPNFQKEAFDRYA---EI
TRV p134  #  121  TRGEKRIHSCCPNLDIRDDQREISRQIFLTAIGDQARSGKRQMSENELWMYDQFRKNIAA
SFV nsP1  #   74  MST-HKYHCVCPMRSAEDPERLDSYAKKLAAASGKVLDREIAGKITDLQTVMATP---DA
Sin nsP1  #   74  FSE-HQYHCVCPMRSPEDPDRMMKYASKLAEKACKITNKNLHEKIKDLRTVLDTP---DA
```

```
                       *
                       *      *                      **
A1MV 1a   #  163  GEADYCMDTFQKCSRRAD---------YAFAIHSTSDLDVGEL
BMV  1a   #  156  EVPNFCLNRAQDCDVQAD---------WAICIHGGYDMGFQGL
CMV  1a   #  162  RQPDFCTKSAAECKVQAH---------FAISIHGGYDMGFRGL
TMV p126  #  174  PEDAVCHNTFQTMRHQPMQ---QSGRVYAIALHSIYDIPADEF
TRV p134  #  181  PNAVRCNNTYQGCTCRGFSDGKKKGAQYAIALHSLYDFKLKDL
SFV nsP1  #  130  ESPTFCLHTDVTCRTAAEVAVYQDVYA-VHAPTSLYHQAMKGV
Sin nsP1  #  130  ETPSLCFHNDVTCNMRAEYSVMQDVY--INAPGTIYHQAMKGV
```

Protein sequences are translated from nucleotide seqeunces of the viral RNAs indicated on the left of the alignment as determined by the following authors: alfalfa mosaic virus[105] (N terminal residue determination from Reference 106), brome mosaic virus,[107] cucumber mosaic virus,[49] tobacco mosaic virus,[16] tobacco rattle virus,[50] Semliki Forest virus,[108] and Sindbis virus.[109] For genetic map locations of the conserved domain, see text, Figure 2. (AlMV and CMV resemble BMV to a first approximation, and TRV resembles TMV in this region of the genetic map.) Residue numbers refer to the first amino acid on each line of the alignment, with respect to the initiator methionine. For key to symbols see Appendix 1.

We note that this domain contains conserved Cys and His residues, in common with a number of proteins which bind nucleic acids in a metal-dependent fashion (see, for example, Reference 110). However, the pattern of conserved Cys and His residues in this domain does not fit any well-characterized nucleic acid binding domain, nor are Cys and His residues by themselves diagnostic of nucleic acid binding domain (see, for example, the Cys protease active site sequences in Appendix 2).

The viruses listed in Table 1 and in alignments 4 to 6 form a more divergent group than the picornaviruses in alignments 1 to 3, and this should be taken into account when using the alignments which make no claim to be definitive.

APPENDIX 5
ALIGNMENT OF THE CONSERVED CORE OF ALPHAVIRUS nsP2 PROTEINS AND THEIR PLANT VIRAL HOMOLOGUES LISTED IN TABLE 1

```
                         * *  * **   *           *
                         * ********* *           * *    *
A1MV 1a   # 816   AHFSVTIVDGVAGCGKTTNIKQIARSSGRDVDLILTSNRSSADELKETIDCS---PLTKL
BMV  1a   # 676   PTCDISMVDGVAGCGKTTAIKDAFR---MGEDLIVTANRKSAEDVRMALFPDTYNSKVAL
CMV  1a   # 704   PTGTISQVDGVAGCGKTTAIKSMFN---PSTDIIVTANKKSAQDVRYALFKST-DSKEAC
TMV p126  # 824   SSAKVVLVDGVPGCGKTKEILSRVN---FDEDLILVPGKQAAEMIRRRANSSG-IIVATK
TRV p134  # 895   VNCSFELVDGVPGCGKSTMIVNSAN---PCVDVVLSTGRAATDDLIERFASKGF-PCKLK
SFV nsP2  # 177   YKTTVVGVFGVPGSGKSAIIKSLV----TKHDLVTSGKKENCQEIVNDVKK----HRGKG
Sin nsP2  # 177   YKVETIGVIGTPGSGKSAIIKSTV----TARDLVTSGKKENCREIEADVLR----LRGMQ
                           ↑    ↑↑↑
```

```
                        **                  **    * *                  **  *
                      * **  *            *  **    * * *                **  **
A1MV 1a   # 873   HYIRTCDSYLMSA---SAVKAQRLIFDECFLQHAGLVYAAATLAGC-SEVIGFGDTEQIP
```

```
BMV  1a    # 733  DVVRTADSAIMHGV----PSCHRLLVDEAGLLHYGQLLVVAALSKC-SQVLAFGDTEQIS
CMV  1a    # 760  AFVRTADSILLNDC----PTVSRVLVDEVVLLHFGQLCAVMSKLHA-VRALCFGDSEQIA
TMV  p126  # 880  DNVKTVDSFMMNFGKSTRCQFKRLFIDEGLMLHTGCVNFLVAMSLC-EIAYVYGDTQQIP
TRV  p134  # 951  RRVKTVDSFLMHCVD-GSLTGDVLHFDEALMAHAGMVYFCAQIAGA-KRCICQGDQNQIS
SFV  nsP2  # 229  TSRENSDSILLNGC---RRAVDILYVDEAFACHSGTLLALIALVKPRSKVVLCGDPKQCG
Sin  nsP2  # 229  ITSKTVDSVMLNGC---HKAVEVLYVDEAFACHAGALLALIAIVRPRKKVVLCGDPMQCG
                                                                       ↑↑  ↑

                                               *
                  *  *                       * *** **   *
AlMV 1a    # 929  FVSRNPSFVFRHH---KLTGKVERKLITWRSPADATYCLEKYFYKNKKP-------VKTN
BMV  1a    # 788  FKSRDAGFKLLHG--NLQYDRRDVVHKTYRCPQDVIAAVNLLKRKCGNRDTKYQ-SWTSE
CMV  1a    # 815  FSSRDASFDMRFS--KLIPDETSDADTTFRSPQDVVPLVRLMATKALPKGTRTKYSDGAQ
TMV  p126  # 939  YINRVSGFPYPAHFAKLEVDEVETRRTTLRCPADVTHYLNRRYEGFVMSTS------SVK
TRV  p134  #1009  FKPRVSQVDLRFSSLVGKFDIVTEKRETYRSPADVAAVLNKYYTGDVRTHNATANSMTVR
SFV  nsP2  # 286  FFNMMQLKVNFNH---NICTEVCHKSISRRCTRPVTAIVSTLHYGGKMRTTNP-----CN
Sin  nsP2  # 286  FFNMMQLKVHFNHPEKDICTKTFYKYISRRCTQPVTAIVSTLHYDGKMKTTNP-----CK

                                                *                         *
                  * *              *           ** **                    * * **
AlMV 1a    # 979  SRVLRSIEVVPINSPVSVERNTNALYLCHTQAEKAVLKAQTHLKGC------DNIFTTHE
BMV  1a    # 845  SKVSRSLTKRRITSGLQVTIDPNRTYLTMTQADKAALQTRAKDFPVSKDWIDGHIKTVHE
CMV  1a    # 873  SKVRKSVTSRAVASVSLVELDPTRFYITMTQADKASLITRAKELNLPKAFYTDRIKTVHE
TMV  p126  # 993  KSVSQEMVGGAAVIN-PISKPLHGKILTFTQSDKEALLSR----------GYSDVHTVHE
TRV  p134  #1069  KIVSKE----------QVSLKPGAQYITFLQSEKKELVNLLALRK-----VAAKVSTVHE
SFV  nsP2  # 338  KPIIIDTTGQTKPKPGDIVLTCFRGWAKQLQLDYR----------------GHEVMTAAA
Sin  nsP2  # 341  KNIEIDITGATKPKPGDIILTCFRGWVKQLQIDYP----------------GHEVMTAAA

                   **     *   *                      *   *               *
                   ** *   *  ** * *   *              ******   * *        *
AlMV 1a    #1033  AQGKTFDNVYFCRLTRTSTSLATGRDPINGPCNGLVALSRHKKTFKYFTIAHDSDDVIYN
BMV  1a    # 905  AQGISVDNVTLVRLKSTKCDLFKHEEY------CLVALTRHKKSFEYCFNGELAGDLIFN
CMV  1a    # 933  SQGISEDHVTLVRLKSTKCDLFKKFSY------CLVAVTRHKVTFRYEYCGVLGGDLIAN
TMV  p126  #1042  VQGETYSDVSLVRLTPTPVSIIAGDSP-----HVLVALSRHTCSLKYYT---VVMDPLVS
TRV  p134  #1114  SQGETFKDVVLVRTKPTDDSIARGREY------LIVALSRHTQSLVYET---VKEDDVSK
SFV  nsP2  # 382  SQGLTRKGVYAVRQKVNENPLYAPASE-----HVNVLLTRTEDRLVWKT---LAGDPWIK
Sin  nsP2  # 385  SQGLTRKGVYAVRQKVNENPLYAITSE-----HVNVLLTRTEDRLVWKT---LQGDPWIK


AlMV 1a    #1093  ACRDAGNTDDSILARSYNHNF*
BMV  1a    # 959  CVK*
CMV  1a    # 987  CIPLV*
TMV  p126  #1094  IIRDLEKLSSYLLDMYKVDAGTQ*
TRV  p134  #1165  EIRESAALTKAALARFFVTETVL*
SFV  nsP2  # 434  VLSNIPQGN-FTATLEEWQEEHDK <---342 residues to C terminus--->
Sin  nsP2  # 437  QPTNIPKGN-FQATIEDWEAEHKG <---348 residues to C terminus--->
```

Protein sequences are translated from nucleotide sequences of the viral RNAs indicated on the left of the alignment. For original references and key to symbols, see legends to Appendixes 4 and 1, respectively. For genetic map locations, see text, Figure 2. Residues are numbered with respect to the initiator methionine of the protein indicated.

Conserved residues marked ↑ are (1) Gly and Gly–Lys–Ser residues in the configuration expected for a purine nucleoside phosphate binding site (see text, Figure 4) and (2) a Gly–Asp containing motif also found in herpes, vaccinia, and adenovirus DNA polymerases.[89]

APPENDIX 6
ALIGNMENT OF THE ALPHAVIRUS nsP4 PROTEINS AND THEIR PLANT VIRAL HOMOLOGUES LISTED IN TABLE 1

```
                    *                   *
                    *                   *      *                    *    *
A1MV 2a   #    301 LP--LHHSIDDLYFQEWVETSDKSLDV-DPCRIDLSVFNNWQSSENCYEPRFKTGALSTR
BMV  2a   #    239 LP--THAYFDDSYHQALVENGDYSMDF-DRIRLKQSDVDWYRDPDKYFQPKMNIGSAQRR
CMV  2a   #    288 LP--THGNYDDSFHQVFVDSADYSTDM-DHVRLRQSDLVAKIPDGGHMLPVLNTGSGHQR
TMV p183    am+ 34 LP--GNSTMMNNFDAVTMRLTDISLNVKDCILDMSKSVAAPKDQIKPLIPMVRTAAEMPR
TRV p194    op+ 35 FP--GNSLRDSSLDGYLVATTDCNLRLDNVTIKSGNWKDKFAEKETFLKPVIRTAMPDKR
MID nsP4    op+138 YPTVTSYQITDEYDA-YLDMVDGSESCLDRAAFCPSKLRSFPKKHSYHRAEIRSAVPSPF
SFV nsP4 #     133 YPTVASYQITDEYDA-YLDMVDGSDSCLDRATFCPAKLRCYPKHHAYHQPTVRSAVPSPF
SIN nsP4    op+138 YPTVASYQITDEYDA-YLDMVDGTVACLDTATFCPAKLRSYPKKHEYRAPNIRSAVPSAM

                          * *  ***
                     *  * * *  ***  ** *                        *
A1MV 2a   #    358 KGTQTEALLAIKKRNMNVPNLGQIYDVNSVANSVVNKLLTTVIDPDKLCMF----PDFIS
BMV  2a   #    296 VGTQKEVLTALKKRNADVPEMGDAINMKDTAKAIAKRFRSTFLNVDGEDCLRAS----MD
CMV  2a   #    345 VGTTKEVLTAIKKRNADVPELGDSVNLSRLSKAVAERFRLSYMNVDALAKSNF-----VN
TMV p183    am+ 92 QTGLLENLVAMIKRNFNAPELSGIIDIENTASLVVDKFFDSYLLKEKRKPNKNV--SLFS
TRV p194    op+ 93 KTTQLESLLALQKRNQAAPDLQENVHA--TVLIEETMKKLKSVVYDVGKIRAD---PIVN
MID nsP4    op+197 QNTLQNVLAAATKRNCNVTQMRELPTLD-SAVFNVECFKKYACNNDYWDEFAQKPIRLTT
SFV nsP4 #     192 QNTLQNVLAAATKRNCNVTQMRELPTMD-SAVFNVECFKRYACSGEYWEEYAKQPIRITT
SIN nsP4    op+197 QNTLQNVLIAAYKRNCNVTQMRELPTLD-SATFNVECFRKYACNDEYWEEFARKPIRITT

                                                           *   *
                                                      *  * *   *         ***  *
A1MV 2a   #    414 EGEVSYFQDYIVGKNPDP---ELYSDPLGVRSIDSYKHMIKSVLKPVEDNSLHLERPMPA
BMV  2a   #    352 VMTKCLEYHKKWGKHMD----LQGVNVAAETDLCRYQHMLKSDVKPVVTDTLHLERAVAA
CMV  2a   #    400 VVSNFHAYMQKWPSSGLS---YDDLPDLHAENLQFYDHMIKSDVKPVVTDTLNVDRPVPA
TMV p183    am+150 RESLNRWLEKQEQVTIGQ---LADFDFVDLPAVDQYRHMYKAQPKQKLDTSIQTEYPALQ
TRV p194    op+148 RAQMERWWRNQSTAVQAKV--VADVRELHEIDYSSYMYMIKSDVKPKTDLTPQFEYSALQ
MID nsP4    op+256 ENITSYVTRLKGPKAAALFAKTYDLKPLQEVPMDRFVVDMKRDVKVTPGTKHTEERPKVQ
SFV nsP4 #     251 ENITTYVTKLKGPKAAALFAKTHNLVPLQEVPMDRFTVDMKRDVKVTPGTKHTEERPKVQ
SIN nsP4    op+256 EFVTAYVARLKGPKAAALFAKTYNLVPLQEVPMDRFVMDMKRDVKVTPGTKHTEERPKVQ

                   **  * *       * *
A1MV 2a   #    471 TITYHDKDIVMSSSPIFLAAAARLMLILRDKITIPSGKFHQLFSIDA---EAFDASFHFK
BMV  2a   #    408 TITFHSKGVTSNFSPFFTACFEKLSLALKSRFIVPIGKISSLELKNV-----RLNNRYFL
CMV  2a   #    457 TITFHKKTITSQFSPLFISLFERFQRCLRERVVLPVGKISSLEMTGF-----SVLNKHCL
TMV p183    am+207 TIVYHSKKINAIFGPLFSELTRQLLDSVDSSRFLFFTRKTPAQIEDFFGDLDSHVPMDVL
TRV p194    op+206 TVVYHEKLINSLFGPIFKEINERKLDAMQ-PHFVFNTRMTSSDLNDRVKFLNTEAAYDFV
MID nsP4    op+316 VIQAAEPLATAYLCGIHRELVRRLNAVLL--PNVHTLFDMSAEDFDAIISEHFRPGDAVL
SFV nsP4 #     311 VIQAAEPLATAYLCGIHRELVRRLNAVLR--PNVHTLFDMSAEDFDAIIASHFHPGDPVL
SIN nsP4    op+316 VIQAAEPLATAYLCGIHRELVRRLTAVLL--PNIHTLFDMSAEDFDAIIAEHFKQGDPVL

                   * *    ***                *
                   * * ******* *            **        *   *      *   *       *
A1MV 2a   #    528 EIDFSKFDKSQNELHHLIQERFLKYLGIPNGFLTLWFNAHRKSRISDSKNGVFFNVDFQR
BMV  2a   #    463 EADLSKFDKSQGELHLEFQREILLALGFPAPLTNWWSDFHRDSYLSDPHAKVGMSVSFQR
CMV  2a   #    512 EIDLSKFDKSQGEFHLMIQEHILNDLGCPAPITKWWCDFHRFSYIKDKRAGVGMPISFQR
TMV p183    am+267 ELDISKYDKSQNEFHCAVEYEIWRRLGFEDFLGEVWKQGHRKTTLKDYTAGIKTCIWYQR
TRV p194    op+265 EIDMSKFDKSANRFHLQLQLEIYRLFGLDEWAAFLWEVSHTQTTVRDIQNGMMAHIWYQQ
MID nsP4    op+374 ETDIASFDKSQDDSLAYTGLMLLEDLGVDQPLLELIEASFGEITSTHLPTGTRFKFGAMM
SFV nsP4 #     371 ETDIASFDKSQDDSLALTGLMILEDLGVDQYLLDLIEAAFGEISSCHLPTGTRFKFGAMM
SIN nsP4    op+374 ETDIASFDKSQDDAMALTGLMILEDLGVDQPLLDLIECAFGEISSTHLPTGTRFKFGAMM

                     *   *   *                         ***
                   ****  *   **                        *****
A1MV 2a   #    588 RTGDALTYLGNTIVTLACLCHVYDLMDPNVKFVVASGDDSLIGTVEEL-PRDQEFLFTTL
BMV  2a   #    523 RTGDAFTYFGNTLVTMAMIAYASDLS--DCDCAIFSGDDSLIISKVK--PVLDTDMFTSL
```

```
CMV  2a    #   572 RTGDAFTYFGNTIVTMAEFAWCYDTD--QFDRLLFSGDDSLAFSKLP--PVGDPSKFTTL
TMV  p183  am+327 KSGDVTTFIGNTVIIAACLASMLPME--KIIKGAFCGDDSLLYFPKGCEFPDVQHSANLM
TRV  p194  op+325 KSGDADTYNANSDRTLCALLSELPLE--KAVMVTYGGDDSLIAFPRGTQFVDPCPKLATK
MID  nsP4  op+492 KSGMFLTLFVNTMLNMTIASRVLEERLTNSKCAAFIGDDNIVHGVKSD--KLLAERCAAW
SFV  nsP4  #   487 KSGMFLTLFINTVLNITIASRVLEQRLTDSACAAFIGDDNIVHGVISD--KLMAERCASW
SIN  nsP4  op+492 KSGMFLTLFVNTVLNVVIASRVLEERLKTSRCAAFIGDDNIIHGVVSD--KEMAERCATW
                     ↑   ↑   ↑                         ↑↑↑

                    * * *             *                            *
                    * * *          *  ** *                   * * * ***
A1MV 2a    #   647 FNLEAKFPHNQ-----PFICSKFLITMPTTSGGKVVLPIPNPLKLLIRLGSKKV----NA
BMV  2a    #   579 FNMEIKVMDPSV----PYVCSKFLVE--TEMGNLVSVP--DPLREIQRLAKRKI--LRDE
CMV  2a    #   628 FNMEAKVMEPAV----PYICSKFYSLM--SLVTRFQS---PTIREIQRLGTKKIPYSDNN
TMV  p183  am+385 WNFEAKLFKKQY----GYFCGRYVIHHDRGCGCIVYY---DPLKLISKLGAKHI---KDW
TRV  p194  op+383 WNFECKIFKYDV----PMFCGKFLLK---TSSCYEFVP--DPVKVLTKLGKKSI---KDV
MID  nsP4  op+492 MNMEVKIIDAVMCERPPYFCGGFIVFDQVTGTCCRVA---DPLKRLFKLGKPLP---AED
SFV  nsP4  #   487 VNMEVKIIDAVMGEKPPYFCGGFIVFDSVTQTACRVS---DPLKRLFKLGKPLT---AED
SIN  nsP4  op+492 LNMEVKIIDAVIGERPPYFCGGFILQDSVTSTACRVA---DPLKRLFKLGKPLP---ADD

                              *
                           *  *                *      **
A1MV 2a    #   698 DIFDEWYQSWIDIIGGFNDHH--VIRCVAAMTAHRYLRRPSLYL--EAALESLGKIFAGK
BMV  2a    #   631 QMLRAHFVSFCDRMKFINQLDEKMITTLCMFVYLKYGKEKPWIF------------EEVR
CMV  2a    #   679 DFLFAHFMSFVDRLKFMDRMSQSCIDQLSIFFELKYKKSGN-----EAALVLGAFKKYTA
TMV  p183  am+435 EHLEEFRRSLCDVAVSLNNCA--YYTQLDDAVWEVHKTAPP------GSFVYKSLVKYLS
TRV  p194  op+431 QHLAEIYISLNDSNRALGNYM--VVSKLSESVSDRYLYKGD------SVHALCALWKHIK
MID  nsP4  op+546 KQDEDRRRALADEAQRWNRVG--IQADLEAAMNSRYEVEGIRNVITALTTLSRNYHNFRH
SFV  nsP4  #   541 KQDEDRRRALSDEVSKWFRTG--LGAELEVALTSRYEVEGCKSILIAMTTLARDIKAFKK
SIN  nsP4  op+546 EQDEDRRRALLDETKAWFRVG--ITGTLAVAVTTRYEVDNITPVLLALRTFAQSKRAFQA

                           *
                           **
A1MV 2a    #   754 TLCKEECLFNEKHE--SNVKIKPRRVKKSHSDARSRARRA*
BMV  2a    #   679 AALAAFSLYSENFLRFSDCYCTEGIRVYQMSDPVCKFKRTT<105 res to C term>
CMV  2a    #   734 NFNAYKELYY------SDRQQCDLVNTFCISEFRV-IRRTT< 72 res to C term>
TMV  p183  am+487 DKYLFRSLFIDGSSC*
TRV  p194  op+483 SFTALCTLFRDENDKELNPAKVDWKKAQRAVSNFYDW*
MID  nsP4  op+604 LRGPVIDLYGGPK*
SFV  nsP4  #   599 LRGPVIHLYGGPRLVR*
SIN  nsP4  op+604 IRGEIKHLYGGPK*
```

Protein sequences are translated from the nucleotide sequences of the viral RNAs indicated on the left of the alignment. For original references and key to symbols, see legend to Appendixes 4 and 1, respectively, except for: alfalfa mosaic virus,[111] cucumber mosaic virus,[48] and Middelburg virus.[112] For genetic map locations, see text, Figure 2. Residues are numbered with respect to the N-terminal methionine (AlMV, BMV, and CMV), the first residue after the suppressible termination codon (TMV, TRV, Sindbis, and Middelburg viruses), or the deduced site of proteolytic cleavage (Semliki Forest virus). Residues thought to be homologous to those in picornaviral P3D polymerase proteins (see Appendix 3 and Reference 54), including the Gly–Asp–Asp sequence found in a wide range of polymerases, are marked ↑.

REFERENCES

1. Recombination at the DNA level, *Cold Spring Harbor Symp. Quant. Biol.*, vol. 49, 1984.
2. **Hirst, G. K.,** Genetic recombination with Newcastle disease virus, poliovirus and influenza, *Cold Spring Harbor Symp. Quant. Biol.*, 27, 303, 1962.

3. **Horiuchi, K.**, Genetic studies of RNA phages, in *RNA Phages*, Zinder, N. D., Ed., Cold Spring Harbor Laboratory, Cold Spring Harbor, N.Y., 1975, 29.
4. **Wittmann, H. G.**, Proteinuntersuchung an Mutanten des Tabakmosaikvirus als Beitrag zum Problem des Genetischen Codes, *Z. Vererbungsl.*, 93, 491, 1962.
5. **Tsugita, A.**, The proteins of mutants of TMV RNA, *J. Mol. Biol.*, 5, 28, 1962.
6. **Flavell, R. A., Sabo, D. L., Bandle, E. F., and Weissmann, C.**, Site-directed mutagenesis: generation of an extracistronic mutation in bacteriophage Qβ RNA, *J. Mol. Biol.*, 89, 255, 1974.
7. **Domingo, E., Flavell, R. A., and Weissmann, C.**, *In vitro* site-directed mutagenesis: generation and properties of an infectious extracistronic mutant of bacteriophage Qβ, *Gene*, 1, 3, 1976.
8. **Holland, J., Spindler, K., Horodyski, F., Grabeau, E., Nichol, S., and vande Pol, S.**, Rapid evolution of RNA genomes, *Science*, 215, 1577, 1982.
9. **Durbin, R. K. and Stollar, V.**, Sequence analysis of the E2 gene of a hyperglycosylated, host restricted mutant of Sindbis virus, and estimation of the mutation rate from frequency of revertants, *Virology*, 154, 135, 1986.
10. **Eigen, M. and Schuster, P.**, The hypercycle, *Naturwissenschaften*, 64, 541, 1977.
11. **Domingo, E., Sabo, D., Taniguchi, T., and Weissmann, C.**, Nucleotide sequence heterogeneity of an RNA phage population, *Cell*, 13, 735, 1978.
12. **Batschelet, E., Domingo, E., and Weissmann, C.**, The proportion of revertant and mutant phage in a growing population as a function of mutation and growth rate, *Gene*, 1, 27, 1976.
13. **Cattaneo, R., Schmid, A., Rebmann, G., Baczko, K., ter Meulen, V., Bellini, W. J., Rozenblatt, S., and Billeter, M. A.**, Accumulated measles virus mutations in a case of subacute sclerosing panencephalitis: interrupted matrix protein reading frame and transcription alteration, *Virology*, 154, 97, 1986.
14. **Sabrino, F., Davila, M., Ortin, J., and Domingo, E.**, Multiple genetic variants arise in the course of replication of foot-and-mouth disease virus in cell culture, *Virology*, 128, 310, 1983.
15. **Steinhauer, D. A. and Holland, J. J.**, Direct method for quantitation of extreme polymerase error frequencies at selected base sites in viral RNA, *J. Virol.*, 57, 219, 1986.
16. **Goelet, P., Lomonossoff, G. P., Butler, P. J. G., Akam, M. E., Gait, M. J., and Karn, J.**, Nucleotide sequence of tobacco mosaic virus RNA, *Proc. Natl. Acad. Sci. U.S.A.*, 79, 5818, 1982.
17. **Schubert, M., Harrison, G. G., and Meier, E.**, Primary structure of the vesicular stomatitis virus polymerase (L) gene: evidence for a high frequency of mutations, *J. Virol.*, 51, 505, 1984.
18. **Strauss, E. G. and Strauss, J. H.**, Replication strategies of the single stranded RNA viruses of eukaryotes, *Curr. Top. Microbiol. Immunol.*, 105, 1, 1983.
19. **Ohno, S.**, Dispensable genes, *Trends Genet.*, 1, 160, 1985.

19a. **Reanney, D. L.**, Genetic error and genome design, *Trends Genet.*, 2, 41, 1986.

20. **Cooper, P. D.**, Genetics of picornaviruses, in *Comprehensive Virology*, Fraenkel-Conrat, H. and Wagner, R. R., Eds., Plenum Press, New York, 1977, 9, 133.
21. **Makino, S., Keck, J. G., Stohlman, S. A., and Lai, M. M. C.**, High frequency RNA recombination of murine coronaviruses, *J. Virol.*, 57, 729, 1986.
22. **Bujarski, J. J. and Kaesberg, P.**, Genetic recombination between RNA components of a multipartite plant virus, *Nature (London)*, 321, 528, 1986.
23. **Perrault, J.**, Origin and replication of defective interfering particles, *Curr. Top. Microbiol. Immunol.*, 93, 151, 1981.
24. **Fields, S. and Winter, G.**, Nucleotide sequences of influenza virus segments 1 and 3 reveal mosaic structure of a small viral RNA segment, *Cell*, 28, 303, 1982.
25. **Monroe, S. S. and Schlesinger, S.**, RNAs from two independently isolated defective interfering particles of Sindbis virus contain a cellular tRNA sequence at their 5′ ends, *Proc. Natl. Acad. Sci. U.S.A.*, 80, 3279, 1983.
26. **Varmus, H. and Swanstrom, R.**, Replication of retroviruses, in *RNA Tumor Viruses*, Vol. 1, Weiss, R., Teich, N., Varmus, H., and Coffin, J., Eds., Cold Spring Harbor Laboratory, Cold Spring Harbor, N.Y., 1984, 369 and supplement.
27. **Jennings, P. A., Finch, J. T., Winter, G., and Robertson, J. S.**, Does the higher order structure of the influenza virus ribonucleoprotein guide sequence rearrangements in influenza viral RNA?, *Cell*, 34, 619, 1983.
28. **King, A. M. Q., Ortlepp, S. A., and Newman, J. W. I., and McCahon, D.**, Genetic recombination in RNA viruses, in *The Molecular Biology of Positive Strand RNA Viruses*, Rowlands, D. R., Mahy, B. W. J., and Mayo, M. A., Eds., Academic Press, New York, 1987, 36, 129.
29. **Rose, J. K., Doolittle, R. F., Anilionis, A., Curtis, P. J., and Wunner, W. H.**, Homology between the glycoproteins of vesicular stomatitis virus and rabies virus, *J. Virol.*, 43, 361, 1982.
30. **Giorgi, C., Blumberg, B. M., and Kolakofsky, D.**, Sendai virus contains overlapping genes expressed from a single mRNA, *Cell*, 35, 829, 1983.
31. **Franssen, H., Leunissen, J., Goldbach, R., Lomonossoff, G., and Zimmern, D.**, Homologous sequences in non-structural proteins from cowpea mosaic virus and picornaviruses, *EMBO J.*, 3, 855, 1984.

32. **Haseloff, J., Goelet, P., Zimmern, D., Ahlquist, P., Dasgupta, R., and Kaesberg, P.**, Striking similarities in amino acid sequence among nonstructural proteins encoded by RNA viruses that have dissimilar genomic organisation, *Proc. Natl. Acad. Sci. U.S.A.*, 81, 4358, 1984.
33. **Shioda, T., Iwasaki, K., and Shibata, H.**, Determination of the complete nucleotide sequence of the Sendai virus genome RNA and the predicted amino acid sequences of the F, HN and L proteins, *Nucleic Acids Res.*, 14, 1545, 1986.
34. **Morgan, E. M. and Rakestraw, K. M.**, Sequence of the Sendai virus L gene, *Virology*, 154, 31, 1986.
35. **Stanley, J., Rottier, P., Davies, J. W., Zabel, P., and van Kammen, A.**, A protein linked to the 5′ termini of both RNA components of the cowpea mosaic virus genome, *Nucleic Acids Res.*, 5, 4505, 1978.
36. **Daubert, S. D., Bruening, G., and Najarian, R. C.**, Protein bound to the genome RNAs of cowpea mosaic virus, *Eur. J. Biochem.*, 92, 45, 1978.
37. **Pelham, H. R. B.**, Synthesis and proteolytic processing of cowpea mosaic virus proteins in reticulocyte lysates, *Virology*, 96, 463, 1979.
38. **Argos, P., Kamer, G., Nicklin, M. J. H., and Wimmer, E.**, Similarity in gene organisation and homology between proteins and animal picornaviruses and a plant comovirus suggest common ancestry of these virus families, *Nucleic Acids Res.*, 12, 7251, 1984.
39. **Stauffacher, C. V., Usha, R., Harrington, M., Schmidt, T., Hosur, M. V., and Johnson, J. E.**, The structure of cowpea mosaic virus at 3.5 Å resolution, submitted.
40. **Domier, L. L., Franklin, K. M., Shahabuddin, M., Hellmann, G. M., Overmeyer, J. H., Hiremath, S. T., Siaw, M. F. E., Lomonossoff, G. P., Shaw, J. G., and Rhoads, R. E.**, The nucleotide sequence of tobacco vein mottling virus RNA, *Nucleic Acids Res.*, 15, 5417, 1986.
41. **Allison, R., Johnston, R. E., and Dougherty, W. G.**, The nucleotide sequence of the coding region of tobacco etch virus genomic RNA: evidence for the synthesis of a single polyprotein, *Virology*, 154, 9, 1986.
42. **Fritsch, C.**, Nucleotide sequence of tomato black ring virus RNA, paper presented at EMBO Workshop on Molecular Plant Virology, Wageningen, Netherlands, July 6 to 10, 1986, 20.
43. **Bloomer, A. C., Champness, J. N., Bricogne, G., Staden, R., and Klug, A.**, Protein disk of tobacco mosaic virus at 2.8 Å resolution showing the interactions within and between subunits, *Nature (London)*, 276, 362, 1978.
44. **Harrison, S. C., Olson, A. J., Schutt, C. E., Winkler, F. K., and Bricogne, G.**, Tomato bushy stunt virus at 2.9 Å resolution, *Nature (London)*, 276, 368, 1978.
45. **Rossmann, M., Arnold, E., Erickson, J. W., Frankenburger, E. A., Griffith, J. P., Hecht, H-J., Johnson, J. E., Kamer, G., Luo, M., Mosser, A. G., Rueckert, R. R., Sherry, B., and Vriend, G.**, Structure of a human common cold virus and functional relationship to other picornaviruses, *Nature (London)*, 317, 145, 1985.
46. **Ahlquist, P., Strauss, E. G., Rice, C. M., Strauss, J. H., Haseloff, J., and Zimmern, D.**, Sindbis virus proteins nsP1 and nsP2 contain homology to nonstructural proteins from several RNA plant viruses, *J. Virol.*, 53, 536, 1985.
47. **Cornelissen, B. J. C. and Bol, J. F.**, Homology between the proteins encoded by tobacco mosaic virus and two tricornaviruses, *Plant Mol. Biol.*, 3, 379, 1984.
48. **Rezian, M. A., Williams, R. H. V., Gordon, K. H. J., Gould, A. R., and Symons, R. H.**, Nucleotide sequence of cucumber mosaic virus RNA 2 reveals a translation product significantly homologous to corresponding proteins of other viruses, *Eur. J. Biochem.*, 143, 277, 1984.
49. **Rezian, M. A., Williams, R. H. V., and Symons, R. H.**, Nucleotide sequence of cucumber mosaic virus RNA 1, *Eur. J. Biochem.*, 150, 331, 1985.
50. **Hamilton, W. D. O., Boccara, M., Robinson, D. J., and Baulcombe, D. C.**, The complete nucleotide sequence of tobacco rattle virus RNA1, *J. Gen. Virol.*, 68, 2563, 1987.
51. **Strauss, E. G. and Strauss, J. H.**, Structure and replication of the alphavirus genome, in *The Togaviruses and Flaviviruses*, Schlesinger, S. and Schlesinger, M., Eds., Plenum Press, New York, in press.
52. **Flanegan, J. B. and Baltimore, D.**, Poliovirus poly(U) polymerase and RNA replicase have the same viral polypeptide, *J. Virol.*, 29, 352, 1979.
53. **Dorssers, L., van der Krol, S., van der Meer, J., van Kammen, A., and Zabel, P.**, Purification of cowpea mosaic virus RNA replication complex: identification of a virus-encoded 110,000-dalton polypeptide responsible for RNA chain elongation, *Proc. Natl. Acad. Sci. U.S.A.*, 81, 1951, 1984.
54. **Kamer, G. and Argos, P.**, Primary structural comparison of RNA-dependent polymerases from plant, animal and bacterial viruses, *Nucleic Acids Res.*, 12, 7269, 1984.
55. **Toh, H., Hayashida, H., and Miyata, T.**, Sequence homology between retroviral reverse transcriptase and putative polymerases of hepatitis B virus and cauliflower mosaic virus, *Nature (London)*, 305, 827, 1983.
56. **Patarca, R. and Haseltine, W. H.**, Sequence similarity among retroviruses — erratum, *Nature (London)*, 309, 728, 1984.

57. **Toh, H., Kikuno, R., Hayashida, H., Miyata, T., Kugimiya, W., Inouye, S., Yuki, S., and Saigo, K.,** Close structural resemblance between putative polymerase of a *Drosophila* transposable genetic element 17.6 and *pol* gene product of Moloney murine leukemia virus, *EMBO J.*, 4, 1267, 1985.
58. **Goldbach, R. W.,** Molecular evolution of plant RNA viruses, *Annu. Rev. Phytopathol.*, 24, 289, 1986.
59. **Murthy, M. R. N.,** Comparison of the nucleotide sequences of cucumber mosaic virus and brome mosaic virus, *J. Mol. Biol.*, 168, 469, 1983.
60. **Cornelissen, B. J. C., Janssen, H., Zuidema, D., and Bol, J. F.,** Complete nucleotide sequence of tobacco streak virus RNA 3, *Nucleic Acids Res.*, 12, 2427, 1984.
61. **Rossmann, M. G., Liljas, A., Branden, C. -I., and Banaszak, L. J.,** Evolutionary and structural relationships among dehydrogenases, in *The Enzymes*, Vol. 11 (Part A), 3rd ed., Boyer, P. D., Ed., Academic Press, New York, 1975, 61.
62. **Walker, J. E., Saraste, M., Runswick, M. J., and Gay, N. J.,** Distantly related sequences in the α- and β-subunits of ATP synthase, myosin, kinases and other ATP-requiring enzymes and a common nucleotide binding fold, *EMBO J.*, 1, 945, 1982.
63. **Halliday, K. R.,** Regional homology in GTP-binding proto-oncogene products and elongation factors, *J. Cyclic Nucleotide Protein Phosphorylation Res.*, 9, 435, 1984.
64. **Gorbalenya, A. E., Blinov, V. M., and Koonin, E. V.,** Prediction of nucleotide binding properties of virus-specific proteins from their primary structure, *Mol. Genet.*, 7, 30, 1985.
65. **Levinson, A. D.,** Normal and activated *ras* oncogenes and their encoded products, *Trends Genet.*, 2, 81, 1986.
66. **Karin, M. and Richards, R. I.,** Human metallothionein genes — primary structure of the metallothionein-II gene and a related processed gene, *Nature (London)*, 299, 797, 1982.
67. **Lemischka, I. and Sharp, P. A.,** The sequences of an expressed rat α-tubulin gene and a pseudogene with an inserted repetitive element, *Nature (London)*, 300, 330, 1982.
68. **Haseloff, J. and Symons, R. H.,** Comparative sequence and structure of viroid-like RNAs of two plant viruses, *Nucleic Acids Res.*, 10, 3681, 1982.
69. **Chu, P. W. G., Francki, R. I. B., and Randles, J. W.,** Detection, isolation and characterisation of high molecular weight double-stranded RNAs in plants infected with velvet tobacco mottle virus, *Virology*, 126, 480, 1983.
70. **Symons, R. H., Haseloff, J., Visvader, J. E., Keese, P., Murphy, P. J., Gill, D. S., Gordon, K. H. J., and Bruening, G.,** On the mechanism of replication of viroids, virusoids and satellite RNAs, in *Subviral Pathogens of Plants and Animals: Viroids and Prions*, Maramorosch, K., Ed., Academic Press, New York, 1986, 235.
71. **Kiberstis, P. A., Haseloff, J., and Zimmern, D.,** 2′ phosphomonoester, 3′-5′ phosphodiester bond at a unique site in a circular viral RNA, *EMBO J.*, 4, 817, 1985.
72. **Prody, G. A., Bakos, J. T., Buzayan, J. M., Schneider, I. R., and Bruening, G.,** Autolytic processing of dimeric plant virus satellite RNA, *Science*, 231, 1577, 1986.
73. **Buzayan, J. M., Gerlach, W. L., Bruening, G., Keese, P., and Gould, A. R.,** Nucleotide sequence of satellite tobacco ringspot virus RNA and its relationship to multimeric forms, *Virology*, 151, 186, 1986.
74. **Guyaux, M., Cornelissen, A. W. C. A., Steinert, M., and Borst, P.,** *Trypanosoma brucei*: a surface antigen mRNA is discontinuously transcribed from two distinct chromosomes, *EMBO J.*, 4, 995, 1985.
75. **Sather, S. and Agabian, N.,** A 5′ spliced leader is added in *trans* to both α- and β-tubulin transcripts in *Trypanosoma brucei*, *Proc. Natl. Acad. Sci. U.S.A.*, 82, 5695, 1985.
76. **Konarska, M. M., Padgett, R. A., and Sharp, P. A.,** *Trans* splicing of mRNA precursors *in vitro*, *Cell*, 42, 165, 1985.
77. **Solnick, D.,** *Trans* splicing of mRNA precursors, *Cell*, 42, 157, 1985.
78. **Zimmern, D.,** Do viroids and RNA viruses derive from a system that exchanges genetic information between eukaryotic cells?, *Trends Biochem. Sci.*, 7, 205, 1982.
79. **Argos, P., Tsukihara, T., and Rossmann, M. G.,** A structural comparison of concanavalin A and tomato bushy stunt virus protein *J. Mol. Evol.*, 15, 169, 1980.
80. **Temin, H.,** Origin of retroviruses from cellular moveable genetic elements, *Cell*, 21, 599, 1980.
81. **Sonigo, P., Barker, C., Hunter, E., and Wain-Hobson, S.,** Nucleotide sequence of Mason-Pfizer monkey virus: an immunosuppressive D-type retrovirus, *Cell*, 45, 375, 1986.
82. **Coffin, J. M.,** Structure, replication and recombination of retrovirus genomes: some unifying hypotheses, *J. Gen. Virol.*, 42, 1, 1979.
83. **Huang, C-C., Hay, N., and Bishop, J. M.,** The role of RNA molecules in transduction of the proto-oncogene *c-fps*, *Cell*, 44, 935, 1986.
84. **Baltimore, D.,** Expression of animal virus genomes, *Bacteriol. Rev.*, 35, 235, 1971.
85. **Michel, F. and Lang, B. F.,** Mitochondrial class II introns encode proteins related to the reverse transcriptase of retroviruses, *Nature (London)*, 316, 641, 1985.

86. **Loeb, D. D., Padgett, R. W., Hardies, S. C., Shehee, W. R., Comer, M. B., Edgell, M. H., and Hutchison, C. A., III,** The sequence of a large L1Md element reveals a tandemly repeated 5′ end and several features found in retrotransposons, *Mol. Cell. Biol.*, 6, 168, 1986.
87. **Hattori, M., Kuhara, S., Takenaka, O., and Sakaki, Y.,** L1 family of repetitive DNA sequences in primates may be derived from a sequence encoding a reverse transcriptase-related protein, *Nature (London)*, 321, 625, 1986.
88. **Dorssers, L., van der Meer, J., van Kammen, A., and Zabel, P.,** Cowpea mosaic virus RNA replication complex and the host-encoded RNA-dependent RNA polymerase-template complex are functionally different, *Virology*, 125, 155, 1983.
89. **Hodgman, T. C.,** An amino acid sequence motif linking viral DNA polymerases and plant virus proteins involved in RNA replication, *Nucleic Acids Res.*, 14, 6769, 1986.
90. **Itoh, H., Kozasa, T., Nagata, S., Nakamura, S., Katada, T., Ui, M., Iwai, S., Ohtsuka, E., Kawasaki, H., Suzuki, K., and Kaziro, Y.,** Molecular cloning and sequence determination of cDNAs for α subunits of the guanine nucleotide-binding proteins G_s, G_i, and G_o from rat brain, *Proc. Natl. Acad. Sci. U.S.A.*, 83, 3776, 1986.
91. **Fry, D. C., Kuby, S. A., and Mildvan, A. S.,** ATP-binding site of adenylate kinase: Mechanistic implications of its homology with *ras*-encoded p21, F_1-ATPase, and other nucleotide-binding proteins, *Proc. Natl. Acad. Sci. U.S.A.*, 83, 907, 1986.
92. **Jurnak, F.,** Structure of the GDP domain of EF-Tu and location of the amino acids homologous to *ras* oncogene proteins, *Science*, 230, 32, 1985.
93. **Moller, W. and Amons, R.,** Phosphate-binding sequences in nucleotide-binding proteins, *FEBS Lett.*, 186, 1, 1985.
94. **Kitamura, N., Semler, B. L., Rothberg, P. G., Larsen, G. R., Adler, C. J., Dorner, A. J., Emini, E. A., Hanecak, R., Lee, J. J., van der Werf, S., Anderson, C. W., and Wimmer, E.,** Primary structure, gene organization and polypeptide expression of poliovirus RNA, *Nature (London)*, 291, 547, 1981.
95. **Racaniello, V. R. and Baltimore, D.,** Molecular cloning of poliovirus cDNA and determination of the complete nucleotide sequence of the viral genome, *Proc. Natl. Acad. Sci. U.S.A.*, 78, 4887, 1981.
96. **Stanway, G., Hughes, P. J., Mountford, R. C., Minor, P. D., and Almond, J. W.,** The complete nucleotide sequence of a common cold virus: human rhinovirus 14, *Nucleic Acids. Res.*, 12, 7859, 1984.
97. **Callahan, P. L., Mizutani, S., and Colonno, R. J.,** Molecular cloning and complete sequence determination of RNA genome of human rhinovirus 14, *Proc. Natl. Acad. Sci. U.S.A.*, 82, 732, 1985.
98. **Skern, T., Sommergruber, W., Blaas, D., Gruendler, P., Fraundorfer, F., Pieler, C., Fogy, J., and Kuechler, E.,** Human rhinovirus 2: complete nucleotide sequence and proteolytic processing signals in the capsid protein region, *Nucleic Acids. Res.*, 13, 2111, 1985.
99. **Carroll, A. R., Rowlands, D. J., and Clarke, B. E.,** The complete nucleotide sequence of the RNA coding for the primary translation product of foot-and-mouth disease virus, *Nucleic Acids Res.*, 12, 2461, 1984.
100. **Palmenberg, A. C., Kirby, E. M., Janda, M. R., Drake, N. L., Duke, G. M., Potratz, K. F., and Collett, M. S.,** The nucleotide and deduced amino acid sequences of the encephalomyocarditis viral polyprotein coding region, *Nucleic Acids. Res.*, 12, 2969, 1984.
101. **Najarian, R., Caput, D., Gee, W., Potter, S. J., Renard, A., Merryweather, J., van Nest, G., and Dina, D.,** Primary structure and gene organisation of human hepatitis A virus, *Proc. Natl. Acad. Sci. U.S.A.*, 82, 2627, 1985.
102. **Lomonossoff, G. P. and Shanks, M.,** The nucleotide sequence of cowpea mosaic virus B RNA, *EMBO J.*, 2, 2253, 1983.
103. **Pincus, S. E., Diamond, D. C., Emini, E. A., and Wimmer, E.,** Guanidine selected mutants of poliovirus: mapping of point mutations to polypeptide 2c, *J. Virol.*, 57, 638, 1986.
104. **Stalhandske, P. O. K., Lindberg, M., and Pettersson, U.,** Replicase gene of coxsackievirus B3, *J. Virol.*, 51, 742, 1984.
105. **Cornelissen, B. J. C., Brederode, F. T., Moormann, R. J. M., and Bol, J. F.,** Complete nucleotide sequence of alfalfa mosiac virus RNA1, *Nucleic Acids Res.*, 11, 1253, 1983.
106. **Lindhout, W. H.,** Regulation of the Translation of Alfalfa Mosaic Virus RNA1, Ph.D. thesis, University of Leiden, Netherlands, 1985.
107. **Ahlquist, P., Dasgupta, R., and Kaesberg, P.,** Nucleotide sequence of the brome mosaic virus genome and its implications for viral replication, *J. Mol. Biol.*, 172, 369, 1984.
108. **Takkinen, K.,** Complete nucleotide sequence of the nonstructural protein genes of Semliki Forest virus, *Nucleic Acids Res.*, 14, 5667, 1986.
109. **Strauss, E. G., Rice, C. M., and Strauss, J. H.,** Complete nucleotide sequence of the genomic RNA of Sindbis virus, *Virology*, 133, 92, 1984.
110. **Miller, J., McLachlan, A. D., and Klug, A.,** Repetitive zinc-binding domains in the protein transcription factor IIIA from *Xenopus* oocytes, *EMBO J.*, 4, 1609, 1985.

111. **Cornelissen, B. J. C., Brederode, F. T., Veeneman, G. H., van Boom, J. H., and Bol, J. F.,** Complete nucleotide sequence of alfalfa mosaic virus RNA 2, *Nucleic Acids Res.*, 11, 3019, 1983.
112. **Strauss, E. G., Rice, C. M., and Strauss, J. H.,** Sequence coding for the alphavirus nonstructural proteins is interrupted by an opal termination codon, *Proc. Natl. Acad. Sci. U.S.A.*, 80, 5271, 1983.
113. **Kirkegaard, K. and Baltimore, D.,** The mechanism of RNA recombination in poliovirus, *Cell*, 47, 433, 1986.
114. **Yusoff, K., Millar, N. S., Chambers, P., and Emmerson, P. T.,** Nucleotide sequence analysis of the L gene of Newcastle disease virus: homologies with Sendai and vesicular stomatitis viruses, *Nucleic Acids Res.*, 15, 3961, 1987.
115. **Miller, R. H.,** Proteolytic self-cleavage of Hepetitis B virus core protein may generate serum e antigen, *Science*, 236, 722, 1987.
116. **Fuller, S. D. and Argos, P.,** Is Sindbis a simple picornavirus with an envelope?, *EMBO J.*, 6, 1099, 1987.
117. **Hodgman, T. C.,** personal communication.

Index

INDEX

A

A. albopictus cells, 174
Actinomycin D, 34
Adhl gene, 90
Agrobacterium, 30, 118—121
Agroinfection, 30, 121
Alfalfa mosaic virus, 220
Alignment of alphavirus nsP4 proteins, 234—235
Alignment of conserved core of alpha-virus nsP2 proteins, 232—233
Alignment of conserved core of picornaviral P2C proteins, 228—229
Alignment of conserved N terminal domain of alphavirus nsP1 proteins, 231—232
Alignment of picornaviral P3D polymerase proteins, 229—231
Alignment of residues around active site of picornaviral P3C protease, 229
Alignment signals, 81
Allogenic parents frequency of recombination between, 152
Alphaviruses, 170—174, 180, 220
Amino acid motifs, 223—224
Amino acid sequence homology, 163
Aminoacylation, 38
Aphid transmission, 26
Aphthovirus, 150
ASBV, see Avocado sunblotch viroid
Autolytic reactions of StobRV RNA, 129—131, 139
Avian leukosis virus, 17—19
Avian myeloblastosis virus, 38
Avocado sunblotch viroid (ASBV), 108, 110, 141

B

Bacteriophage SP6 RNA polymerase, 130
Bacteriophage T4 RNA ligase, 140
Bacteriophage T4 RNA p2Spl, 130
Baculovirus, 75
B104, 70, 73, 75, 80, 90, 93
Brome mosaic virus, 121, 158, 1621—163, 220
bsl, 70, 74—75, 80, 82
Budding, 44

C

Calf intestinal alkaline phosphatase, 135
CaMV, see Cauliflower mosaic virus
Capsid assembly, 26, 51
Carbon source, 85, 88
CARNA 5, 132
Carnation etched ring virus, 26
Catalytic RNA, 129
CAT gene, see Chloramphenicol acetyltransferase
Cauliflower mosaic virus (CaMV), 24—28, 31, 34—38, 53—55, 119, 141, 159, 220
Caulimoviruses, 24
cccDNA, 51, 55
cDNA of SFV, 171
Cell culture, 44
c-erbB, 17
Chimeric viroid cDNAs, 116—117
Chloramphenicol acetyltransferase (CAT) gene, 173
Chrysanthemum stunt viroid, 110
cinl, 70, 73, 75, 80
Circle junction, 70
Circularization reaction, 132—133, 138—140, 142
Circular DNA forms, 72, 83—84
Circular RNAs, 129—130, 132, 135, see also Viroids
cis-acting sequences, 79—83
Citrus exocortis viroid, 110
Cleavage, 14, 18
c-myc, 18
Coconut cadang-cadang viroid, 110
Coding sequence, 89—91
Columnea latent, 110
Complementary RNA, 130
Complementary termini, 175, 177
Complementation, 161
3′ Conserved molecules, 179
5′ Conserved molecules, 179
Conserved sequences, 26, 28—30, 170
Copia, 64, 69—70, 73, 75, 78, 80, 82—84, 94
Copyback structures, 176, 178
Copy choice recombination, 159—160, 215, 225—226
Coronaviruses, 150, 158, 215
Cowpea mosaic virus, 217—219
Crossovers, 152—160
  genetic exchanges between distantly related picornaviruses, 154—157
  homologous recombination, 158—160
  minimum length of sequence homology, 157
  occurence of recombination at many sites in picornavirus genome, 152—154
  recombination as general phenomenon, 157—158
Crown gall, 120
Cucumber mosaic virus, 128, 132
Cucumber pale fruit viroid, 110
Cucumoviruses, 220
Cyclic phosphodiester, 131
2′: 3′-Cyclic phosphodiester, 132, 133, 140—141
Cyclization, 141
Cytidine-2′: 3′-cyclic phosphodiester, 130
Cytopathic retroviruses, 18—19
Cytoplasmic phase, 32—33
Cytoplasmic polyhedrosis virus, 200

D

Dahlia mosaic virus, 26
Defective-interfacing (DI) genomes, 173—174, 179
Defective-interfacing RNAs, 128, 167—185, 204, 215, 225
  mechanisms of interference, 179—180

negative strand RNA viruses, 175—179, see also specific viruses
positive strand RNA viruses, 168—175, see also specific virus
Deletion mapping, 31, 171—172
Deletion mutants, 187—206
double-stranded RNAs associated with hypovirulence conversion in *Endothia parasitica*, 200—201
mechanism repsonsible for generation of remnant dsRNAs, 201—203
remnant dsRNAs as experimental tools, 203—204
wound tumor virus, 188—196, see also Wound tumor virus
yeast killer viruses, 188, 196—200
Developmental regulation, 88
Dictyostelium, 72, 79, 85
Dimeric RNA, 10, 129—130, 132, 141
DIRS1, 70, 72—73, 75, 79, 85—87, 92
Distribution, 70—76
DNA polymerase, 12, 34, 44—46, 64
DNA synthesis, 32—33
Double-stranded RNA, 187—206, see also Deletion mutants
experimental tools, 203—204
hypovirulence-associated, 200—201
mechanisms responsible for generation of, 201—203
Drosophila, 63, 75—79, 84, 92—94
Duck hepatitis B virus, 28, 46—48
Dysgenesis, 93—94

E

Electrophoresis of RNA, 133
Encapsidation, 38, 65, 67—70, 128, 172
Encephalomyocarditis virus, 169
Endogenous proviruses, 19
Endonuclease, 64
Endoplasmic reticulum, 16, 69
Endothia parasitica, 200—201, 204
Enhancers, 13, 17, 51, 82
Env gene, 7, 9, 15—16, 54, 64—65, 75, 79
Env reading frame, 76—78
Enzymatic questions, 34—36
Enzymic ligation of sobemovirus satellite RNAs, 140—141
Error correction, 214
Escherichia coli, 36, 77
Evolution, 211—228
genetic material, 212
homologous recombination, 161
interviral homologies, 215—225
general routes of variation, 215—216
homologies between diverse virus groups, 216—220
recombination of functional modules, 223—225
modular construction of RNA virus genomes, 216—225
nonhomologous recombination, 163
point mutation, 212—214
protovirus hypothesis, 225—227
recombination, 212, 214—215, 223—225
relationships, 55, 171
retroviruses, 19—20
RNA genome variation, mechanisms and routes of, 212—213
selection pressure, 220—223
specific mechanisms, 213—215
Expression, 13—15

F

Figwort mosaic virus, 26
Fingerprinting, 153, 155
Flea, 73, 80, 92
Foot-and-mouth disease virus, 121, 150—151
412, 73, 75, 80, 93—94
Frameshift mutations, 15, 27, 76, 78, 88, 89

G

Gag gene, 7—9, 27, 54, 78
Gag open reading frame, 35, 84—85
Gag-pol readthrough, 76
Gag proteins, 15, 64—65
GALI/Ty fusion, 84
Gene arrangements, 216
Gene conversion, 63, 93
Gene expression, 36—38, 49—51, 216
Gene organization, 47, 51, 54
General recombination, 215
Genetic recombination, 150—165, see also Positive strand RNA viruses
Genome-linked viral protein (VPg), 8—10, 212, 214, 217
Genomic RNA, 27, 37, 64, 128—129, 132, 162
Glucocorticoid-responsive element, 51
Glued locus, 90
Gly-Asp-Asp motif, 219, 227
Glycosylation, 16
Golgi apparatus, 16
Grapevine, 110
Guanidine, 150
Gypsy, 70, 73, 75, 80, 92—95

H

Hairpin dsDNAs, 11, 32
HBeAg, 48
HBV, see Hepatitis B virus
Heat shock, 85, 88
Hepadnaviruses, 79, 220
Hepatitis B virus (HBV), 43—58
budding, 44
cell culture, 44
direct repeats in genome, 48—49, 51—52
DNA, 44—46, 48—49
enhancer, 51
evolutionary relationship, 55
gene expression, control of, 49—51
gene organization, 47, 54

life cycle compared with other retroviruses and retrotransposons, 53—55
minus strand DNA, 48—49, 51—52
nucleic acid in infected liver cells and virions, 45—49
plus strand DNA, 48—49, 52, 53
pregenome production, 51
promoters, 49, 54
proteins, 44—45, 52—53
related viruses in animals, 44
replication cycle, 46, 51—53
reverse transcription, 44
RNA, 46, 48
sequence homology with region of reverse transcriptase of retroviruses, 53
splicing, 54
structure, 44—45
transcriptional control, 49—51
transcriptional maps, 46—48
translational control, 51
Heterogeneity, 65
HIS4, 91, 94—95
HMS Beagle, 73, 75, 80, 91
Homologous recombination, 92, 150—161
complementation, 161
coronaviruses, 158
detection, 150
evolution, 161
frequency, 151—152
intertypic, 154—157
intratypic, 152
isolation, 150
mapping, 151
mechanism of, 159—160
mutation frequency, 161
in nature, 160
other RNA viruses, 158—159
picornaviruses, 150—158, see also other subtopics hereunder
plant viruses, 158—159
poliovirus vaccines, 156—157
reasons for, 160—161
reversiviruses, 159
RNA repair, 161
selection of recombinants, 150
sites, 152—154, 157—158
tissue culture, 154—155
tobraviruses, 158—159
Hop stunt viroid, 110
Horseradish latent virus, 26
Host cell recognition, 44
Host genes, 60
Host-pathogen interaction, 108
Host-range, 26—27
Host target site duplications, 82
*Hpa*II, 70, 72—73, 80, 92
hsRTVL, 80
hsRTVL-H, 74, 76
Human immunodeficiency virus, 4, 13, 16, 18—20
Human reovirus, 202
Hybrid dysgenesis, 93—94
5′-Hydroxyl group, 140—141
Hypovirulence-associated dsRNAs, 200—201, 204

I

IAP, see Intracisternal A-particle
Ilarviruses, 220
IL-2, 18
Inclusion bodies, 24, 32—33
Indonesian tomato, 110
Influenza virus, 178—180, 215
Initial events, 10—12
Insect vector, 26, 35, 188
Insertion suppressor, 60, 63, 90, 94—95
Integrase, 12, 64
Integration, 5—6, 9, 12—13, 70, 81, 92
Interference, mechanisms of, 179—180
Intermolecular DI RNAs, 30, 215
Internal deletions, 175—179
Intertypic recombinants, 155—157
Interviral homologies, 215—225, see also Evolution
Intracisternal A-particle (IAP), 64, 69, 71, 74—75, 79—80, 82
Introns, 85, 91—92, 112, 129, 141
Isogenic parents, 151—152

J

Junction-containing RNAs, 141
Junction region, 170

L

Lentiviruses, 4
Lepidoptera, 74—75
Life cycle, 53—55, 62
Ligase, 12
Ligated RNA, 135
Long terminal repeat (LTR), 5—6, 61
domain of, 67
retrotransposons, 55, 79—81
retrovirus replication, 12—15
solo, 61, 63, 93—94
transposons, 73—74
U3-R-U5, 10, 67
Lor, 16
LTR, see Long terminal repeat
LTR-LTR recombination, 63, 93
Lucerne transient streak virus, 131—132
LYS2, 92

M

Mammals, 75—76
Mammary tumor virus, 17
MAT locus, 88
mdgl, 73, 75, 80, 84
mdg3, 73, 80, 84, 89
mdg4, 73
Mengovirus, 169
Minichromosomes, 26—28

Minus strand DNA, 48—49, 51—52
Minus strand synthesis, see Minus strand DNA
Mirabilis mosaic virus, 26
Modular construction of RNA virus genomes, 216—225, see also Evolution
Molds, 72, 75
Moloney murine leukemia virus (MuLV), 17, 76, 94
Mouse elements, 63, 79, 158
mRNA, 48
Multimeric RNA, 129—130, 132, 139, 141
MuLV, see Moloney murine leukemia virus
MuRRS, 74, 76, 80
Mutants, 26, 28—30
Mutation rates, 161, 213, 223
Mys, 74, 76, 80

N

Nascent strand, 6
Negative strand RNA viruses, 175—179, 216
Nepoviruses, 128, 132, 141
Neutralizing antibody, 150
Nicked genome, 6
Nicotiana clevelandii, 32
5′ Noncoding region, 91
Nonenzymic ligation of STobRV RNA, 133—139
Nonhomologous recombination, 162—163
Notch locus, 91
Nuclear cccDNA, 46
Nuclear Factor I, 50
Nuclear phase, 32—33
Nuclease digestions, 129
Nucleic acid, 25, 45—49
Nucleotide sequence homology, 128

O

Oligoribonucleotides, 135—138
Oncogenes, 17—18, 54
Open reading frames, 8, 64—65, 76—79
Overlapping reading frames, 49

P

Packaging signals, 82
Pancreatic ribonuclease A, 135, 138
Paramyxoviruses, 216—217
Partially double-stranded circular DNA, 44
Partially single-stranded cicular DNA, 52
Persistent infections, 177, 204, 214
2′-Phosphate, 132, 140, 141
5′-Phosphate group, 140
Phosphodiester bonds, 129, 135
Phosphomonoesterase, 133
Phosphorylation, 27, 88
Physarum, 72
Phytoreoviruses, 188—189
Picornaviruses, 150—158, 168—169, 215, 217—219, see also Homologous recombination
pim-1, 18
Plant viruses, 75, 158—159, 162, 171
Plus strand DNA, 48—49, 52, 53
Point mutation, 212—214
Polyadenylation by-pass, 37
Polarity, 37, 53
Pol gene, 7—9, 15, 54, 88, 226
Poliovirus, 121, 156—157, 169, 179—180, 218
Pol open reading frame, 35, 64—65, 76—78, 84—85
Pol proteins, 15
Polyadenylation, 14
Polycistronic genomic RNA, 216
Polymerase activity, 45, 53
Polynucleotide kinase, 138
Polypeptides, 24
Polypurine tract, 10, 28—29, 53
Positive strand RNA viruses
  defective-interfering RNAs derived from, 168—175, see also specific viruses
  recombinant in, 149—165, see also Homologous recombination; Nonhomologous recombination
Posttranslational regulation, 88—89
Potato spindle tuber viroid (PSTV), 108—125, 128, 141—142, see also Viroids
Potyviruses, 219
Pregenome production, 51
Primer, 48—49, 51, 53, 81—82, 133, 160
Processed pseudogenes, 70, 88
Promoters, 5, 17, 37, 49, 54, 89
Protease domain, 35, 65, 79, 217
Proteins
  cauliflower mosaic virus, 26—27
  hepatitis B virus, 44—45, 52—53
  homology, 35
  kinase, 27
  processing, 37, see also Protein synthesis and processing
  products, 78—79
  synthesis and processing, 15—16, 37
Proteolytic processing, 24, 27, 78
Protooncogene, 17
Prototypic retrovirus isolates, 4
Protovirus hypothesis, 225—227
Provirus, 5, 12—13, 18, 30

Q

Qβ, 174—175, 213

R

RAD52, 93
ras, 18
Rate of sequence divergence (drift), 222—223
Readthrough, 76
Receptor-mediated endocytosis, 10
Reciprocal recombination events, 93
Recombination events, 10, 12, 30—31, 65, 92—94, 150—151, 212, 214—217, 223—225
Redundancy, 48, 52—53
Regulation, 82, 85—89
Relay race translation, 38

Remnant dsRNAs, see Double-stranded RNAs
Repair, 28
Replicase, 214, 217, 226—227
Replication
  cycle, 4—5, 9, 46
  intermediates, 32, 36, 112, 121
  model, 24—30, 36—38, 51—53, 139—140
  recombination, 30
Retroid elements, 33, 37
Retrotransposition, 65—70, 83—89
Retrotransposons, 24, 33, 55, 59—103, 159
  alignment signals, 81
  circular DNA forms, 72
  cis-acting sequences, 79—83
  Class I, 89—91
  Class II, 91
  Class III, 91—92
  Class IV, 92
  coding sequence, 89—91
  DIRS1, 79
  distribution, 70—76
  DNA, 64
  Drosophila elements, 76—79, 92
  encapsulation, 65, 67—70
  enhancers, 82
  env reading frame, 76—78
  evolutionary relationship, 55
  gene conversion, 93
  gene organization, 54
  hepatitis B virus life cycle compared with, 53—55
  heterogeneity, 65
  host target site duplications, 82
  insertions, 60, 89—92, 94, 95
  integration, 70, 81, 92—94
  introns, 91—92
  life cycle, 62
  long terminal repeat sequences, 55, 73—74, 79—81
  LTR-LTR recombination, 93
  model, 65—70
  mouse elements, 79
  neighboring genes, effects on, 89—92
  5′ noncoding region, 91
  open reading frames, 64—65, 76—79
  packaging signals, 82
  pol reading frame, 76—78
  primer binding sites, 81—82
  properties, 61—64
  protein products, 78—79
  reciprocal recombination events, 93
  recombination between, 92—94
  retrotransposition, see Retrotransposition
  reverse transcription, 70
  RNA, 64
  sequences, 80
  silencers, 83
  structural comparison, 76—83, see also other subtopics hereunder
  structure, 64—65
  target duplication, 73—74
  target gene, 92
  transcription, 65, 79, 81
  Ty elements, 92
  yeast elements, 76—79
Retroviruses, 3—24, 159, 220, 227
  evolution and genetics, 19—20, 55
  gene organization, 54
  genome, 8—10
  hepatitis B virus life cycle compared with, 53—55
  importance of, 4
  life cycle, 62
  open reading frames, 8
  replication, 3—22
    consequences of, 17—19
    cycle, 4—5, 9
    cytopathic interactions, 18—19
    expression, 13—15
    initial events, 10—12
    integration, 12—13
    protein synthesis and processing, 15—16
    RNA processing, 15
    transactivation, 16
    transformation and oncogenes, 17—18
  virion, 4—8
Reverse transcriptase, 6, 10, 38, 53, 61, 64, 69, 84—85, 89, 133, 135, 220, 226
  cauliflower mosaic virus, 34—36
  dimeric STobRV(+)RNA, 129
  strand switching, 215
Reverse transcription, 18, 48, 61, 217
  cauliflower mosaic virus, 23—38, see also other subtopics hereunder
  conserved sequences, 28—30
  cytoplasmic phase, 32—33
  deletions, 31
  direct evidence for, 32—36
  enzymatic questions, 34—36
  genetic evidence for the replication model, 28—30
  hepatitis B virus, 44
  in vivo replication intermediates, 32
  intracellular sites of DNA synthesis, 32—33
  jumps, 11
  mutants, 28—30
  nuclear phase, 32—33
  recombination events, 30—31
  retrotransposons, 70
  retrovirus, 9
  RNA pregenome, 46, 48, 51
  steps of, 71
  terminally redundant RNA species, 53
  transport problems, 32—33
  viral replicative complexes, 33—34
Reversion rate, 213
Reversiviruses, 159
Revertants, 151
Rhabdoviruses, 216—217
Ribonucleases, 129, 133, 135—138
Ribosome frame shifting, 35, 51, 53—55
RNA, 46, 48, 64, 108, 110, 121
  dependent DNA polymerase, 35, 112, 219, 226
  genome variation, 212—213
  polymerase II, 64

pregenome, 44, 46, 48, 51, 54—55
processing, 15, 37
recombination, 121, 150, see also Homologous recombination; Nonhomologous recombination
repair, 161
synthesis, 37
terminal repeat, 27
transcripts, 5
viral genomes, 222, 223
RNase H activity, 28, 33, 51—53, 65
Rodents, 75—76
Rolling circle transcription, 139, 142
roo, 73, 93
Rous sarcoma virus (RSV), 76, 82

S

Saccharomyces cerevisiae, 72, 196—197
Satellite RNAs, 225, see also Small satellite RNAs
Saturation mutagenesis, 118
Secondary hairpins, 114, 116—117
Selection, 213—214, 216, 220—223
Semliki Forest virus, 169—172
Sendai virus, 176—178, 180
17.6, 73, 75, 80, 92
1731, 73
Short sequence motif, 227
Sindbis virus, 162—163, 169—174, 180, 213, 215, 219
Single-stranded interruption, 24
Site-specific mutagenesis, 115—116
Slime molds, 72, 75
Small satellite RNAs, 127—145, see also STobRV RNA
Snap-back structures, 176
Sobemovirus satellite RNAs, 128, 131—132, 140—141
Solanum nodiflorum mottle virus, 131—132
Solo LTRs, 61, 63, 93—94
Soybean chlorotic mottle virus, 26
Specificity, 12, 14
Splicing, 31, 37, 54
Springer, 73, 80, 92
SP6 bacteriophage DNA-dependent RNA polymerase, 172
SPT3, 94—95
STobRV RNA, 128—140
autolytic processing, 139
circular forms, 129—130, 132
complementary, 130
dimeric form, 129—130, 132
multimeric forms, 129—130, 132, 139
nonenzymic ligations, 133—139
(-) RNA, 133—138
(+) RNA, 138—139
replication, 129, 139—140
transcription, 128
translation, 128
Strong-stop DNA, 11, 28, 30—32, 81—82, 84, 159
Structure, see specific viruses
Subgenomic RNA, 219
Subterraneum clover mottle virus, 132
su(Hs), 94—95
su(Hw), 94
Supporting virus, 128
Suppressor mutations, 63
Suppressors of retrotransposon insertions, 60, 94—95
SV40, 27, 49, 171

T

Target duplication, 73—74, 92
Tat, 16
TATA box sequence, 50
T-DNA, 119—120
TED, 70, 73, 75, 80
Temperature, 89
Template switch, 6, 31, 52—53, 84, 140, 159
Terminator insertion, 18
THE-1, 74, 76, 80
Three strand helix configuration, 25
3518, 73
Ti plasmid, 118—119
Tissue culture, 154—155
Tobacco plants, 132
Tobacco rattle virus, 159
Tobacco ringspot virus (TobRV), 128—129
Tobamoviruses, 220
Tobraviruses, 158—159, 220
Togaviruses, 169—174
Tomato apical stunt viroid, 110
Topological knot, 24
Transactivation, 16
Transcription, 9, 27, 49—51, 65, 85, 88, 128, 139, 203—204
Transcriptional maps, 46—48
Transcriptional signals, 79, 81
Transfection, 44, 172
Transformation and oncogenes, 17—18, 132
Translation, 37—38, 51, 88, 128, 219
Transport problems, 16, 32—33
Transposable elements, 60, 220, 226
Transposition, 61, 83—85, 94
trans-Splicing, 225
tRNA, 6, 10, 28, 77, 81—82, 84, 88—89, 140, 170—171
297, 73, 75, 80, 92
Ty elements, 65, 70, 72, 73, 75, 80, 82—85, 91—94

U

ura3, 89—90
UV light, 85, 88

V

Velvet tobacco mottle virus, 132
Vesicular stomatitis virus (VSV), 173, 175—177, 180
Viral DNA synthesis, 27—28, 46, 51

Viral genome, 24—26
Viral protein sequence homologies, 216—220
Viral replication site, 32—34
Virions, 4—8, 15, 24, 45—49
Viroids, 107—125, 128, 132
  cDNAs, 112—118
  cross protection, 117
  defined, 108
  development of assays for specific functions, 118—119
  emerging approaches, 118—122
  infections as analytical tools, 121—122
  potential mRNAs, 112
  processing, 114
  replication cells transformed by *Agrobacterium*, 112, 119—121, 141—142
  sequence variants, significance of, 110—111
  small satellite RNAs contrasted, 128
  structure, 108—110
  virulence-modulating regions, 111
Viroplasms, 24, 32—34
Virus-like particles (VLPs), 65, 67—69, 78, 82, 84—85, 188
Virusoids, 132
VL30, 72, 74, 76, 80
VPg, see Genome-linked viral protein

W

w^{a}, 91
w^{hf}, 91
Wheat germ RNA ligase, 141
w^{hd}, 90
Wound tumor virus, 204
  deletion mutants, 188—196
    discovery and characterization of, 189—192
  genome remnant dsRNAs, 189—192
  sequence organization of genome, 192—196
w^{sp}, 91
w^{zm}, 91

Y

Yeast, 63, 72, 76—79, 188, 196—200, 204
  distribution, 72
  structural comparison, 76—79
Yellow gene, 95
Yield test, 150—151